ENVIRONMENTAL STATISTICS, ASSESSMENT, AND FORECASTING

EDITED BY

C. RICHARD COTHERN

CENTER FOR ENVIRONMENTAL STATISTICS DEVELOPMENT STAFF
U. S. ENVIRONMENTAL PROTECTION AGENCY
WASHINGTON, D. C.

N. PHILLIP ROSS

ENVIRONMENTAL STATISTICS AND INFORMATION DIVISION
U. S. ENVIRONMENTAL PROTECTION AGENCY
WASHINGTON, D. C.

LEWIS PUBLISHERS
Boca Raton   Ann Arbor   London   Tokyo

Library of Congress Cataloging-in-Publication Data

Environmental statistics, assessment, and forecasting / edited by C. Richard Cothern and
N. Phillip Ross.
     p.   cm.
   Includes bibliographical references and index.
   ISBN 0-87371-936-0
   1. Environmental sciences — Statistical methods. 2. Environmental policy. I. Cothern,
   C. Richard. II. Ross, N. Phillip.
GE45.S75E58   1993
363.7'0072--dc20   93-25598
                                                      CIP

Preface*

Environmental statistics are generated, analyzed, integrated, presented, and interpreted by many different people that practice many different disciplines, have a multitude of conflicting concerns, and work in many different circumstances. The communication between these multivarious individuals is sporadic, incomplete, and often indifferent. We organized the symposium reported in this volume to focus specifically on the problems that the above characteristics bring to the attempts to describe the state of the environment. We called this symposium "Environmental Statistics, Assessment, and Forecasting" to emphasize these three aspects of the overall effort. We assembled for a full day of presentations and discussions at the National Meeting of the American Chemical Society (ACS) in Washington, D.C., on August 26, 1992. We organized this symposium as part of the overall program of the Environmental Division of the ACS. We were aided in organizing this symposium by the following who acted as the chairmen for the various sections: Joseph Abe, U.S. EPA; Daniel Krewski, Health Canada; Rick Linthurst, U.S. EPA; Barry Nussbaum, U.S. EPA; and Alvin Pesachowitz, U.S. EPA.

This volume displays the diversity and complexity that characterizes the field of environmental statistics. From the symposium and the volume we can observe at least two common themes. One concept that is shared by almost everyone in this field is that of quantitative uncertainty. Some call this error, others variability, and yet others distributions. There are as many different concepts of uncertainty as there are workers in environmental statistics. And, yet, all would agree on its importance in describing their contribution. Another common theme is that of incompleteness. Our current picture of the state of the environment is like an unfinished jigsaw puzzle for which many of the parts are missing. There is much work to be done before a credible, coherent, and understandable picture emerges of the state of the environment. Our volume is one of several steps that are needed to achieve the goal of describing the state of the environment.

C. Richard Cothern
N. Phillip Ross
Washington, D.C.

---

*The thoughts and ideas expressed in this volume are those of the authors and are not necessarily those of the U.S. Environmental Protection Agency.

About the Editors

C. Richard Cothern, Ph.D., is presently with the U.S. Environmental Protection Agency's Center For Environmental Statistics Development Staff. He has served as the Executive Secretary of the Science Advisory Board at the U.S. EPA and as their National Expert on Radioactivity and Risk Assessment in the Office of Drinking Water. In addition, he is an Associate Professorial Lecturer in the Chemistry Department of the George Washington University. Dr. Cothern has authored over 70 scientific articles including many related to public health, the environment, and risk assessment. He has written and edited ten books involving such diverse topics as science and society, energy and the environment, trace substances in environmental health, lead bioavailability, risk assessment, and radon and radionuclides in drinking water. He received his B.A. from Miami University (Ohio), his M.S. from Yale University, and his Ph.D. from the University of Manitoba.

N. Phillip Ross, Ph.D., is Chief Statistician of the U.S. Environmental Protection Agency and Chairman of the Organization for Economics and Cooperative Development (OECD) Group on the State of the Environment. He is presently Director of the Environmental Statistics and Information Division in the EPA Office of Policy, Planning, and Evaluation. He has been a prime mover in the Federal initiative to create a National Center for Environmental Statistics. For his work in this area, he has received EPA's highest award for Management Excellence. Dr. Ross has been Chair of the American Statistical Association Section on Statistics and the Environment. He represents the statistical community on a number of Federal and nongovernmental panels and task forces involving statistics and the environment. He enjoys teaching and the outreach and has been an Adjunct Professor of Mathematics and Statistics at the American University in Washington, D.C.

Roster

Symposium on Environmental Statistics, Assessment, and Forecasting

Organizers

C. Richard Cothern
Center For Environmental Statistics
  Development Staff
PM-223
U.S. Environmental Protection Agency
Washington, D.C. 20460

N. Phillip Ross, Director
Environmental Statistics and
  Information Division
PM-222B
U.S. Environmental Protection Agency
Washington, D.C. 20460

Participants

Joseph Abe
OPPE
PM-222A
U.S. Environmental Protection Agency
Washington, D.C. 20460

Asit Basu
Department of Statistics
328 Math Sciences Building
University of Missouri
Columbia, MO 65211

Ralph Baumgardner
Atmospheric Research and Exposure
  Laboratory
U.S. Environmental Protection Agency
MD-80A
Research Triangle Park, NC 27711

Barbara A. Beard
Office of Air Quality Planning and
  Standards
U.S. Environmental Protection Agency
MD-14
Research Triangle Park, NC 27711

Peter Blair
Manager, Energy and Materials
  Division
Office of Technology Assessment
U.S. Congress
Washington, D.C. 20510-6336

Mark Boroush
Senior Economist and Policy
  Analyst
The Futures Group
Glastonbury, CT 06033

Susan Brunenmeister
Computer Sciences Corporation
Chesapeake Bay Program
  Office
410 Severn Ave., Suite 113
Annapolis, MD 21403

Eliot J. Christian
U.S. Geological Survey
802 National Center
Reston, VA 22092

Michael Cole
Department of Agronomy
University of Illinois
1102 South Goodwin
Urbana, IL 61801

Noel Cressie
Department of Statistics
Iowa State University
Ames, IA 50011

Carroll Curtis
Center for Environmental Statistics
  Development Staff
PM-222B
U.S. Environmental Protection Agency
Washington, D.C. 20460

James Daley
Office of Policy, Planning, and
  Evaluation
PM-223Y
U.S. Environmental Protection Agency
Washington, D.C. 20460

H. T. David
Department of Statistics
304B Snedecor Hall
Iowa State University
Ames, IA 50011–1210

Thomas G. Dewald
Office of Information Resources
  Management
3405R
U.S. Environmental Protection Agency
Washington, D.C. 20460

T. E. Dorfler
Arthur D. Little, Inc.
35-321A Acorn Park
Cambidge, MA 02140-2390

Steven Edland
University of Washington
Department of Enviromental Health
Blakely Building, SC-34
Seattle, Washington 98195

Mark Ellersieck
Agricultural Experiment Station
University of Missouri
Columbia, MO 65211

Paul Feder
Battelle
Room 11-7-060
505 King Street
Columbus, OH 43201

Warren P. Freas
Office of Air Quality Planning and
  Standards
MD-14
Research Triangle Park, NC 27711

Timothy L. Gauslin
U.S. Geological Survey
802 National Center
Reston, VA 22092

David Gaylor
Biometry Staff
National Center for Toxicological
  Research
U.S. Food and Drug Administration
Jefferson, AR 72079

Richard O. Gilbert
Battelle, Pacific NW Laboratories
PO Box 999, MSN K7-34
Richland, WA 99352

M. J. Goddard
Environmental Health Centre
Health Canada
Room B9
Tunney's Pasture
Ottawa, Ontario, Canada K1A 0L2

S. D. Gore
Center for Statistical Ecology and
  Environmental Statistics
The Pennsylvania State University
University Park, PA 16802

Tim Haas
School of Business Administration
University of Wisconsin - Milwaukee
P.O. Box 742
Milwaukee, WI 53201

David Holland
Atmospheric Research and Exposure
    Assessment Laboratory
MD–56
U.S. Environmental Protection Agency
Research Triangle Park, NC 27711

Gary F. Krause
Agricultural Experiment Station
University of Missouri
Columbia, MO 65211

Daniel Krewski
Environmental Health Centre
Health Canada
Room 109
Tunney's Pasture
Ottawa, Ontario, Canada K1A 0L2

Herbert Lacayo
Environmental Statistics and
    Information Division
PM-222B
U.S. Environmental Protection Agency
Washington, D.C. 20460

Eleanor Leonard
Center for Environmental Statistics
    Development Staff
PM-222B
U.S. Environmental Protection Agency
Washington, D.C. 20460

Rick Linthurst
Atmospheric Research and Exposure
    Assessment Laboratory/ORD
MD–75
U.S. Environmental Protection Agency
Research Triangle Park, NC 27711

Tom Mace
OIRM
PM-211
U.S. Environmental Protection Agency
Washington, D.C. 20460

David A. Marker
Westat, Inc.
1650 Research Boulevard
Rockville, MD 20850–3129

Foster J. Mayer, Jr.
Environmental Research Laboratory
U.S. Environmental Protection Agency
Gulf Breeze, FL 32561

Tom Mazzuchi
Department of Operations Research
Staughton Hall
The George Washington University
707 22nd Street, NW
Washington, D.C. 20052

Neerchal K. Nagaraj
Deptment of Mathematics and
    Statistics
UMBC
Baltimore, MD 21228

Brand Niemann
The Center for Environmental
    Statistics
Environmental Statistics and
    Information Division
PM-222B
U.S. Environmental Protection Agency
Washington, D.C. 20460

Barry D. Nussbaum, Chief
Statistical Analysis and Computing
    Branch
PM-222B
Environmental Statistics and
    Information Division
U.S. Environmental Protection Agency
Washington, D.C. 20460

Tom O'Connor
National Oceanic and Atmospheric
    Administration
N/ORCA 21
1305 East-West Highway
Silver Spring, MD 20910

Gary Oehlert
Department of Applied Statistics
University of Minnesota
St. Paul, MN 55108

G. P. Patil
Professor of Mathematical Statistics
and Director of the Center for
Statistical Ecology and
Environmental Statistics
421 Classroom Building
The Pennsylvania State University
University Park, PA 16802

Alan M. Pesachowitz
Office of Information Resources
Management
PM-208
U.S. Environmental Protection
Agency
Washington, D.C. 20460

David W. Rejeski
OPPE
PM 222A
U.S. Environmental Protection Agency
Washington, D.C. 30852

Svetlana Ryaboy
Westat, Inc.
1650 Research Boulevard
Rockville, MD 20850-3129

Joel Schwartz
OPPE
PM 221
U.S. Environmental Protection Agency
Washington, D.C. 20460

Jeanne C. Simpson
Batelle, Pacific NW Laboratories
P.O. Box 999, MSN K7-34
Richland, WA 99352

A. K. Sinha
Center for Statistical Ecology and
Environmental Statistics
The Pennsylvania State University
University Park, PA 16802

David South
Argonne National Laboratory
EID/900
9700 South Cass Avenue
Argonne, IL 60439-4832

Kai Sun
Department of Statistics
University of Missouri
Columbia, MO 65211

Charles Taylor
Strategic Futurist
Strategic Studies Institute
U.S. Army War College
Carlisle Barracks, PA 17103

Gerald van Belle
Department of Environmental
Health/SC–34
University of Washington
Seattle, WA 98195

William J. Warren-Hicks
The Cadmus Group
Executive Park
1920 Highway 54
Durham, NC 27713

Robert L. Wolpert
Institute of Statistics and Decision
Sciences
Duke University
Durham, NC 27708-0251

Seongmo Yoo
Electronics and Telecommunications
Research Institute
P.O. Box 8, Daedog Science Town
Daejeon, Korea

Yiliang Zhu
Department of Epidemiology and
Biostatistics
University of South Florida
Tampa, FL 33612-3805

Reviewers

Kenneth G. Brown, Chris Campany, Ed Calabrese, Tom Curran, J. Clarence
(Terry) Davies, Jerzy Filar, George Flatman, Chapman Gleason, Jerry Glen,
Richard Hayes, Rolf Hartung, Herbert A. (Pepi) Lacayo, David Lesh, Dean
Neptune, Brand Niemann, Robert O'Brien, Brenda Odom, Milton Siegel, Bimal
Sinha, Tom Starr

We, the editors, dedicate this volume to our wives,
Margaret Fogt Cothern and Nina Ross
in appreciation of their
patience, support, and interest
in our involvement

Contents

Section III
Models and Data Interpretation
Introduction by Daniel Krewski

Section IV
Future Environmental Management
Introduction by Joseph M. Abe

Overview

Uncertainties in Assessing the State of the Environment: An Overview of Environmental Statistics, Assessment, and Forecasting*

C. Richard Cothern and N. Phillip Ross

INTRODUCTION

Clear, complete, statistically accurate, and understandable information is essential to making informed decisions concerning the state of the environment. To develop a coherent picture of national and regional environmental trends and conditions requires: collecting quality data; assembling the existing data which is scattered among many governmental, industrial, academic, and environmental organizations; statistically analyzing and integrating the information; and providing complete, accurate, and understandable presentations. Unfortunately, in most cases much of the data do not exist or are not available and, in only a few cases, are they fully analyzed and integrated. In general, they are constantly changing and, thus, the overall the picture of the environment is unclear. Our picture of the state of the environment is like a partially assembled jigsaw puzzle where many of the pieces are missing, only parts are put together and the picture is incomplete and potentially misleading of what is actually there.

There is often a difference between what people perceive as the state of the environment and what it actually is (i.e., see the report *Reducing Risk*[1]). For example, many feel that human beings are causing the weather to be hotter, but whether this is true is not clearly known. In

---

*This chapter is an overview of the other chapters in this volume as well as an overview of the Symposium on Environmental Statistics, Assessment, and Forecasting held by the Environmental Division of the American Chemical Society in Washington, D.C. on August 26, 1992. The thoughts and ideas expressed in this chapter are those of the authors and are not necessarily those of the U. S. Environmental Protection Agency.

another dimension, many feel that hazardous waste sites are dangerous while indoor air concentrations of radon are unimportant. In reality, many orders of magnitude more die from radon than hazardous waste sites. One of the inputs needed to bring the different views of the scientific community and the general public together is a more complete and more understandable quantitative description of the state of the environment. The objective of this volume is to examine the problems in assembling, analyzing, and presenting this information in a form to be understandable and useful to various groups interested in the environment: print and electronic media, policy makers, planners, researchers, the public, environmental groups, academe, industry, etc.

Perhaps the most important characteristic of the existing environmental data and information is its uncertainty. This characteristic has numerous meanings and all are germane to the present discussion, which makes this concept a useful thread for understanding what is happening in the environment. The concepts of uncertainty and the incompleteness of the data and its analysis are useful organizing tools in understanding and putting the current volume into perspective.

The objectives of the current volume are to:

- Assess some aspects involved in the state of assessing the environment, including discussions of uncertainties, data gaps, assessment techniques, and modeling efforts
- Examine the current proposed frameworks for assessing the state of the environment and future needs in this area
- Examine some ways to use raw data to produce information such as forecasting and assessments
- Discuss other potential uses for the existing and future data
- Examine some problems in harmonizing and integrating data from the Federal, state, and local levels, as well as that from other sources

These objectives are discussed in the following chapters in this volume in sections entitled: Combining Data and Data Uncertainty, Spatial Chemostatistics, Models and Data Interpretation, Future Environmental Management, Geographic Information Systems (GIS), and Environmental Statistics.

DATA UNCERTAINTY AND VARIABILITY

Webster's Dictionary defines uncertain as: "not known or established, doubtful; not determined, undecided; not having sure knowledge; subject to change, variable; unsteady, fitful." This definition covers a wide range of possible situations and it illustrates the current state of the data and information available describing the state of the environment. There are basically two kinds of uncertainty: no information and quantitative. In all too many cases there is no information or large data gaps where the data and information is not known or is not available. In this case, decision makers are called on to develop models and then policy to fill the gaps. For example, the effect of low level ionizing radiation on human health is not known. For this example the adopted policy, which errs on the safe side, is to make a linear extrapolation of the dose-response curve down to zero dose. Quantitative uncertainty involves a quantifiable inherent error or variability in the data which, in a few cases, can be as little as 10 or 20% (few situations require this good an error). In some cases the variability can be quite large, e.g., some model predictions can span six to twelve orders of magnitude. It is not always necessary for the required range of variability or error to be small. The acceptability of uncertainty range depends on the decision and what requirements that imposes. For example, if the estimated lifetime individual risk level is in the range of 10^{-20} to 10^{-10}, and the acceptable level is anything less than 10^{-6}, then even though the error range is ten orders of magnitude, it is clear that the range is acceptable.

The current volume focuses on quantitative uncertainty which comes in a range of possible forms:

- No information or data
- Incomplete or missing data or some pieces missing
- Incompatible data due to different sampling methods, differing analytical methods, or different times or spaces
- Anecdotal data—this is usually the case and involves such problems as nonstatistical sampling, grab samples, little information concerning where, how, and when the sample was taken, and so on
- Variability—involving the inherent error and such concepts as precision, accuracy, and Heisenberg's uncertainty principle
- Error, as in the statistical error where it helps to know the distribution
- Temporal or time variations—in order to determine time trends one needs to know the background level; predicting or estimating into the future by use of models involves unique uncertainty problems, mainly large uncertainties; the data involved is usually a snapshot in time
- Spatial variations—at the local, regional, national, and global levels
- Estimation—human health (extrapolating from animals to humans and from high doses to low doses); predicting the future involves enormous uncertainties

Another way to focus on the different aspects of uncertainty is to ask a series of questions concerning uncertainty in assessing the state of the environment such as

- What do we know with a high degree of accuracy and confidence—little, if anything?

- What do we think we know?
- What has been estimated—with high or low accuracy and confidence?
- What has been conjectured on the basis of judgment?
- Is the judgment scientific, professional, or just a guess?
- What do we not know?
- What is planned?
- What is needed?

The broad range of contributions to the variability in environmental data need to be considered in the effort to describe the state of the environment.

CHARACTERISTICS AND PHILOSOPHY OF ENVIRONMENTAL DATA

Conceptually, environmental processes can be characterized as: complex, multifaceted, chaotic, and dynamic. Complexity can be seen in such diverse areas as global warming, species diversity, synergisms, and antagonisms. Another example of complexity is thresholds such as those in dose response curves for health endpoints. The chaotic element is the instabilities in environmental phenomena introduced by the activities of humans from automobiles, power plants, large cities, industry, by nature, and other activities. The study of such environmental processes as those mentioned here must involve many different disciplines and skills. Technical, scientific (biological, physical, and social), and engineering skills are needed involving such disciplines as: biology (including such specialties as microbiology, toxicology, medicine, physiology, and pathology), chemistry (environmental and analytical), physics (environmental and health), statistics, civil engineering, environmental engineering, economics, political science, law, communications, to mention most of the major ones. Many of these disciplines were represented at the symposium.

The establishment of a sound basis for determining the state of the environment requires a strong philosophical basis. A proper framework is needed and could involve the following aspects:[2]

- Establishment of the Existential Base
  - Proper sampling, statistically planned to assure obtaining a credible and sufficiently complete, accurate, and precise sets of data that describe reality
- Establishment of Data Quality
  - Includes description of the metadata to determine completely the details of data collection and the variability, error, or distribution of the data

- Establishment of Observation Standards
  - Determination of what may be inferred or deduced from the data (and what cannot) — that is, what has actually been observed?
- Establishment of Evidential Relevance
  - Determination of what assumptions are required for use of the data, what confidence can be placed in this data, and what range of predictions can be made from this data
- Establishment of Coherence and Summarization
  - How rational is the prediction, confidence distribution, and results of the analysis of the data

This framework can be functionally expressed in the data quality objectives. These need to be established before the data are collected. In some respects this procedure boils down to the scientific method.

All of these aspects, characteristics, and philosophies need to be considered in any attempt to describe the state of the environment.

UNCERTAINTY IN ENVIRONMENTAL STATISTICS ANALYSIS

One of the best ways to describe statistically quantitative uncertainty is by using distributions. There has been, for several years now, a common thought among most people involved in the analysis of the environment that most variables follow a lognormal distribution. This oral tradition is based on some real data and in most cases it appears that the lognormal distribution fits better than the normal distribution. Professor G. P. Patil, the keynote speaker at the Environmental Statistics, Assessment, and Forecasting Symposium, observed that we are beyond the age of lognormality. He noted that many of the data sets describing the state of the environment have more skewness than that exhibited by the lognormal distribution. However, more detailed and careful analysis is needed even if the lognormal distribution was the perfect model. It may not be necessary in all cases to know the actual distribution and even knowing it we do not know all that we need to know. Nonparametric procedures circumvent the need to know the distribution in some situations. Patil recommended that this implies that more detailed and careful analysis needs to be done in the effort to describe the state of the environment.

Professor Patil suggested that the determination of the state of the environment requires more careful and complete analysis of the data and that this means care from the beginning. The first step in collecting data concerning the environment is the design of the sampling methods. An attitude here that needs to be overcome is the thought expressed by many that "I don't need to take a second measurement." The complete and careful design of sampling takes extra effort, but the alternative is that the data collected may be of limited value if the sampling plan is not statistically valid.

There is also a general problem due to the large variability and soft data where some errors are known and some are not. This led Patil to ask if the public is ready for environmental statistics? And, are the decision makers ready? Patil described the need for a consortium for outreach to the public, decision makers, and planners to provide environmental statistics involved with a cross-disciplinary aspect.

In the current volume there are several aspects of collecting and analyzing environmental data that require more attention. These include sampling procedures, such as composite sampling, cluster sampling and transect sampling, combining data from different sources such as that produced in the laboratory and the field, dealing with data that are below the detection limit, multivariate data, data that has spatial and temporal variations, and the general overall quality of the data.

Composite Sampling and Combining Data

The more data one has the more complete is the picture. With less data there could be uncertainty. However, there is a conflict in that there are limited resources for sampling and collecting data. One alternative, composite sampling, is discussed in some detail in Chapter 3 by Patil et al. and Chapter 2 by Edland and Van Belle. These chapters include a review of the literature in such areas as ranked set sampling, cluster sampling, and transect sampling. The general idea is to preserve limited resources and provide the best and most complete data where the cost is too large to sample for all possible situations individually. Thus, what is needed is a balance between cost and uncertainty. This process is described as trying to achieve observational economy and the authors feel that perhaps these techniques are under utilized. Problems encountered here include rapidly changing technology and the need for a broad view of the life stages of living things (plants, humans, microlife, etc.) and of the contaminants (source, transport, exposure, and fate).

A new and more generalized algorithm for estimating population mean and variance from composite samples is offered by Edland and Van Belle. Their algorithm allows for an arbitrary, potentially unequal number of field samples in a composite sample. The idea is to reduce the cost without substantially reducing the quality of the information determined.

Often data are collected from sources that are not necessarily or easily compatible. Combining such data and information can lead to increased uncertainty. One example of trying to combine such data is given in Chapter 4 by Warren-Hicks and involves combining data collected in the laboratory and in the field. The example he uses is laboratory bioassay and field observations of fish responses to acidification. He combines the data using classical inference by Bayesian techniques that can also combine data from expert opinion.

Dealing with Data Below the Detection Level

One of the more common uncertainties in dealing with environmental data is the existence of values that are below the detection level. Nagaraj and Brunenemeister, in Chapter 5, review some of these methods and conclude that most methods involve substituting the "best value" for the censored usually being zero, half the detection limit, or the detection limit. They propose a weighted regression approach which incorporates the relationship between the detection limit and the precision of the data. They illustrate this method using water quality monitoring data collected by the Chesapeake Bay Program. The results show that their weighted regression approach is much less sensitive to the method of substitution of the "best value." They list three advantages for their approach: lowered sensitivity to the level of censoring, less sensitivity to the choice of data substitution (zero, half the detection limit, or the detection limit) and less complicating effects of autocorrelation. The reader should also note that in Chapter 6, by Cressie, there is also a discussion of the usefulness of detection limits in modern spatial statistics.

Multivariate Data

The area of multivariate data contains many potential contributions to handling the uncertainty in the data. Spatial data can be censored, incomplete, or simply incompatible.

Problems relating to air pollution data are discussed in Chapter 7 by Holland et al. They discuss the methodology for monitoring trends in air depositions to test the effect of the Clean Air Act Amendments of 1990 for wet sulfate deposition. They investigated spatial and temporal status and trends in aquatic and terrestrial ecosystems using a rural monitoring network. Their analysis led them to recommend some additional stations.

The early pioneers in our country recognized that the quality of vegetation could distinguish between good and poor quality soil. However, as Cole points out in Chapter 12, current data on soil quality involves uncertainty as it includes several indicators such as physical, chemical, topographical, and those involving biota. The problem is how to combine this information in a meaningful and useful way to reduce the uncertainty concerning the quality of the soil. In addition, the available data have problems such as being collected by different disciplines and investigators for different purposes, involving different sampling strategies, sampling processing, and analytical methods. The botanical surveys involve complex plant taxonomy and many subdisciplines in invertebrate biology to characterize the microbes. Also involved are the complications of microscopic and macroscopic spatial variability as well as temporal variations. These complications all contribute to the overall uncertainty.

The overall quality of environmental data bases are affected by many sources of variability and bias. These general contributions are discussed in Chapter 18 by Marker et al. The biggest roadblock to successful development of a list of tools describing data quality is the language differences between statisticians, chemists, and other environmental scientists. He outlined the six elements of data quality: precision, bias, mean square error, representativeness, comparability, and completeness. Another dimension of this problem is discussed in Chapter 1, by Gilbert and Simpson, relating to the quality of historical data, especially the lack of expert opinion for events that happened some time ago. Other problems involved in the methods for spatial interpolation of environmental data are discussed in Chapter 21 by David and Yoo.

Problems encountered in trying to develop a data base to represent a national view are discussed by O'Connor, in Chapter 19, in the context of monitoring chemical contamination in the coastal waters of the U.S.

Presentation of Data: Dealing with Quantitative Illiteracy and Subjectivity

One of the important dimensions of uncertainty in environmental data is the problem of presenting the data clearly and with the ranges of variability, missing data, and conflicting values as well as the different layers of data. The use of geographic information systems (GIS) can speak to some of these problems and are discussed in Chapter 16 by Pesachowitz and Chapter 17 by Christian and Gauslin. They point out that protecting the environment is a job that is inherently geographic in nature and involves understanding the spatial relationships of contaminants and natural resources. These systems can be used to visualize and understand the interactions between media and to highlight the areas of environmental interest and concern. The use of advanced computer technology to manage, analyze, and display large volumes of spatially related data can provide rapid information retrieval and can display multiple data layers to help analysts identify spatial relationships. However, a bridge is needed between library science and information service so that information can be accurately and completely intercommunicated. This will require a standardization for such factors as language, feedback of information to further refine a search, consistency, format, search methodology, publicity and access, as well as hardware and software requirements.

Other aspects of uncertainty involve credibility and subjectivity. In Chapter 1, Gilbert and Simpson deal with these problems as well as that of assigning probability distributions using professional judgement. They are also concerned about statistical illiteracy in attempts to communicate with the public.

An organization that could centrally speak to many of the problems involved in uncertainties, and describing the state of the environment, is an effort currently happening at the U.S. EPA. It is called the Center for Environmental Statistics and is described in some detail in Chapter 22 by Niemann et al. The topics discussed there include a framework (a modification of the pressure-state-response framework) and an information system and sample products (including the *Guide to Selected National Environmental Statistics in the U.S. Government*, a *Sourcebook of Environmental Statistics* and an *Atlas of Selected Environmental Databases* and *Cross-Media Issues and Integrated Environmental Statistics Fact Sheets*).

UNCERTAINTY IN QUANTITATIVE RISK ASSESSMENTS AND MODELING

Quantitative risk assessment involves the introduction of numerous uncertainties into the description of the state of the environment. All models require some assumptions and too often the variability and error introduced in these estimations are large and not well documented and described.

Goddard et al. use Bayesian shrinkage estimators in the threshold determination in their discussion, Chapter 10, of measures of carcinogenic potency. They also include a historical review. Gilbert and Simpson deal with uncertainties of reconstructing doses from exposure to non-ionizing radiation in the 1940s using modeling based on the Monte Carlo probabilistic approach. Basu et al., in Chapter 20, addresses the problem of how to establish a safe level of exposure for cancer-causing substances or a standard safe dose in a standardized way.

Developmental bioassay data is extremely important, as pointed out in Chapter 9 by Gaylor. Determining the safe level for developmental toxicants is more difficult than that for carcinogens. Developmental effects are complicated by the chemical effects that happen in the pregnant mother and it is not clear if there is a threshold or not.

Schwartz, in Chapter 11, asks if we are worrying about the right things in the area of air pollution? Further, he discusses the importance of contrasting short- and long-term determinations of concentration and how to decide on which is best to focus. He thinks that perhaps it would be more cost effective to concentrate on the long term — we know a lot about the acute effects but need more information concerning the chronic effects. He observes that we are spending the least on preventing high levels of particulates, but that they are the most important from the human health perspective.

UNCERTAINTY IN GAUGING FUTURE ENVIRONMENTAL TRENDS

The development of future scenarios about the state of the environment is discussed in Chapters 13 through 15 by Boroush, Taylor, and Rejeski, respectively. Boroush characterized the specific uncertainty involved in this area by describing prediction of the future as a softer subject. He advocates a continuing outlook process involving trend scanning and monitoring (these can be both judgment and guesses), issue analysis, and scenario forecasting. In order to make decisions concerning the environment, knowledge of the statistical information and data is not enough, as we need to also know policy, public opinion, resources available, or, in general, the operating environment of the environmental community. He emphasized devopment of strategic vision, not master plans. Rejeski adds a broader vision of the use of future scenarios, including the ideas that the public wants to participate in policy developments and that planners are goal-oriented and not visionary.

Rejeski, in his chapter on images of the future, observes that quantitative risk assessment models and their predictions are not enough. We need to go the next step and combine this with visions of the future, which represent the ultimate uncertainty. He calls for a vision into the risk and priority process and further observes that risk and the circle of blame focuses on the negative outcomes—this encourages risk aversion and blaming someone. What is needed is to focus on the positive consequences.

CONCLUSIONS

Uncertainty in the assessment of the state of the environment is widespread and has a variety of meanings. The extent and meaning of uncertainty depends on who is involved, what their objectives are, what disciplines they represent, and what their viewpoint is. This variety is a characteristic of the field of environmental statistics and must be considered in any evaluation of the state of the environment.

One main goal in environmental statistics, assessment, and forecasting should be to reduce the uncertainties. Uncertainty, variability, and bias will always be present in any measurement of physical phenomena. They cannot be completely eliminated. However, by focusing on selective reductions in uncertainty a clearer picture can be achieved.

The other central characteristic problem that clearly emerges from reading the chapters in this volume is its incompleteness. This will only be addressed and solved with time and resources. Both will be needed to sample and collect data in a statistically meaningful way, analyze and integrate data that is often disparate, and present it in a manner that can be understood and used by a wide variety of people. The current volume will be a useful step in that process.

REFERENCES

1. U. S. Environmental Protection Agency, *Reducing risk: setting priorities and strategies for environmental protection*, Science Advisory Board, SAB-EC-90–021, Washington, D.C., September 1990.
2. Crawford-Brown, D., University of North Carolina, Department of Environmental Sciences and Engineering, Chapel Hill, NC, private communication.

SECTION I

Combining Data and Data Uncertainty

N. Phillip Ross

INTRODUCTION

Data and information concerning the environment come in a multitude of forms which all too often are incompatible with each other, anecdotal or fragmentary, and nonrepresentative of the time or space they purport to describe. In addition, the data often involve the use of inadequate measuring devices that leave too many values below the detection level or of questionable accuracy. Some data are so old that the methodology cannot be reconstructed. Too often the data were collected without a sampling plan or with one that is seriously flawed. A myriad of problems arise in the process of combining such data and information, and each one requires considerable thought and development of techniques to combine and deal with the wide range of uncertainties.

There are several methodologies and approaches that can be used or further developed to deal with the diversity of data and information that exists concerning the environment. However, it should be emphasized that any environmental measurement must be conducted under a statistically developed sampling plan to be fully credible. To develop such a plan the data quality objectives (DQOs) first must be described. These include what accuracy and precision are required, how complete and representative the data need to be, what level of statistical error must be achieved, and what confidence level is required. The DQOs are determined by the use to which the data are to be put, the funds available, and the feasibility of the measurements. Perhaps the most important problem involved in determination of the state of the environment is that too much of the existing data have been collected under a sampling plan that cannot ensure its quality and compatibility with other measurements.

Two major elements needed in any attempt to describe the state of the environment are the methods for combining data and the techniques for dealing with the uncertainty in the data and information. Uncertainty

13

comes in many shades. There is uncertainty in simply not knowing. For example, Chapter 1, the problem of reconstructing radiation doses received by people in the 1940s and 1950s is addressed, and that is so long ago that many of the details are forgotten. In many cases concerning environmental statistics the data are not known and must be guessed at by using professional judgment. Uncertainty also exists in a quantitative way in that all measurements have a limit to the accuracy and precision achievable. Often the quality achieved is adequate for immediate needs but presents a problem when data are used later for another purpose. Several of the characteristics and problem areas in the determination of the quantitative uncertainty are described in Chapter 1. In addition uncertainty generally includes a subjective element, although its determination is essential for credible data. The importance of the parameters involved can be tested by sensitivity analysis. However, presenting the data in an understandable form involves some unique problems, especially when dealing with the media and the public.

Cost is a significant problem in the development of sampling protocols and the collection of data. In an effort to reduce costs and not substantially reduce the quality of data determined, methods have been developed to combine data. Some of these techniques are described in Chapters 2 and 3. Several problem areas are explored including: the impact of rapidly changing technology, the need for a broader view over life changes of living things (humans, plants, microlife) and contaminants (source, transport, and fate), the large variability in data form and quality, the prevalence of soft data, the multitude of disciplines involved (and hence conflicting nomenclature and meanings of words), and a public that is not educated in the quantitative skills needed to understand environmental statistics. The several approaches described include: compositing data and cluster, composite, and transect sampling.

Combining data presents a range of difficult problems, and some of these are discussed in Chapter 4. The authors describe the use of the Bayesian approach in combining data from the laboratory and field measurements, and the addition of verification data and expert opinion. This methodology can also be used to project into areas where there are no data.

One of the most contentious areas encountered in combining data is the existence of data below the detection level. The most common technique is to use the "best value" which can be zero, one half the detection limit, the detection limit, or some other fraction of the detection limit. In Chapter 5 this area is reviewed and a new approach is presented that involves the use of a weighted regression method that incorporates the relationship between detection limit and precision of the data. The authors illustrate this method using water quality monitoring data collected by the Chesapeake Bay program. Their unique approach uses the relationship between the uncertainty and the detection limit.

The problem areas and methods described in this section are only some of the major ones involved in combining data in order to describe the state of the environment. In any attempt to combine all the data it quickly becomes clear that little of it was collected for that purpose. In order to achieve the overall goal of describing the state of the environment, methods and techniques will have to be developed to make maximum use of the existing data and to encourage that future measurements are conducted in a manner to be of the greatest value.

CHAPTER 1

Uncertainty Issues of the Hanford Environmental Dose Reconstruction Project

Richard O. Gilbert and Jeanne C. Simpson

ABSTRACT

The staff of the Hanford Environmental Dose Reconstruction (HEDR) project is estimating (reconstructing) the doses received by persons exposed to past emissions of radioactive materials in air and water from the Hanford site nuclear facilities in southeastern Washington State. A major objective of the project is to quantify the uncertainties of the reconstructed doses. In this chapter the statistical and uncertainty issues associated with the HEDR project will be discussed, particularly with regard to reconstructing doses to the thyroids of residents who were exposed to iodine-131 in air effluents from the Hanford site in 1945–1947 (the time period when the largest releases of iodine-131 occurred). Reconstructing doses that occurred up to 47 years ago is fraught with problems such as (1) validating spatial-temporal environmental transport and dose models when few historical data exist for that purpose, (2) developing data quality objectives for a retrospective study, (3) relying on expert opinion (subjective, "soft" information) to specify model parameter values and their uncertainties, (4) evaluating uncertainties about models themselves, (5) conducting research in an open public forum, and (6) communicating complex uncertainty results to the public.

INTRODUCTION

The goal of this book is to examine the current state of environmental data and their statistical analysis, as well as the modeling of environmental systems. This chapter addresses this goal from the perspective of a large computer-simulation modeling effort included in the Hanford

Environmental Dose Reconstruction (HEDR) project. The staff of the HEDR project is estimating (reconstructing on the basis of historical information) the doses received by specific individuals and groups of demographically similar individuals who were exposed to radionuclides in air and/or water effluents from the Hanford site plutonium production facilities in southeastern Washington State beginning in 1944. The HEDR project began in 1987 after data became available in 1986 from the U.S. Department of Energy that indicated large amounts of iodine-131 (a by-product of plutonium production) had been released into the air at Hanford, particularly in the period from 1945 to 1947. The releases were large enough to suggest that residents who were exposed to iodine-131, particularly infants and small children, could have received large radiation doses to the thyroid by consuming fresh milk from cows that ate pasture or feed contaminated by iodine-131. Radionculides were also released into the Columbia River via the water used to cool the nuclear reactors. The largest quantities of radionuclides probably entered the river during the 1960s when plutonium production was at a peak.

The HEDR project and a companion study, the Hanford Thyroid Disease study (HTDS),[1-3] were initiated primarily in response to public pressure to assess potential adverse health effects from exposures to radionuclides in air and water emissions from Hanford. The HTDS — which is being conducted by the Fred Hutchinson Cancer Research Center in Seattle, Washington, in collaboration with the U.S. Centers for Disease Control — is evaluating whether the incidence of thyroid disease has increased among persons exposed as infants and young children to releases of Hanford iodine-131 into the air in 1945–1957. In support of the HTDS, the HEDR project is reconstructing the doses to the thyroid that resulted from exposure to iodine-131. However, the scope of the HEDR project is larger than estimating thyroid doses, and includes estimating doses to the body from other important radionuclides released from Hanford into the air and the Columbia River.

Battelle, Pacific Northwest Laboratories is conducting the HEDR project under the direction of an 18-member independent Technical Steering Panel (TSP), which is composed of scientists and engineers as well as representatives of the states of Washington, Oregon, and Idaho, native Americans; and the public. Quarterly scientific public meetings are held in different cities throughout the study region. Members of the general public and organized public interest groups regularly participate in the meetings.

RECONSTRUCTING IODINE-131 DOSES FOR THE AIR PATHWAY

Estimating doses to the thyroid from iodine-131 that reached persons via the air pathway will be accomplished using the air pathway dose computer model HEDRIC (Hanford Environmental Dose Reconstruction Integrated Codes). The HEDRIC model consists of four separate models (computer codes) that have well-defined links for transferring information from one model to another.[4] The models, which must be executed in sequence, are the source term model (STRM), atmospheric transport model (RATCHET), environmental accumulation model (DESCARTES), and the dose model (CIDER). The geographic domain of the RATCHET model consists of 20 counties in eastern Washington State, 7 counties in north central Oregon, 6 counties in northern Idaho, and the portions of the Nez Perce Indian Reservation that project beyond these counties. Currently, plans are to estimate doses for a 19-county area that is more or less in the center of the RATCHET model domain.

The source term model (STRM), which has about 10 uncertain parameters, is used to reconstruct on an hourly basis the historic releases of iodine-131 into the air from Hanford separations facilities. The iodine was produced along with other fission products (such as krypton-85) and plutonium when uranium fuel rods were irradiated in Hanford nuclear reactors. When the irradiated fuel rods were dissolved, the iodine was released from solution and discharged to the air through high stacks. The estimates of hourly iodine-131 releases, as well as hourly meteorologic data such as wind speed and wind direction, are used as input to the atmospheric Lagrangian-trajectory, Gaussian puff dispersion model called RATCHET.[5] This model is used to estimate (reconstruct) the concentration of iodine-131 in air and the amount deposited at each of approximately 2100 locations spaced 6 mi apart on a square grid pattern over the study domain. RATCHET has about 10 parameters with values that change over space and time.

The outputs of the atmospheric transport model are inputs to the environmental accumulation model DESCARTES (Dynamic EStimates of Concentrations and Accumulated Radionuclides in Terrestrial Environments)[6] which is used to calculate time-dependent estimates of iodine-131 concentrations in soil, plants, and animal products. This model, which operates on a daily time step and has about 20 uncertain parameters, is used to estimate the historic amount of iodine-131 in significant environmental media such as milk and leafy vegetables at the appropriate grid nodes. These spatially and temporally varying environmental iodine-131 concentrations correspond to the air and surface ground concentrations obtained using the air transport model. A commercial foods submodel accounts for the fact that agricultural products produced at one location may be transported to different locations before they are consumed by humans or fed to livestock. Time delays between harvest and consumption are important because iodine-131 has a short (8-day) half-life, i.e., half of the iodine-131 present at any given moment decays away in 8 days.

The CIDER dose model has about 20 uncertain parameters, many of which have values that change over time. Dose estimates will be made at a maximum of 2100 grid nodes, the same nodes at which RATCHET estimates ground depositions and air concentrations of iodine-131 and at which DESCARTES calculates the concentration in plants and animal products. However, dose estimates for a given individual may be made at fewer than 2100 grid nodes because it is unlikely that an individual resided or traveled in the vicinity of all grid nodes. Also, during screening calculations, it may be shown that dose estimates at nearby nodes in the outer regions of the study region are very similar, in which case, dose estimates at all nodes may not be computed.

Current plans are to repeat the above modeling process 100 times to generate 100 realizations of daily environmental concentrations of iodine-131 at each grid point during the 1945–1949 time period, which includes the period of largest iodine-131 releases. More than 100 realizations may be used if the time and cost required to generate realizations can be sufficiently reduced. For each realization, a new value for each parameter in the model is randomly generated from its specified probability density function (pdf). These parameter pdfs, which represent the uncertainty in our knowledge of the true parameter values, are being developed from either the scientific literature, via elicitation of expert opinion and/or from actual measurements of the parameter. The parameter values are generated using simple random sampling or Latin Hypercube sampling.[7-9] For each iteration of the process, all information from each model is saved and passed on directly to the next model. The information from each model is not summarized in any way so that all spatial and temporal associations are preserved. Hence, the explicit and implicit correlations among the models and model parameters are retained and propagated through all four models.

The set of 100 realizations of environmental concentrations will be generated only once and stored for repeated use. Each realization is interpreted as a potential (plausible) set of environmental iodine-131 concentrations (over time and space) to which specific individuals were exposed in varying degrees depending on what their lifestyles were and on when and where they lived in the area. For each specific individual living in the study area, each realization of environmental concentrations is used as input to the dose computer code, Calculation of Individual Doses from Environmental Radionuclides (CIDER). In this way, 100 possible (plausible) estimates of thyroid dose are obtained for each specified individual. These 100 dose estimates can be used to quantify the uncertainty in the individual's dose resulting from a lack of knowledge about true parameter values. Uncertainty in dose estimates resulting from lack of knowledge about the mathematical form of the model is assessed using model validation and other techniques, as discussed below. Sensitivity analyses are also being conducted to determine which pathways and model parameters contribute the most uncertainty (resulting from lack of knowledge) about the true dose, as discussed below.

The 100 estimates of dose will be summarized by computing standard summary statistics such as the mean, median, geometric mean, range, standard deviation, and geometric standard deviation of the estimates, as well as various percentiles. In addition, the set of 100 dose estimates will be plotted as a pdf and the associated cumulative distribution function (cdf). The interpretation of this cdf, making use of the computed 5th and 95th percentiles for illustration purposes, is as follows: there is a subjective 90% chance that the true value of the iodine-131 air-pathway dose to the thyroid of this specific individual is between the 5th and 95th percentiles.

In addition, one-sided and two-sided distribution-free statistical tolerance limits[9] for specified percentiles will be constructed using the minimum and maximum of the 100 dose estimates.

The dose cdf is subjective because elicitation of expert opinion was required to specify values of many of the model parameters and because the various models are based on the best professional judgment. The deterministic predicted (DP) dose (the dose obtained when the central value, usually the mean or median, of each parameter is used in the model) will also be computed and plotted on the cdf. The ratios (DP dose):(5th percentile) and (95th percentile):(DP dose) may be computed and used as multiplicative measures of lack of knowledge in the dose received by the individual.

STATISTICAL AND UNCERTAINTY ISSUES

The remainder of this chapter discusses some of the uncertainty and statistical issues associated with the approach outlined above. These issues are grouped under the following headings: technical, policy, communication, and legal.

TECHNICAL ISSUES

Data Quality Objectives

It is well known that data quality objectives (DQOs),[10-13] as well as criteria for assessing their attainment, should be developed during the planning stages of environmental field studies for which measurements of pollutant concentrations in soil, air, water, or biota will be made. DQOs specify the quality of the data required. They are typically described in terms of accuracy, precision, representativeness, completeness, and comparability.[13] Even though the HEDR project staff is not collecting and measuring new environmental samples, the TSP has mandated that DQOs must be specified in all project work plans. Unfortunately, it is not always clear how DQOs should be specified for retrospective studies such as the HEDR project. To reconstruct doses, the emphasis is on collecting data and information by searching historic records and by talking to people who can supply information about diets, lifestyles, and farming practices of the 1940s.

Consider the "accuracy" DQO. The accuracy of reconstructed doses for specific individuals among the general public in the mid-1940s cannot be directly assessed because their thyroids were not measured for iodine at that time (although some radiation measurements of Hanford workers were made[14]). Nevertheless, it is vital that efforts be made to evaluate or validate the accuracy of reconstructed doses. Some validation methods being used by the HEDR project staff are discussed below. The accuracy of dose reconstruction that can actually be achieved by the HEDR project will not be known until the project is essentially completed.

The "precision" DQO for the dose received by a specific individual is captured in part by the width of the subjective 90% interval stated above, which depends on the lack-of-knowledge uncertainty in model parameter values. However, it is not meaningful at the beginning of the HEDR project to specify the width of the 90% interval that should be achieved, because the quantity and quality of historical data and information are not known at that time. The width of the interval will decrease as the uncertainty of each parameter value decreases. This decrease in uncertainty may occur for some parameters if additional historic information is uncovered. However, some data and information needed to reduce parameter uncertainties (e.g., wind speed and direction data at critical locations) may never have been collected. Also, there are limits to the amount and quality of information that can be obtained by eliciting information from experts or by surveying persons at the present time. Toward the end of the project, the pdfs of important parameters can be changed and the uncertainty analyses repeated to see how much the width of the dose 90% interval changes. This exercise may provide information about the potential benefits that could be gained by conducting additional search of historic documents.

For the HEDR project, the DQOs for representativeness, completeness, and comparability apply to historic data, data searches, computer codes, and models.[15] The DQO for representativeness of historic data can be expressed in terms of acceptable sampling designs, analytical measurement methods, detection limits, and data analysis procedures. The DQO for completeness of the data/information record should take into consideration the possible effect on the accuracy and precision of reconstructed doses (and important intermediate model predictions) of alternative completeness levels. However, this evaluation is highly uncertain itself. The degree of completeness required of critically important parameters should, in general, be greater than that for less important parameters. The DQO for comparability is a qualitative statement expressing the degree of agreement among sets of data, parameter values, computer codes, or other components of the modeling effort. The attainment of the DQOs for representativeness, completeness, and comparability is assessed via independent peer reviews by acknowledged experts.

Eliciting and Analyzing Expert Opinion ("Soft" Information)

For many models and model parameters, the HEDR project staff must elicit information from experts to supply missing data and information. Unfortunately, there are few such experts, they are sometimes difficult to identify and locate, and lapses of memory result in missing or biased information. The methods used to elicit information must be carefully selected and the results properly analyzed to ensure that reliable information is obtained.[16-18] A consequence of using expert opinion is that the dose cdf and 90% intervals are subjective. Hence, the credibility of the dose cdf depends on the credibility of experts and the methods used to elicit and analyze that information.

There are several methods for eliciting information. The HEDR project staff relies primarily on self-assessment and/or informal solicitation[17] of experts not connected with the project to supplement information from published reports and journal papers. However, for critical parameters (those for which increased uncertainty substantially affects the variability of the dose estimates), probability encoding[17,19] may be used to elicit—in a structured and self-consistent manner—a technical expert's assessment of the pdf of a parameter. This pdf represents the expert's lack-of-knowledge uncertainty about the historic value of the parameter. Although probability encoding is the most systematic and defensible approach to developing subjective probability assessments, it is also the most expensive.

Evaluating the Validity, Uncertainty, and Sensitivity of Dose Estimates

A model is valid when model predictions agree sufficiently well with actual measurements of the predicted quantity.[9,20-22] Validation efforts will focus on validating the source term, air transport, and environmental accumulation models (STRM, RATCHET, and DESCARTES, respectively). A complication is that the air pathway dose model (CIDER) predicts doses for many different combinations of lifestyles, times, and residence histories. It is perhaps unrealistic to expect the model to be equally valid for all of these combinations.

Thus far, only a few data sets have been found that would be useful for validation purposes. One historic data set is gross beta measurements of sagebrush leaves and twigs collected throughout the Hanford environs beginning in 1945. The spatial patterns of these measurements, after they are converted to iodine-131 equivalents (with uncertainty), can be compared with vegetation iodine-131 concentration patterns that are predicted by the environmental accumulation model DESCARTES.[6] Also, the dispersion part of the air transport model (RATCHET) can be validated using isotopes such as krypton-85 (a fission product produced in the nuclear reactors) for which more recent data exist.[23] However, this effort does not validate the deposition portion of the model. Consideration is also being given to partially validating the STRM, RATCHET, and DESCARTES models using environmental iodine-131 measurements that were made during Hanford iodine-131 releases to the air that occurred in 1949 and 1951.

The project is also participating in the Validation of Models for Radionuclide Transfer in the Terrestrial, Urban, and Aquatic Environments (VAMP) model-intercomparisons study of the International Atomic Energy Agency.[24] A valuable supplement to formal validation efforts are peer reviews by acknowledged experts. Such reviews have been, and will continue to be, an important component of the HEDR project.

The approach being used to conduct the parameter uncertainty analyses for specific individuals was described above. Sensitivity analyses will be conducted to identify the most important parameters and pathways. Present plans are to conduct sensitivity analyses first for the dose model and then for the environmental accumulation model, the air transport model, and source term model, in that order. Analyses will be conducted for different combinations of lifestyles, age groups, and residence locations because the important parameters may change for the different combinations. For model outputs that are monotonic in response to inputs, partial correlation coefficients and partial rank correlation coefficients obtained using regression/correlation methods[25] will be used to rank parameters with respect to their contribution to the uncertainty in the model. The sensitivity of model segments that give nonmonotonic responses will be evaluated using methods that are not yet developed. In addition, methods to rank parameters of models that give nonmonotonic responses need to be developed. Sensitivity analyses will also be conducted to assess the amount and type of change in an individual's dose cdf that occurs when models and pdfs of model parameters are changed.

Meaning of Dose pdfs and cdfs

A potential problem with the Monte Carlo (computer simulation) parameter uncertainty analysis described above is that the meaning of the resulting pdf and cdf of dose may not be well understood or explained. Their meaning depends on the question being asked.[9,26] If the question relates to the dose that a specific individual received, then using the central value for each parameter in the model provides a deterministic answer (the calculated deterministic predicted dose) to that question. The cdf of dose obtained by a parameter uncertainty analysis for this case gives our uncertainty in the true dose for the individual that results from lack of knowledge about the correct parameter values to use. This interpretation is appropriate for the cdf of dose obtained for specific individuals by the HEDR project.[9]

If the question being asked refers to doses for a population group, then the answer is expressed in the form of a distribution, not a single deterministic value. This distribution of doses, which is obtained from computer simulations, arises because of differences in the value of input parameters from person to person within the group. The distribution estimates the variability in dose among persons of the group. For this case, the specified pdfs of the model parameters must characterize this among-person variability.

POLICY ISSUES

Two policy questions associated with assessing uncertainty of model predictions are: "to what extent should lack-of-knowledge uncertainty or variability be taken into account?" and "when is it appropriate to conduct uncertainty and sensitivity analyses of a complex or lengthy model using a relatively simple analytical approach (such as truncated Taylor Series) rather than the more accurate Monte Carlo computer simulation approach?" Some of the factors that are considered by the Technical Steering Panel and HEDR project staff when considering these questions are project goals, DQOs, public expectations, credibility, and cost.

COMMUNICATION ISSUES

To build and maintain credibility with members of the public, the TSP takes care to keep the public informed of progress, to invite their participation at all stages of work, and to provide results in language that is understandable by a general technical audience. Effective communication of project methods and results is crucially important to establishing and maintaining the public's confidence. Effective communication requires that technical words and jargon be effectively explained and interpreted. Even communicating with other scientists can be a challenge if they are not familiar with statistical terminology and methods. Conducting science in a public forum can also raise unnecessary concerns about the quality of the effort if the public does not understand that midcourse corrections are part of the normal course of events in complex research projects.

Another issue is the appropriate role for statisticians in large multidisciplinary study teams. For the HEDR project, statisticians play a significant role in making day-to-day technical decisions. Although the Statistics Task was not established for several months after the project began, the task has evolved over a 4-year period to the point where statisticians are involved in all technical aspects of the project, with particular emphasis on assessing model reliability (validity, uncertainty, and sensitivity).

LEGAL ISSUES

Statisticians should have some experience dealing with legal matters pertaining to the use and interpretation of statistics and uncertainty analyses, particularly for environmental studies for which litigation is a common occurrence. Indeed, Battelle is currently under subpoena to retain and supply on demand all HEDR project records for use in litigation already initiated by persons who consider their health problems to have been caused by exposure to radiation released from Hanford. The possibility of litigation motivates increased attention to quality control in computer code development and implementation, statistical analyses, records management, and documentation of what was done and why.

SUMMARY

This chapter has briefly discussed some of the statistical and uncertainty issues that are important to the HEDR project and to the wider range of environmental sampling programs. We believe that statisticians must prepare themselves to be effective leaders and communicators of statistical concepts, methods, and interpretations in multidisciplinary environmental studies. The statistician needs to develop and assess the attainment of statistically meaningful DQOs; to ensure that expert opinion is properly elicited, interpreted, and used; to guide other researchers in the type of validation, uncertainty, and sensitivity analyses that are appropriate to use given the questions being addressed; to ensure that computer simulation results are properly interpreted and communicated, and to be prepared to give legal testimony regarding statistical and uncertainty aspects of the study. These needs are present for current environmental sampling programs, just as much as they are for retrospective studies such as the HEDR project.

ACKNOWLEDGMENT

The HEDR project is funded by the U.S. Centers for Disease Control by a grant from the U.S. Department of Energy.

REFERENCES

1. Cate, S., A. J. Ruttenber, and A. W. Conklin. "Feasibility of an Epidemiologic Study of Thyroid Neoplasia in Persons Exposed to Radionuclides from the Hanford Nuclear Facility Between 1944 and 1956," *Health Phys.* 59:169–178 (1990).
2. David, S., K. J. Kopecky, T. E. Hamilton, and B. Amundson. "Hanford Thyroid Disease Study Protocol," Fred Hutchinson Cancer Research Center, Seattle, WA (April 1990).
3. Gilbert, R. O., J. C. Simpson, B. A. Napier, H. A. Haerer, A. M. Liebetrau, A. J. Ruttenber, and S. Davis. "Statistical Aspects of the Hanford Environmental Dose Reconstruction Project and the Hanford Thyroid Disease Study," *Radiat. Res.*124(3):354–355 (1990).
4. Napier, B. A., J. C. Simpson, T. A. Ikenberry, R. A. Burnett, and T. B. Miley. "Final Design Specification for the Environmental Pathways and Dose Model," Pacific Northwest Laboratory, Richland, WA, PNL-7844 HEDR (1992).
5. Ramsdell, J. V., Jr. and K. W. Burk. "Regional Atmospheric Transport Code for Hanford Emission Tracking (RATCHET)," Pacific Northwest Laboratory, Richland, WA, PNL-8003 HEDR (February 1992).
6. Ikenberry, T. A., R. A. Burnett, B. A. Napier, N. A. Reitz, and D. B. Shipler. "Integrated Codes for Estimating Environmental Accumulation and Individual Dose from Past Hanford Atmospheric Releases," Pacific Northwest Laboratory, Richland, WA, PNL-7993 HEDR (February 1992).
7. McKay, M. D., R. J. Beckman, and W. J. Conover. "A Comparison of Three Methods for Selecting Values of Input Variables in the Analysis of Output from a Computer Code," *Technometrics* 21(2):239–245 (1979).
8. Iman, R. L. and M. J. Shortencarier. "A FORTRAN 77 Program and User's Guide for the Generation of Latin Hypercube and Random Samples for Use with Computer Models," Sandia National Laboratories, Albuquerque, NM, NUREG/CR-3624 (1984).
9. IAEA. "Evaluating the Reliability of Predictions Made Using Environmental Transfer Models," International Atomic Energy Agency, Vienna, Safety Series No. 100 (1989).

10. "Data Quality Objectives for Remedial Response Activities: Development Process," U.S. EPA Report EPA/540/G-87/003 (March 1987).
11. "Data Quality Objectives for Remedial Response Activities: Example Scenario," U.S. EPA Report EPA/540/G-87/004 (March 1987).
12. Neptune, D., E. P. Brantly, M. J. Messner, and D. I. Michael. "Quantitative Decision Making in Superfund: A Data Quality Objectives Case Study," *Hazardous Mater. Control* 3(3):18–27 (1990).
13. Barth, D. S., B. J. Mason, T. H. Starks, and K. W. Brown. "Soil Sampling Quality Assurance User's Guide," 2nd ed., U.S. EPA Report EPA 600/8–89/046 (March 1989).
14. Ikenberry, T. A. "Evaluation of Thyroid Radioactivity Measurement Data from Hanford Workers, 1944–1946," Pacific Northwest Laboratory, Richland, WA, PNL-7254 HEDR (May 1991).
15. Shipler, D. B. "FY 1992 Task Plans for the Hanford Environmental Dose Reconstruction Project," Pacific Northwest Laboratory, Richland, WA, PNL-7757 HEDR Rev. 1 (April 1992).
16. Hora, S. C. and R. L. Iman. "Expert Opinion in Risk Analysis: The NUREG-1150 Methodology," *Nucl. Sci. Eng.* 102:323–331 (1989).
17. Roberds, W. J. "Methods for Developing Defensible Subjective Probability Assessments," Transportation Research Record, No. 1288, Transportation Research Board, National Research Council, Washington, D.C. (1990).
18. Meyer, M. A. and J. M. Booker. *Eliciting and Analyzing Expert Judgement, A Practical Guide*, Academic Press, New York (1991).
19. SRI. "SRI Probability Encoding Manual," SRI International, Menlo Park, CA (1979).
20. Brier, G. W. "Statistical Questions Relating to the Validation of Air Quality Simulation Models," Environmental Protection Agency, Research Triangle Park, NC, EPA-650/4–75–010 (March 1975).
21. Londergan, R. J. and N. E. Bowne. "Validation of Plume Models, Statistical Methods and Criteria," Electric Power Research Institute, Palo Alto, CA, EA-1673-SY, Research Project 1616–1 (January 1981).
22. Gilbert, R. O. "Assessing Uncertainty in Pollutant Transport Models," *TRANSTAT: Statistics for Environmental Studies*, Number 27, Pacific Northwest Laboratory, Richland, WA, PNL-SA-12523 (September 1984).
23. Ramsdell, J. V., Jr. "Atmospheric Transport and Dispersion Modeling for the Hanford Environmental Dose Reconstruction Project," Pacific Northwest Laboratory, Richland, WA, PNL-7198 HEDR (July 1991).
24. IAEA. "Progress Report Number 3 on the IAEA/CEC Coordinated Research Programme on Validation of Models for the Transfer of Radionuclides in Terrestrial, Urban and Aquatic Environments and Acquisition of Data for that Purpose," International Atomic Energy Agency, Vienna (1991).
25. Iman, R. L., M. J. Shortencarier, and J. D. Johnson. "A FORTRAN 77 Program and User's Guide for the Calculation of Partial Correlation and Standardized Regression Coefficients," Sandia National Laboratories, Albuquerque, NM, NUREG/CR-4122 (1985).
26. Hofer, E. "On Some Distinctions in Uncertainty Analysis," *Proceeding of the CEC/USDOE Workshop on Uncertainty Analysis*, Pacific Northwest Laboratory, Richland, WA, PNL-SA-18372, pp. 29–30 (1990).

CHAPTER 2

Decreased Sampling Costs and Improved Accuracy with Composite Sampling

Steven D. Edland and Gerald van Belle

ABSTRACT

Composite sampling (i.e., pooling samples before making measurements) reduces the cost of a sampling program in terms of time and/or laboratory costs without substantially reducing the effective sample size. For many sampling programs, the resources freed by this reduction in cost can be used to collect more field samples prior to compositing, increasing the effective sample size achieved and improving the accuracy of population estimates and the overall effectiveness of the sampling program. This important sampling method has applications in field ecology, ecosystem monitoring and assessment, and hazardous waste site assessment. This chapter reviews the composite sampling literature with regard to these applications. A more general algorithm for estimating the population mean and variance from composite samples that allows for an arbitrary (potentially unequal) number of field samples per composite sample is presented, and hypothesis tests and power calculations for composite sampling plans are described.

INTRODUCTION

Composite sampling, defined as the pooling of field samples prior to measurement or laboratory analysis, is a simple and straightforward method of enhancing sampling programs. However, composite sampling methods have seen only limited application in the environmental sciences, perhaps because their treatment in statistical and environmental sampling texts has (with notable exceptions[1-3]) been sparse. Much of the statistical literature on composite sampling deals instead with agricultural and industrial applications and their attendant problems. Examples include sampling shipments of bales of wool for quality in Australia[4,5] and sampling the continuous output of industrial and mining processes.[6] More recently we have seen published applications of composite sampling in the field of environmental science, including sampling for plankton densities in aquatic samples[7,8] and sampling for contamination at hazardous waste sites[9] and in groundwater supplies.[10] Rohde[3] includes a review of published examples of composite sampling.

The intent of this chapter is to review composite sampling from the perspective of the environmental sciences, and to provide some motivation for why and how composite sampling can improve the efficiency of an environmental sampling program. Special attention is given to the simple but important case of sampling to estimate the mean and variance. The general applicability of this estimation problem is suggested by the fact that almost every environmental sampling program is concerned to some extent with the estimation of the mean and variance of one or more environmental parameters. Statistical aspects of this application are also relatively easy to describe, making it a good example for providing some motivation for the composite sampling approach. It is hoped that this presentation will lead to greater understanding and accessibility of this potentially important, and perhaps underutilized, approach to sample design.

Some Useful Nomenclature and Notation

Composite sampling involves the pooling of smaller samples into larger samples on which measurements are to be made. The samples to be pooled have been referred to in the literature variously as experimental units, elements, increments, members, scoops, or other names depending on the application being discussed or the historic perspective of the authors. The pooled samples have been referred to variously as bulk, group, or composite samples. For simplicity and for clarity in regard to the environmental applications, we will refer throughout this chapter to the samples to be pooled as **field samples** and to the samples on which measurements are to be made as composite **laboratory samples** or simply as composite samples. Typically we are interested in an (unobserved) property of the field samples: X, say, based on measurements of this property, Y, say, made on the laboratory samples. This will be characterized formally in the section to follow. Unless otherwise noted, we assume throughout that the laboratory samples are measured with negligible error. We also assume that the field samples are independent and are the same size.

COMPOSITE SAMPLING FOR THE MEAN AND VARIANCE

Composite Sampling Defined

Suppose that X_{ij}, $i = 1, 2, \ldots, I$, $j = 1, 2, \ldots, n_i$, $(E(X) = \mu_x$, $V(X) = \sigma_x^2)$ are independent, identically distributed outcomes from a population of interest. X_{ij} refers to the potentially unrealized values of field samples in an environmental survey. In many environmental applications the goal of a sampling program is to estimate the mean and variance. Suppose, however, that the X_{ij} are not observed, but composite values

$$Y_i = \frac{1}{n_i} \sum_{j=1}^{n_i} X_{ij}$$

are observed instead. As an example, if X_{ij} were the concentrations of a trace metal in Σn_i field samples drawn at random from a body of water and pooled into I composite samples, then Y_i, $i = 1, 2, \ldots, I$, would be the concentrations of trace metal in the resulting composite samples. We now describe estimators for the population parameters of the distribution of X based on the observed values Y_i. General algorithms allowing for an arbitrary, potentially unequal, number of field samples per composite sample are presented. This level of generality is required when the compositing ratio is disrupted for some reason, for example, when field samples are lost or damaged prior to compositing, or when samples are being composited in the field and field procedures need to be simplified.

Estimating μ_x and σ_x^2

It is convenient to present estimators of μ_x and σ_x^2 using the results of linear regression analysis (e.g., Reference 11, pp. 42–65). Because

$$EY_i = \frac{1}{n_i} \sum_{j=1}^{n_i} EX_{ij} = \mu_x$$

and

$$Var(Y_i) = \frac{1}{n_i^2} \sum_{j=1}^{n_i} Var(X_{ij}) = \frac{1}{n_i}\sigma_x^2,$$

we can write $Y_i = \mu_x + \epsilon_i$,

where μ_x = expectation of X, and
              ϵ_i = the "error" term containing the random variation of
              Y_i about μ_x.

The distribution of ϵ_i is dependent on σ_x^2 and n_i. In particular, $E(\epsilon_i) = 0$ and $Var(\epsilon_i) = \sigma_x^2/n_i$. The least-squares estimator of μ_x is obtained by the weighted least-squares regression procedure with weights n_i, $i = 1, 2,$ ..., I. The weighted least-squares estimator of μ_x is

$$\hat{\mu}_x = \frac{\sum\limits_{i=1}^{I} n_i Y_i}{\sum\limits_{i=1}^{I} n_i}.$$

For $n_i \equiv n$, $\hat{\mu}_x$ further reduces to

$$\hat{\mu}_x = \frac{1}{I} \sum\limits_{i=1}^{I} Y_i.$$

We note that

$$E\hat{\mu}_x = \frac{\sum\limits_{i=1}^{I} n_i E Y_i}{\sum\limits_{i=1}^{I} n_i} = \mu_x.$$

By Theorem 3.2 and Theorem 3.6, corollary 2, of Seber,[11] $\hat{\mu}_x$ is the best linear unbiased estimator of μ_x. In particular, it is better than the naive estimator obtained by the unweighted average of the Y_i's. Kussmaul and Anderson[12] and Edelman[13] develop similar estimators for multiple-stage nested designs.

The estimator

$$\hat{\sigma}_x^2 = \frac{1}{I-1} \sum\limits_{i=1}^{I} n_i (Y_i - \hat{\mu}_x)^2$$

is the corresponding unbiased estimator of $\sigma_x{}^2$. $\hat{\sigma}_{\bar{x}}^2$ estimates the population variance relative to the field sampling unit used, and should always be reported in this way. The same caution holds when interpreting variance estimates. For example, increasing the size of the field sampling unit decreases the sample to sample variability, $\sigma_x{}^2$, so that care should be taken to ensure that the same field sampling units have been used when comparing $\hat{\sigma}_x{}^2$ from study to study. Numerical examples estimating μ_x and $\sigma_x{}^2$ are presented in the next section.

Unlike estimates of μ_x, estimates of $\sigma_x{}^2$ are not always improved by increasing the number of field samples through composite sampling. The variance of estimates of $\sigma_x{}^2$ is unaffected by compositing when sampling from a normal distribution. It is improved for leptokurtic distributions (e.g., the log distribution) and is made worse for platykurtic distributions. This is because the variance of $\hat{\sigma}_x{}^2$ involves third- and fourth-order moments. For the normal distribution these are 0. (See Appendix 1.)

Additional sources of variance may be suspected in different applications. For example, industrial and agricultural applications have been concerned with modeling additional sources of variability in a sample population, such as shipment-to-shipment variability, or "lot-to-lot" variability. This could correspond to lake-to-lake or site-to-site variability in environmental applications. Approaches along the lines of variance components models have been applied to these situations (for discussion see, e.g., References 4, 5, 7, 12, and 13).

Example Calculation of $\hat{\mu}_x$ and $\hat{\sigma}_x{}^2$

This section presents two examples of composite sampling suggestive of the range of possible applications of this technique. Sample data illustrating the calculation of $\hat{\mu}_x$ *and* $\hat{\sigma}_x{}^2$ are provided.

Example 1: Field Measurements

A random sample of smolting pink salmon fry are captured in a weir trap. We are interested in estimating the distribution of the mass (in grams) of the smolting fry, but it is impractical to weigh individual fry for several reasons: excessive handling may stress the fry, it is especially tedious to weigh individual fry under field conditions, and field measurement equipment may not be sensitive in the range of values expected for individual fry (smolting pink salmon fry weigh from 0.2 to 0.5 g). By compositing the fry into groups of approximately 100 prior to measurement, we substantially reduce handling of fry and field effort; and we bring the mass of the composite samples to within the effective range of field equipment. Sample data are given in Table 1. In this example, Y_i would correspond to the mean mass for the ith composite sample. We

Table 1.   Mean Mass of Composite Samples of Pink Salmon Fry

Composite sample no.	n_i	Composite sample T_i(g)	Y_i (g)
1	92	26.864	0.292
2	105	31.395	0.299
3	110	30.800	0.280
4	102	32.232	0.316

note that in this example the totals $n_iY_i = T_i$ are measured instead of the composite sample means. This simplifies calculations, as

$$\hat{\mu}_x = \frac{\sum\limits_{i=1}^{I} T_i}{\sum\limits_{i=1}^{I} n_i},$$

and

$$\hat{\sigma}_x^2 = \frac{1}{I-1}\left[\sum_{i=1}^{I}\left(\frac{T_i^2}{n_i}\right) - \frac{\left(\sum\limits_{i=1}^{I} T_i\right)^2}{\sum\limits_{i=1}^{I} n_i} \right].$$

Sampling for the mean mass is an example where compositing is especially easy. Weighing small specimens in the field can be tedious or problematic, and compositing can greatly facilitate the measurement process. In this example, we estimate the mean mass, μ_x, and variability, σ_x^2, of the salmon fry as 0.297 g and 0.024 g^2, respectively.

Example 2: Trace Metal Concentrations

A random sample of 15 fish is taken from a fishery contaminated with copper by mining operations. However, there are resources for only three laboratory analyses. To utilize all available sample material, tissue samples of equal mass are taken from each fish and randomly composited into three laboratory samples, with five fish represented in each sample. Based on the laboratory results (Table 2), we estimate μ_x as 3.25 and σ_x^2 as 1.16.

Table 2. **Mean Concentration of Copper on Composite Tissue**

Composite sample no.	n_i	Y_i (mg/kg dry weight)
1	5	2.75
2	5	3.71
3	5	3.28

Confidence Interval Calculation

The section "Estimating μ_x and σ_x^2" above described some properties of $\hat{\mu}_x$. If we assume further that the field samples X_{ij} are normally distributed, it follows that $\hat{\mu}_x$ is normally distributed with mean μ_x and variance $\sigma_x^2/\Sigma n_i$. Furthermore, (by, for example, Theorem 4.1 (*iii*), Seber[11]) under the null hypothesis H_N: $\mu_x = \mu_o$, we have:

$$t = \frac{\hat{\mu}_x - \mu_o}{\hat{\sigma}_x/\sqrt{\Sigma n_i}} \sim t_{I-1}. \tag{1}$$

That is, t is distributed as the Student's t distribution with $(I - 1)$ degrees of freedom. Hence $(1 - \alpha)100\%$ confidence intervals for μ_x are determined by

$$\hat{\mu}_x \pm t_{I-1,1-\alpha/2} \cdot \hat{\sigma}_x/\sqrt{\Sigma n_i}, \tag{2}$$

where α is the probability that the confidence interval does not straddle the true population mean.

From Equation 2, we can see how increasing the number of field samples through composite sampling can improve the precision of $\hat{\mu}_x$. For example, for the case of simple random sampling without compositing ($n_i \equiv 1$), Equation 2 becomes

$$\hat{\mu}_x \pm t_{I-1,1-\alpha/2} \cdot \hat{\sigma}_x/\sqrt{I},$$

the familiar formula for a $(1 - \alpha)100\%$ confidence interval for the mean. By doubling the number of field samples per laboratory sample, Equation 2 becomes

$$\hat{\mu}_x \pm t_{I-1,1-\alpha/2} \cdot \hat{\sigma}_x/\sqrt{2I}.$$

That is, the expected size of the confidence interval is decreased by a factor of $1/\sqrt{2}$ which translates to a 29% decrease in the expected size of the confidence interval. Tripling the number of field samples decreases the expected confidence interval by 42%, and increasing the number of field samples per composite sample by four cuts the expected size of the confidence interval in half. These improvements in the precision of point estimates are also reflected in the power of tests based on composite samples. The effect of compositing on the power of statistical tests will be described in "Power Calculations" below.

We also see from Equation 2 that increasing the number of laboratory samples, I, decreases the expected size of the confidence interval through $t_{I-1, 1-\alpha/2}$. This effect can be dramatic when the initial size of I is small. However, as I increases $t_{I-1, 1-\alpha/2}$ approaches its asymptotic limit—the critical value for the standard normal distribution—so that there is very little return in terms of improved precision of $\hat{\mu}_x$ for increasing the number of laboratory samples beyond about 30; after this point the return becomes negligible. To illustrate this with an example, consider two sample designs consisting of either 60 composite samples with 2 field samples per composite or 30 composite samples with 4 field samples per composite sample. The total number of field samples (120) is the same for each design, so that the gain in precision is only derived from the increased degrees of freedom associated with the t statistic. With $\alpha = 0.05$, the critical value for the first design is $t_{59} = 2.05$; for the second design the critical value is $t_{29} = 2.00$. Hence the first design with half the laboratory cost is only 2.5% less precise then the second design. This effect is also reflected in the power of statistical tests, and will be discussed in greater detail in the "Power Calculations" section.

Hypothesis Tests

One- and two-sided hypothesis tests follow from Equations 1 and 2. For the one-sided alternative $H_A: \mu_x \geq \mu_o$, we reject H_O if

$$t > t_{I-1, 1-\alpha}.$$

For the two-sided alternative $H_A: \mu_x - \mu_o$, we reject H_O if

$$|t| > t_{I-1, 1-\alpha/2}.$$

For the two sample case, we test the null $H_O: \mu_{x1} - \mu_{x2} = c$ against alternatives with the test statistic

$$t = \frac{\hat{\mu}_{x1} - \hat{\mu}_{x2} - c}{\hat{\sigma}_{pooled} \sqrt{(1/\Sigma n_{1i} + 1/\Sigma n_{2i})}} \sim t_{I+J-2}, \tag{3}$$

where

$$\hat{\sigma}_{pooled} = \sqrt{\frac{(I-1)\hat{\sigma}_{x1}^2 + (J-1)\hat{\sigma}_{x2}^2}{I + J - 2}}$$

and $\hat{\mu}_{x1}$, $\hat{\mu}_{x2}$, $\hat{\sigma}_{x1}^2$, and $\hat{\sigma}_{x2}^2$ are the respective one sample estimators as described in the section "Estimating μ_x and σ_x^2".

The test statistic t is compared against the appropriate critical value for the t distribution with $(I + J - 2)$ degrees of freedom, where I and J are the number of laboratory analyses for the respective populations. For example, for the one-sided alternate, H_A: $\mu_{x1} - \mu_{x2} \leq c$, we use for the critical value $t_{I+J-2, 1-\alpha}$. For the two-sided alternate, H_A: $\mu_{x1} - \mu_{x2} \neq c$, we use $t_{I+J-2, 1-\alpha/2}$.

Equations 1 and 3 actually depend on only the relaxed assumption that the Y_i, $i = 1, 2, \ldots, I$, are normally distributed. Hence by the Central Limit theorem Equations 1 and 3, and the confidence intervals and hypothesis tests described above, hold asymptotically in n_i, $i = 1, 2, \ldots, I$, regardless of the distribution of X_{ij}.

Estimation in the Presence of Subsampling

Another statistical issue related to composite sampling involves problems introduced by the subsampling of composite samples prior to laboratory measurement (e.g., References 4, 7, 14). Such sampling programs, which we will call **composite sampling/subsampling** plans, arise when, due to reasons of convenience or necessity, field samples are grouped into one or possibly more large composite samples. Each composite sample is subsequently mixed and subsampled one or more times to obtain laboratory samples of the appropriate volume. This can be contrasted with **simple composite sampling** described in this chapter and Brumelle et al.,[15] where it is assumed that the mean for the entire composite sample can be determined (see Figure 1).

In composite sampling/subsampling, an unknown proportion, p_{ijk} say, of field sample X_{ij} winds up in composite subsample Y_{ik}, where $k = 1, 2, \ldots, s_i$ indexes subsamples taken from composite sample i. Estimation of μ_x remains fairly straightforward, but extracting estimates of σ_x^2 from the observed variability in the Y_{ik}'s is more involved because of the variability introduced to the Y_{ik}'s by the unknown p_{ijk}'s. The proportions, p_{ijk}, can be variable if there is imperfect mixing, or "blending," of the composite samples so that field samples are represented disproportionately in the laboratory samples. Hence, in composite sampling/subsampling plans, estimation of σ_x^2 requires assumptions about the distribution of the unknown proportions. More involved sampling programs and estimators are also required. This has been approached in several ways. Brown and Fisher[4] discuss an application where, given an independent estimate of the variance of p_{ijk} from experiments performed prior to the sampling program and given various assumptions about the joint distribution of the proportions, an estimate of σ_x^2 can be made. Rohde[7]

Composite Sampling/Subsampling (e.g., Rohde, 1976) ⋮ Simple Composite Sampling (e.g., Brumelle, 1984)

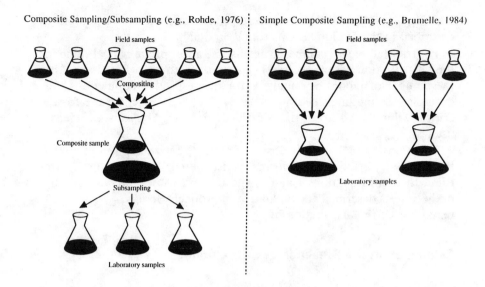

FIGURE 1. Composite sampling/subsampling (e.g.,[7]) and simple composite sampling compared (e.g.,[15]).

discusses modeling the proportions with the Dirichlet distribution and describes an estimator of σ_x^2 achieved by reserving half of the laboratory analyses for estimation of the portion of the variability in the Y_{ik}'s that was introduced by the random p_{ijk}'s. Elder et al.[14] take these models a step further by developing a model and estimators that includes a variance term for the within field sample inhomogeneity translated to subsamples.

The appropriateness of the distribution assumptions regarding the p_{ijk}'s in environmental sampling applications has not been discussed in detail in the literature, but may be of concern. For some applications, such as aquatic sampling, completely homogeneous mixing of the composite sample is to be expected. If there is complete, homogeneous mixing of the composite, then subsampling proportions are not random, but are fixed constants. This would imply, for example, that no additional information would be contained in multiple subsamples from a single composite (laboratory effort wasted). If there were only one composite sample, then σ_x^2 could not even be estimated. In environmental sampling applications, the situation may often be even worse—homogeneous mixing followed by chelation, settling, or other physical processes may introduce additional variability unrelated to the variances we are interested in.

Given the above discussion, simple composite sampling may be preferred over composite sampling/subsampling when an estimate of σ_x^2 is desired. The examples in the "Example Calculation of $\hat{\mu}_x$ and $\hat{\sigma}_x^2$" section suggest some of the applications in environmental sampling where simple composite sampling is the "natural" and correct approach to sampling. These are applications where the complications imposed by a composite sampling/subsampling study design are unnecessary and unwarranted. Investigators contemplating a composite sampling/subsampling program may want to consider the simple composite sampling options. If field sampling units can be redefined (made smaller) or analytical methods revised (made to accommodate larger samples) a simple composite sampling plan may be used instead.

POWER CALCULATIONS

Introduction

Power calculations are useful for reviewing a proposed study or for designing a new study. For a proposed study, power calculations tell us whether a study is feasible — studies with low power being ill-advised because they have a low probability of providing meaningful results. More practically, power calculations can be used in designing a study to find the combination of field sampling and laboratory analytic effort that most efficiently meets study objectives or that optimizes the use of available resources.

In all cases covered in the earlier section on hypothesis testing, the test statistic t is distributed as the noncentral t under the alternate hypothesis. Hence, the noncentral t distribution can be used to find the power of a test, that is, the probability that the test will detect a significant effect if it is there. For example, consider the one sample case with two-sided alternate. For this case, the test statistic t is distributed as the noncentral t with $(I - 1)$ degrees of freedom and noncentrality parameter

$$\delta = \frac{\sqrt{\Sigma n_i}\,\Delta}{\sigma_x},$$

where Δ is the minimum shift in the mean that one desires to detect. The power of a test increases as the number of degrees of freedom $(I - 1)$ increases, and as the noncentrality parameter δ increases. In terms of the number of laboratory samples and field samples, we see that the power of a test based on composite samples increases in I through the degrees of freedom, and in Σn_i through the noncentrality parameter. Hence, increasing the number of laboratory samples or increasing the number of field samples increases the power of a study. As with the confidence intervals discussed in "Example Calculations of $\hat{\mu}_x$ and $\hat{\sigma}_x^2$" increasing the number of laboratory samples increases the power of a study (through the critical value $t_{I-1,\ 1-\alpha/2}$) when the initial I is small; however, this effect diminishes as I increases and $t_{I-1,\ 1-\alpha/2}$ asymptotically approaches the critical value for the standard normal. Sample power calculations illustrating the relative effect of increasing Σn_i and I are described in the next section. Optimizing a sampling program using power calculations is discussed and some practical considerations in the design of a sampling program are suggested. The assumption that laboratory error is negligible compared to σ_x^2 should be carefully considered before applying these methods.

Using Power Calculations

Power calculations based on the noncentral t can be used to compare various sampling strategies in environmental surveys. In the following, we present graphically summarized power tables and illustrate the uses of power calculations for the one sample two-sided hypothesis case. Power calculations were performed using the PowerPack software package.[16] Graphically summarized power calculations are useful for comparing the relative influence of Δ, I, and Σn_i. More complete power tables for a range of α, Δ, I, and Σn_i are provided in Appendix 2.

Figures 2 and 3 provide power calculations for various I, Σn_i, and Δ under the one sample, two-sided hypothesis test scenario. For comparison, these figures also include power calculations for the simple random sampling plan (that is, $n_i \equiv 1$). All power calculations are based on an α error level of 0.05. To use the figures, first identify the minimum shift in means, Δ, you wish to detect; the number of field samples, Σn_i, to be gathered; and the number of laboratory samples, I, to be analyzed. Δ is typically expressed in units of σ_x using the best estimate of σ_x available at the time of the study design.

Consulting the appropriate figure (Figure 2: $\Delta = 0.5\sigma_x$ or Figure 3: $\Delta = \sigma_x$), locate the number of laboratory samples on the x axis, and move upward until you cross the power curve corresponding to the correct number of field samples (expressed on the power table in multiples of I). For example, to detect a shift of $0.5\sigma_x$ with a simple random sample of size 15, (Figure 2: I = 15 and Σn_i = I), the power is 0.44. For a simple composite sampling example, to detect a $0.5\sigma_x$ shift with 30 field samples and 15 composite laboratory samples, (Figure 2: I = 15, Σn_i = 2I), the power is 0.72.

Effect of Increasing Field Effort

Increasing the number of field samples always increases the power of a study. This is particularly useful if field sampling costs are low or if laboratory costs are restrictively high. For example, if extreme analytical

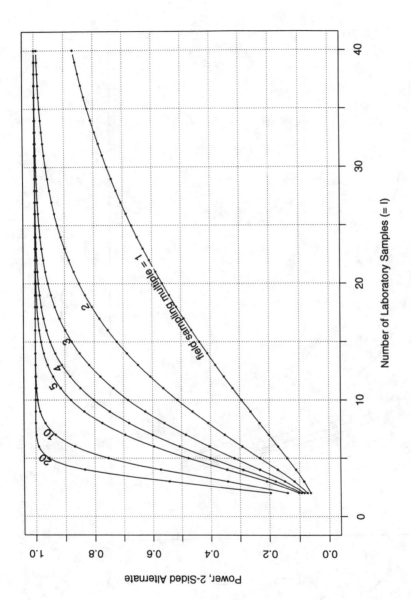

FIGURE 2. Power curves for the one sample two-sided hypothesis test $\mu_x \neq \mu_0$, $\alpha = 0.05$, and $\Delta = \sigma_x$.

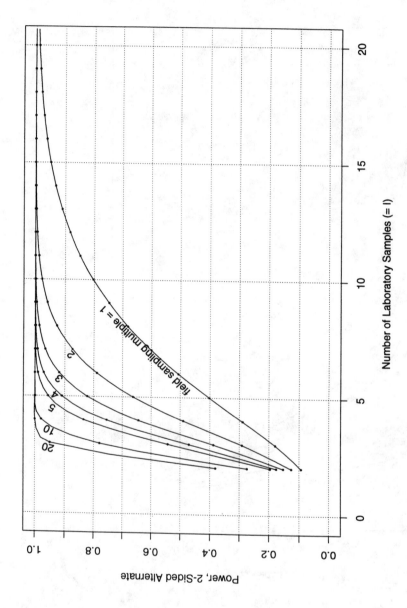

FIGURE 3. Power curves for the one sample two-sided hypothesis test $\mu_x \neq \mu_0$, $\alpha = 0.05$, and $\Delta = 0.5\sigma_x$.

costs fix the maximum number of laboratory samples possible, then the power curves can be used to determine the minimum number of field samples needed to ensure an adequate power. For the example stated above, doubling the field sampling effort from $\Sigma n_i = 15$ to $\Sigma n_i = 30$ improved the power from 0.44 to 0.72. Tripling the field sampling effort ($\Sigma n_i = 3I = 45$) would result in a further increase in power to 0.88.

Effect of Decreasing Laboratory Effort

Given an available number of field samples, Σn_i, we may want to see how much power we lose by pooling samples prior to laboratory analysis. If a substantial decrease in laboratory cost can be achieved while maintaining adequate power, then compositing is suggested. For example, compare the power of a simple random sample of size 40 (Figure 2: power = 0.87) with a simple composite sample based on 40 field samples and 20 laboratory samples (Figure 2: power = 0.85). This example dramatically illustrates the potential savings that can be realized by composite sampling; laboratory expenditures are cut in half with essentially no effect on the power or utility of the study.

Effect of Varying Δ

From Figures 2 and 3, we can also see the importance of Δ in power calculations. Consider, for example, the case of 20 field samples composited into 10 laboratory samples. The power is only 0.51 for this example when Δ is equal to $0.5 \, \sigma_x$, but jumps to 0.97 if Δ is equal to σ_x. The sensitivity of power calculations to the value of Δ underscores the importance of using conservatively small values of Δ when planning a study. In practice, an important consideration is the magnitude of Δ relative to the magnitude of preliminary estimates of σ_x. Hence we should also be careful when estimating σ_x. If σ_x is underestimated, Δ/σ_x is artificially inflated and power calculations can dramatically overstate the power of a proposed study.

Optimizing a Sampling Program

Optimizing a sampling plan involves calculating the power of conceivable sampling plans based on different combinations of laboratory and field samples and choosing the plan that most effectively meets the study objectives. Study objectives can be expressed in terms of the minimum targeted power, typically 0.80 or higher, or in terms of the available resources. Often it is practical to simply calculate the power of all reasonable sampling plans and to choose from among them. For example,

Neptune et al.[9] present a useful ad hoc procedure for designing a composite sampling program. The authors first established the maximum compositing ratio that could be achieved given field sampling, compositing, and laboratory analytic constraints. The samples were soil samples from a hazardous waste site, and physical constraints such as the number of field samples that could be effectively mixed into a composite sample placed practical limitations on the range of sampling plans that could be considered. Power calculations were performed for each plan defined by this range. The sampling plan that most efficiently met their study objectives was selected for implementation. Instead of using power calculations based on the assumption of normally distributed Y_i such as provided in Figures 2 and 3, Neptune et al.[9] estimated the power using Monte Carlo simulations. Such precautions are especially recommended when the distribution of field sample concentrations is highly skewed and the number of field samples per composite sample is small. This is a special concern in most applications involving sampling for pollutants, and especially in hazardous waste site sampling where highly skewed distributions (e.g., the lognormal distribution) are the norm.

Some Practical Sample Size Considerations

As noted earlier in the text, the change in the critical values for the t distribution is small beyond 30 degrees of freedom. In fact, a common practical rule is to assume a standard normal distribution if there are 30 degrees of freedom or more; that is, the argument is that $t_{30} = z$. Since the number of degrees of freedom is determined by the number of composite samples, there is on the basis of statistical considerations little payoff in considering more than a total of about 30 composite samples. We postulate that 30 is an upper limit on the total number of composite samples that should ordinarily be considered. Beyond this, the samples should be composited. In the context of a two-group comparison, 16 composite samples per group will ensure at least 30 degrees of freedom in the two-sample test. This frees resources for increased field sampling. Of course, there may be other considerations for taking more composite samples than this, but the reason cannot be statistical aspects of precision.

In terms of sample size considerations, suppose that the number of composite samples has reached this limit of 30. This simplifies matters, because we can use the usual normal distribution formula for sample size calculation. For a one-sample test in the context of composite sampling (with I composite samples and $n_i \equiv n$ field samples per composite sample) this is

$$I = \frac{(z_{1-\alpha/2} + z_{1-\beta})^2 \, \sigma_x^2/n}{\Delta^2}.$$

Suppose the usual situation is envisaged with a power of 0.80 and an α error level of 0.05. Then

$$(z_{1-\alpha/2} + z_{1-\beta})^2 = 7.85 \qquad (=8, \text{ say}).$$

Substituting and solving for n, this leads to:

$$n = \frac{8\sigma_x^2}{I\,\Delta^2}$$

as a general formula for the minimum number of field samples per composite sample when I is at least 30. From this formula we can also see that, for I greater than 30, compositing is of interest only if

$$\frac{\Delta}{\sigma_x} < \sqrt{\frac{8}{I}} < 0.52.$$

For I below 30, we begin to see the effect of the t distribution in the power calculations. The key explanation is that the estimation of σ_x^2 requires a reasonable number of composite samples to determine it accurately. After this has been achieved, the noncentrality parameter takes over. We note that if σ_x^2 is known from previous studies, or a priori, compositing should be conducted as much as practically feasible.

OTHER APPLICATIONS OF COMPOSITE SAMPLING

Sampling for Presence or Absence, e.g., of Hazardous Waste

There may be situations where we want to test simply for the presence or absence of something. For example, we may wish to test whether a certain contaminant is present in a suspected hazardous waste site. Because of the potential patchy spatial distribution of contaminants in hazardous waste sites, it is very likely that a contaminant would be missed if only a few randomly located samples are taken. In this situation we may want to consider taking a large number of field samples and pooling them into a small number of samples for laboratory analysis. This situation is essentially equivalent to composite sampling for the mean, as described earlier, except that care must be taken to ensure that composite samples are only so large that contamination levels of concern at the field sample level are not diluted to below analytical detection limits by the pooling process. The extra expense of analytical procedures

with smaller analytical detection limits may be warranted in this application.[10]

Isolating Hot Spots

If the hazardous waste site is divided into quadrants and composite
samples taken from each quadrant instead of from the site as a whole, we
can use the composite information to map contaminated quadrants.
Neptune et al.,[9] for example, present a thorough and informative description of the development of such a composite sampling plan. Their
plan maps and creates decision rules regarding the cleanup of a Superfund site suspected of contamination with polyaromatic hydrocarbons.
A simple extension of this sampling plan is to split each field sample
prior to compositing, reserving part of each field sample for future analysis. If a composite sample shows contamination, then the reserved field
samples can be analyzed separately, providing a more detailed mapping
of contaminated quadrants and potentially isolating hot spots. This situation is analogous to the so-called group screening problem, and has a
well-developed literature concerned with optimizing the search for all
contaminated field samples.[17-25] Rajagopal and Williams[10] present an
example, with an analysis of the economic benefits, involving isolating
contaminated wells in a groundwater monitoring network.

Compliance Monitoring

In compliance monitoring, we are often concerned with detecting peak
values or violations, for example, during ongoing monitoring of a point
source or of ambient conditions. Composite sampling may be considered
for compliance monitoring, but would rely in varying degree on distribution assumptions about the underlying values X_{ij}. Sobel and Tong,[26] for
example, develop distribution quartile estimates based on composite
samples assuming an underlying normal distribution. These estimates
suggest the probability that a violation has occurred, but do not defini-
tively document the violation. A more conservative variation of this
approach that avoids this problem somewhat is to split the field samples
prior to compositing. If a composite sample suggests the possibility of a
violation, the reserved field sample splits can be analyzed for a more
definitive answer. Similarly, Casey et al.[27] explore efficient compositing
algorithms for finding the maximal measurement from among a set of
serially autocorrelated field samples.

These methods are similar in that they rely on assumptions about the distribution of the X_{ij} and in that they accept some small probability that violations captured by field samples may go undetected. This is not to say that they these methods are not useful—more efficient sampling programs allow more field samples to be taken and should in general increase the overall probability that noncompliance periods are detected. It does suggest though that the sensitivity of these methods to various possible failures of distribution assumptions must be reviewed before they are implemented. Breiman[28] provides a model for testing the performance of compliance monitoring rules under a range of possible underlying distributions using Monte Carlo methods. Composite sampling plans that are robust to expected distributions should lead to compliance monitoring standards that improve the overall effectiveness of a monitoring program.

DISCUSSION/CONCLUSIONS

This chapter has described and provided some motivation for simple composite sampling study designs. We have reviewed the statistical literature on composite sampling and presented a more generalized approach, allowing an arbitrary, potentially unequal, number of field samples per composite sample. Based on power calculations, we conclude that the utilization of additional field samples by composite sampling will always improve the results of a study. The accuracy of environmental parameter estimates is improved, and the power to detect effects or trends in the data is increased by composite sampling. Alternatively, composite sampling can be used to reduce laboratory expenditures without significantly reducing the information obtained by a study.

Another application where composite sampling is suggested is in the sampling for the simple presence or absence, e.g., of a contaminant, in environmental surveys. Applications analogous to this have appeared in medical and other research dating to the 1940s, and can be extended naturally to the problem of developing optimum strategies for isolating "contaminated" subjects by taking advantage of composite sampling. Other applications, such as composite sampling for peak concentrations or extremes in compliance monitoring, require assumptions about the distribution of values being monitored and at this point should be considered with caution.

In conclusion, simple composite sampling is an important, and perhaps underutilized, sampling methodology for environmental applications. Composite sampling is suggested whenever field samples are readily available and/or laboratory costs are prohibitively high. Given the high cost of laboratory analyses in many environmental applications, compositing procedures may be expected to become increasingly prevalent and should merit increasing attention from environmental scientists and statisticians. Some of the outstanding issues to be addressed include designing more efficient sampling programs to map and prioritize cleanup of Superfund sites, investigation of the properties of decision rules based on composite samples in compliance monitoring, and incorporation of quality control data on measurement error into the variance estimators based on composite sampling. Statisticians must carefully qualify the characteristics and limitations of composite sampling, and environmental scientists must weigh these considerations against the various needs of the investigative and regulatory communities. Given carefully defined statistical procedures, it is felt that composite sampling data may be adapted to many regulatory and assessment problems beyond the examples outlined in this chapter.

ACKNOWLEDGMENT

The authors gratefully acknowledge Douglas J. Martin and Jonathan P. Houghton, Pentec Environmental, Inc., Edmonds, WA, for originally suggesting this problem. This project is supported in part by NIEHS grant ES 04696.

REFERENCES

1. Boswell, M. T., Burnham, K. P., and Patil, G. P. "Role and use of composite sampling and capture-recapture sampling in ecological studies." In *Handbook of Statistics,* Vol. 6, P. R. Krishnaiah and C. R. Rao (Eds.) Amsterdam: Elsevier Science Publishers B. V., 469–488 (1988).
2. Mack, G. A. and Robinson P. E. "Use of composited samples to increase the precision and probability of detection of toxic chemicals." American Chemical Society Symposium *Environmental Applications of Chemometrics*, J. J. Breen and P. E. Robinson (Eds.) Philadelphia, 174–183 (1985).
3. Rohde, C. A. "Batch, bulk, and composite sampling," in *Sampling Biological Populations*, R. M. Cormack, G. P. Patil, and D. S. Robson (Eds.) Fairland, MD: International Co-operative Publishing House, 365–377 (1979).
4. Brown, G. H. and Fisher, N. I. "Subsampling a mixture of sampled material." *Technometrics* 14(3), 663–668 (1972).
5. Cameron, J. M. "The use of components of variance in preparing schedules for sampling baled wool." *Biometrics* 7, 83–96 (1951).
6. Duncan, A. J. "Bulk sampling: problems and lines of attack." *Technometrics* 4, 319–344 (1962).
7. Rohde, C. A. "Composite sampling." *Biometrics* 32, 272–282 (1976).

8. Heyman, U., Ekbohm. G., Blomqvist, P., and Grundström, R. "The precision of abundance estimates of plankton from composite samples." *Water Res.* 16, 1367–1370 (1982).

9. Neptune, D., Brantly, E. P., Messner, M. J., Michael, D. I. "Quantitative decision making in Superfund: a data quality objectives case study." *Hazardous Mat. Control*, May/June, 19–27 (1990).

10. Rajagopal, R. and Williams, L.R. "Economics of sample compositing as a screening tool in ground water quality monitoring." *GWMR* Winter 186–192 (1989).

11. Seber, G. A. F. *Linear Regression Analysis*. New York: John Wiley & Sons, (1977).

12. Kussmaul, K. and Anderson, R. L. "Estimation of variance components in two-stage nested designs with composite samples." *Technometrics* 9(3), 373–389 (1967).

13. Edelman, D. A. "Three-stage nested designs with composited samples." *Technometrics* 16(3), 409–417 (1974).

14. Elder, R. S., Thompson, W. O., and Myers, R. H. "Properties of composite sampling procedures." *Technometrics* 22(2), 179–186 (1980).

15. Brumelle, S., Nemetz, P., and Casey, D. "Estimating means and variances: the comparative efficiency of composite and grab samples." *Environ. Monitoring Assessment* 4, 81–84 (1984).

16. Lenth, R. V. *PowerPack, Version 2.2 User's Guide*. Iowa City: Russell V. Lenth (1987).

17. Feller, W. *An Introduction to Probability Theory and Its Applications*, Vol. I., 3rd ed., New York: John Wiley & Sons (1957).

18. Sobel, M. and Groll, P. A. "Group testing to eliminate efficiently all defectives in a binomial sample." *Bell Syst. J.* 38, 1179–1252 (1938).

19. Watson, G. S. "A study of the group screening method." *Technometrics* 3, 371–388 (1961).

20. Finucan, H. M. "The blood-testing problem." *Appl. Statistics* 13, 43–50 (1964).

21. Hwang, F. K. "A method of detecting all defective members in a population by group testing." *J. Am. Stat. Assoc.* 67, 605–608 (1972).

22. Hwang, F. K. "A generalized binomial group testing problem." *J. Am. Stat. Assoc.* 70, 923–926 (1975).

23. Kumar, S. and Sobel, M. "Finding a single defective in binomial group-testing." *J. Am. Stat. Assoc.* 66, 824–828 (1971).

24. Garey, M. R. and Hwang, F. K. "Isolating a single defective using group testing." *J. Am. Stat. Assoc.* 69, 151–153 (1974).

25. Pfeifer, C. G. and Enis, P. "Dorfman-type group testing for a modified binomial model." *J. Am. Stat. Assoc.* 73, 588–592 (1978).

26. Sobel, M. and Tong, Y. L. "Estimation of a normal percentile by grouping." *J. Am. Stat. Assoc.* 71, 189–192 (1976).

27. Casey, D., Nemetz, P. N., and Uyeno, D. "Efficient search procedures for extreme pollutant values." *Environ. Monitoring Assessment* 5, 165–176 (1985).

28. Breiman, L. "Robust confidence bounds for extreme upper quantiles." *J. Stat. Computation Simulations* 37, 127–149 (1990).
29. Snedecor, G. W. and Cochran, W. G. *Statistical Methods*, 7th ed., Ames, Iowa:Iowa State University Press (1980).

APPENDIX 1

Estimates of μ_x are always improved by composite sampling. However, an interesting result is that for platykurtic distributions estimates of σ_x^2 are actually made worse. This result has been previously demonstrated by Brumelle et al.[15] or see Seber,[11] Theorem 3.4. The following is a simplified presentation of their results. For simplicity we assume that $n_i \equiv n$.

Recall that for $n_i \equiv n$ the estimator of $\hat{\sigma}_x^2$ is simply

$$\hat{\sigma}_x^2 = n\,\hat{\sigma}_Y^2,$$

where $\hat{\sigma}_Y^2$ is the sample variance of Y. Hence

$$V\,(\hat{\sigma}_x^2) = n^2 V\,(\hat{\sigma}_Y^2)$$

$$= n^2 \cdot \frac{2\,\sigma_Y^4}{(I-1)}\left[1 + \frac{(I-1)}{I}\,\frac{\gamma_2(Y)}{2}\right]$$

(e.g., Equation 5.14.3, Snedecor and Cochran[29]), where $\sigma_Y^4 = \sigma_x^4/n^2$ and γ_2, the kurtosis of a distribution, is defined as

$$\gamma_2 = \frac{\mu_4}{\sigma^4} - 3.$$

γ_2 equals 0 for normal distributions, $\gamma_2 < 0$ defines a platykurtic distribution, and $\gamma_2 > 0$ defines a leptokurtic distribution. By Equation 5.14.2, Snedecor and Cochran,[29] $\gamma_2(Y) = \gamma_2(X)/n$. Substituting terms in X and reducing, we have:

$$V\,(\hat{\sigma}_x^2) = \frac{2}{(I-1)}\,\sigma_x^4 + \frac{\gamma_2(X)}{I \cdot n}\,\sigma_x^4. \tag{1}$$

Thus, the precision of $\hat{\sigma}_x^2$ improves as a function of n for leptokurtic distributions and declines as a function of n for platykurtic distributions. The precision of $\hat{\sigma}_x^2$ is independent of n for normal distributions.

Note: Equation 1 is equal to the Brumelle et al.[15] equation for $D^2(\hat{\sigma}^2)$ when n = 1, and their equation for $D^2(\hat{\sigma}^2)$ for the general n.

APPENDIX 2A. Power table for the one sample two-sided hypothesis test $\mu_x \neq \mu_o$, $\alpha = 0.05$

$\Delta = 2\sigma_x$ $\sum n_i =$								I	$1.25I$	$1.5I$	$2I$
$\Delta = \sigma_x$ $\sum n_i =$											
$\Delta = .5\sigma_x$ $\sum n_i =$	I	$1.5I$	$2I$	$2.5I$	$3I$	$3.5I$	$4I$	$5I$	$6I$	$8I$	
$I = 2$	0.0620	0.0676	0.0730	0.0782	0.0832	0.0881	0.0928	0.1018	0.1102	0.1258
3	0.0841	0.1007	0.1170	0.1330	0.1487	0.1641	0.1793	0.2087	0.2371	0.2909
4	0.1113	0.1418	0.1721	0.2020	0.2315	0.2604	0.2888	0.3435	0.3953	0.4900
5	0.1405	0.1861	0.2313	0.2757	0.3190	0.3610	0.4014	0.4773	0.5461	0.6625
6	0.1707	0.2316	0.2915	0.3494	0.4048	0.4574	0.5068	0.5960	0.6722	0.7891
7	0.2013	0.2773	0.3507	0.4204	0.4855	0.5456	0.6005	0.6948	0.7701	0.8737
8	0.2321	0.3224	0.4080	0.4873	0.5594	0.6239	0.6808	0.7738	0.8425	0.9268
9	0.2627	0.3665	0.4627	0.5494	0.6257	0.6917	0.7480	0.8349	0.8941	0.9587
10	0.2932	0.4093	0.5144	0.6061	0.6843	0.7496	0.8031	0.8811	0.9300	0.9772
11	0.3233	0.4506	0.5627	0.6575	0.7355	0.7981	0.8475	0.9153	0.9544	0.9876
12	0.3528	0.4902	0.6075	0.7035	0.7796	0.8384	0.8829	0.9404	0.9706	0.9934
13	0.3819	0.5280	0.6489	0.7445	0.8174	0.8714	0.9107	0.9584	0.9813	0.9965
14	0.4102	0.5638	0.6869	0.7806	0.8494	0.8983	0.9324	0.9712	0.9882	0.9982
15	0.4379	0.5977	0.7215	0.8124	0.8763	0.9200	0.9491	0.9802	0.9926	0.9991
16	0.4649	0.6296	0.7530	0.8400	0.8989	0.9374	0.9619	0.9865	0.9954	0.9995
17	0.4910	0.6596	0.7814	0.8641	0.9177	0.9512	0.9716	0.9908	0.9972	0.9998
18	0.5164	0.6877	0.8070	0.8848	0.9332	0.9621	0.9790	0.9938	0.9983	0.9999
19	0.5409	0.7139	0.8300	0.9027	0.9460	0.9707	0.9845	0.9959	0.9990	0.9999
20	0.5645	0.7383	0.8506	0.9180	0.9564	0.9775	0.9886	0.9972	0.9994	1.0000
21	0.5873	0.7610	0.8689	0.9311	0.9650	0.9827	0.9916	0.9982	0.9996	1.0000
22	0.6092	0.7820	0.8852	0.9422	0.9719	0.9868	0.9939	0.9988	0.9998	1.0000
23	0.6302	0.8014	0.8997	0.9517	0.9776	0.9899	0.9956	0.9992	0.9999	1.0000
24	0.6504	0.8193	0.9125	0.9597	0.9821	0.9923	0.9968	0.9995	0.9999	1.0000
25	0.6697	0.8358	0.9238	0.9664	0.9858	0.9942	0.9977	0.9997	1.0000	1.0000
26	0.6882	0.8510	0.9337	0.9721	0.9887	0.9956	0.9983	0.9998	1.0000	1.0000
27	0.7058	0.8649	0.9424	0.9768	0.9911	0.9967	0.9988	0.9999	1.0000	1.0000
28	0.7227	0.8776	0.9501	0.9808	0.9929	0.9975	0.9991	0.9999	1.0000	1.0000
29	0.7387	0.8893	0.9568	0.9841	0.9944	0.9981	0.9994	0.9999	1.0000	1.0000
30	0.7540	0.8999	0.9626	0.9869	0.9956	0.9986	0.9996	1.0000	1.0000	1.0000

| . | . | . | I | 1.5I | 2I | 2.5I | 3I | 3.5I | 4I |
| 2.5I | 3I | 3.5I | 4I | 6I | 8I | 10I | 12I | 14I | 16I |
10I	12I	14I	16I	24I	32I	40I	48I	56I	64I
0.1398	0.1527	0.1646	0.1757	0.2142	0.2464	0.2743	0.2993	0.3220	0.3428
0.3409	0.3874	0.4306	0.4708	0.6050	0.7052	0.7799	0.8357	0.8774	0.9085
0.5726	0.6435	0.7040	0.7550	0.8879	0.9502	0.9782	0.9906	0.9960	0.9983
0.7528	0.8211	0.8718	0.9089	0.9781	0.9950	0.9989	0.9998	1.0000	1.0000
0.8677	0.9186	0.9507	0.9705	0.9966	0.9996	1.0000	1.0000	1.0000	1.0000
0.9330	0.9654	0.9825	0.9913	0.9995	1.0000	1.0000	1.0000	1.0000	1.0000
0.9674	0.9860	0.9941	0.9976	0.9999	1.0000	1.0000	1.0000	1.0000	1.0000
0.9847	0.9946	0.9981	0.9994	1.0000	1.0000	1.0000	1.0000	1.0000	1.0000
0.9930	0.9980	0.9994	0.9998	1.0000	1.0000	1.0000	1.0000	1.0000	1.0000
0.9969	0.9993	0.9998	1.0000	1.0000	1.0000	1.0000	1.0000	1.0000	1.0000
0.9986	0.9997	0.9999	1.0000	1.0000	1.0000	1.0000	1.0000	1.0000	1.0000
0.9994	0.9999	1.0000	1.0000	1.0000	1.0000	1.0000	1.0000	1.0000	1.0000
0.9997	1.0000	1.0000	1.0000	1.0000	1.0000	1.0000	1.0000	1.0000	1.0000
0.9999	1.0000	1.0000	1.0000	1.0000	1.0000	1.0000	1.0000	1.0000	1.0000
1.0000	1.0000	1.0000	1.0000	1.0000	1.0000	1.0000	1.0000	1.0000	1.0000
1.0000	1.0000	1.0000	1.0000	1.0000	1.0000	1.0000	1.0000	1.0000	1.0000
1.0000	1.0000	1.0000	1.0000	1.0000	1.0000	1.0000	1.0000	1.0000	1.0000
1.0000	1.0000	1.0000	1.0000	1.0000	1.0000	1.0000	1.0000	1.0000	1.0000
1.0000	1.0000	1.0000	1.0000	1.0000	1.0000	1.0000	1.0000	1.0000	1.0000
1.0000	1.0000	1.0000	1.0000	1.0000	1.0000	1.0000	1.0000	1.0000	1.0000
1.0000	1.0000	1.0000	1.0000	1.0000	1.0000	1.0000	1.0000	1.0000	1.0000
1.0000	1.0000	1.0000	1.0000	1.0000	1.0000	1.0000	1.0000	1.0000	1.0000
1.0000	1.0000	1.0000	1.0000	1.0000	1.0000	1.0000	1.0000	1.0000	1.0000
1.0000	1.0000	1.0000	1.0000	1.0000	1.0000	1.0000	1.0000	1.0000	1.0000
1.0000	1.0000	1.0000	1.0000	1.0000	1.0000	1.0000	1.0000	1.0000	1.0000
1.0000	1.0000	1.0000	1.0000	1.0000	1.0000	1.0000	1.0000	1.0000	1.0000
1.0000	1.0000	1.0000	1.0000	1.0000	1.0000	1.0000	1.0000	1.0000	1.0000
1.0000	1.0000	1.0000	1.0000	1.0000	1.0000	1.0000	1.0000	1.0000	1.0000
1.0000	1.0000	1.0000	1.0000	1.0000	1.0000	1.0000	1.0000	1.0000	1.0000

APPENDIX 2B. Power table for the one sample two-sided hypothesis test $\mu_x \neq \mu_o$, $\alpha = 0.01$

$\Delta = 2\sigma_x$ $\Sigma n_i =$	.	.	.	.	.	.	.	.	.	.
$\Delta = \sigma_x$ $\Sigma n_i =$	.	.	.	.	.	.	I	1.25I	1.5I	2I
$\Delta = .5\sigma_x$ $\Sigma n_i =$	I	1.5I	2I	2.5I	3I	3.5I	4I	5I	6I	8I
$I=2$	0.0124	0.0135	0.0146	0.0157	0.0167	0.0177	0.0186	0.0204	0.0221	0.0253
3	0.0174	0.0210	0.0247	0.0283	0.0319	0.0355	0.0391	0.0463	0.0534	0.0674
4	0.0248	0.0326	0.0408	0.0492	0.0578	0.0667	0.0757	0.0941	0.1131	0.1518
5	0.0342	0.0479	0.0626	0.0781	0.0943	0.1111	0.1284	0.1641	0.2008	0.2754
6	0.0453	0.0664	0.0895	0.1140	0.1399	0.1667	0.1942	0.2506	0.3075	0.4184
7	0.0578	0.0877	0.1205	0.1556	0.1924	0.2303	0.2688	0.3460	0.4212	0.5586
8	0.0717	0.1113	0.1551	0.2016	0.2498	0.2989	0.3478	0.4430	0.5315	0.6810
9	0.0867	0.1370	0.1923	0.2505	0.3100	0.3693	0.4273	0.5359	0.6314	0.7789
10	0.1027	0.1643	0.2315	0.3013	0.3712	0.4392	0.5041	0.6207	0.7171	0.8521
11	0.1196	0.1930	0.2721	0.3527	0.4317	0.5067	0.5761	0.6953	0.7876	0.9041
12	0.1373	0.2228	0.3134	0.4040	0.4904	0.5702	0.6419	0.7591	0.8436	0.9394
13	0.1558	0.2533	0.3550	0.4542	0.5464	0.6289	0.7006	0.8121	0.8868	0.9627
14	0.1748	0.2844	0.3964	0.5028	0.5989	0.6822	0.7522	0.8554	0.9193	0.9774
15	0.1944	0.3157	0.4371	0.5493	0.6476	0.7299	0.7967	0.8899	0.9433	0.9866
16	0.2144	0.3472	0.4768	0.5934	0.6921	0.7721	0.8346	0.9171	0.9607	0.9922
17	0.2348	0.3785	0.5153	0.6347	0.7325	0.8090	0.8664	0.9382	0.9730	0.9955
18	0.2555	0.4095	0.5524	0.6732	0.7688	0.8408	0.8929	0.9543	0.9817	0.9975
19	0.2763	0.4401	0.5878	0.7088	0.8012	0.8681	0.9147	0.9665	0.9877	0.9986
20	0.2973	0.4701	0.6215	0.7414	0.8298	0.8914	0.9325	0.9756	0.9918	0.9992
21	0.3184	0.4994	0.6534	0.7712	0.8549	0.9109	0.9469	0.9824	0.9946	0.9996
22	0.3395	0.5279	0.6834	0.7983	0.8768	0.9273	0.9584	0.9874	0.9965	0.9998
23	0.3605	0.5555	0.7115	0.8227	0.8958	0.9410	0.9676	0.9910	0.9977	0.9999
24	0.3815	0.5822	0.7377	0.8446	0.9122	0.9523	0.9749	0.9936	0.9985	0.9999
25	0.4023	0.6079	0.7621	0.8642	0.9262	0.9615	0.9806	0.9955	0.9990	1.0000
26	0.4229	0.6326	0.7846	0.8816	0.9382	0.9691	0.9851	0.9968	0.9994	1.0000
27	0.4432	0.6563	0.8055	0.8971	0.9485	0.9753	0.9886	0.9978	0.9996	1.0000
28	0.4633	0.6788	0.8246	0.9108	0.9571	0.9803	0.9913	0.9985	0.9998	1.0000
29	0.4831	0.7003	0.8422	0.9228	0.9644	0.9844	0.9934	0.9989	0.9998	1.0000
30	0.5026	0.7208	0.8583	0.9334	0.9706	0.9876	0.9950	0.9993	0.9999	1.0000

.	.	.	I	1.5I	2I	2.5I	3I	3.5I	4I
2.5I	3I	3.5I	4I	6I	8I	10I	12I	14I	16I
10I	12I	14I	16I	24I	32I	40I	48I	56I	64I
0.0281	0.0308	0.0332	0.0354	0.0434	0.0501	0.0560	0.0613	0.0662	0.0708
0.0812	0.0948	0.1082	0.1214	0.1723	0.2203	0.2655	0.3081	0.3482	0.3859
0.1912	0.2306	0.2695	0.3077	0.4491	0.5686	0.6660	0.7435	0.8043	0.8515
0.3491	0.4197	0.4860	0.5472	0.7386	0.8562	0.9234	0.9601	0.9796	0.9897
0.5204	0.6104	0.6875	0.7520	0.9090	0.9694	0.9903	0.9970	0.9991	0.9997
0.6732	0.7636	0.8323	0.8829	0.9753	0.9954	0.9992	0.9999	1.0000	1.0000
0.7915	0.8681	0.9187	0.9509	0.9944	0.9995	1.0000	1.0000	1.0000	1.0000
0.8741	0.9313	0.9637	0.9813	0.9989	0.9999	1.0000	1.0000	1.0000	1.0000
0.9275	0.9662	0.9848	0.9934	0.9998	1.0000	1.0000	1.0000	1.0000	1.0000
0.9598	0.9841	0.9940	0.9978	1.0000	1.0000	1.0000	1.0000	1.0000	1.0000
0.9785	0.9929	0.9977	0.9993	1.0000	1.0000	1.0000	1.0000	1.0000	1.0000
0.9888	0.9969	0.9992	0.9998	1.0000	1.0000	1.0000	1.0000	1.0000	1.0000
0.9943	0.9987	0.9997	0.9999	1.0000	1.0000	1.0000	1.0000	1.0000	1.0000
0.9972	0.9995	0.9999	1.0000	1.0000	1.0000	1.0000	1.0000	1.0000	1.0000
0.9986	0.9998	1.0000	1.0000	1.0000	1.0000	1.0000	1.0000	1.0000	1.0000
0.9994	0.9999	1.0000	1.0000	1.0000	1.0000	1.0000	1.0000	1.0000	1.0000
0.9997	1.0000	1.0000	1.0000	1.0000	1.0000	1.0000	1.0000	1.0000	1.0000
0.9999	1.0000	1.0000	1.0000	1.0000	1.0000	1.0000	1.0000	1.0000	1.0000
0.9999	1.0000	1.0000	1.0000	1.0000	1.0000	1.0000	1.0000	1.0000	1.0000
1.0000	1.0000	1.0000	1.0000	1.0000	1.0000	1.0000	1.0000	1.0000	1.0000
1.0000	1.0000	1.0000	1.0000	1.0000	1.0000	1.0000	1.0000	1.0000	1.0000
1.0000	1.0000	1.0000	1.0000	1.0000	1.0000	1.0000	1.0000	1.0000	1.0000
1.0000	1.0000	1.0000	1.0000	1.0000	1.0000	1.0000	1.0000	1.0000	1.0000
1.0000	1.0000	1.0000	1.0000	1.0000	1.0000	1.0000	1.0000	1.0000	1.0000
1.0000	1.0000	1.0000	1.0000	1.0000	1.0000	1.0000	1.0000	1.0000	1.0000
1.0000	1.0000	1.0000	1.0000	1.0000	1.0000	1.0000	1.0000	1.0000	1.0000
1.0000	1.0000	1.0000	1.0000	1.0000	1.0000	1.0000	1.0000	1.0000	1.0000
1.0000	1.0000	1.0000	1.0000	1.0000	1.0000	1.0000	1.0000	1.0000	1.0000
1.0000	1.0000	1.0000	1.0000	1.0000	1.0000	1.0000	1.0000	1.0000	1.0000

CHAPTER 3

Environmental Chemistry, Statistical Modeling, and Observational Economy

G. P. Patil, S. D. Gore, and A. K. Sinha

ABSTRACT

Environmental sampling differs from classical theory of sampling in that the former may entail sampling of different types of material and the sampled materials often influence the sampling procedure. Therefore, while determining an optimal sampling design, the physical (chemical or biological) characteristics of the material to be sampled need to be taken into consideration. In this chapter, we discuss several environmental sampling and statistical modeling situations to illustrate how composite sampling and ranked set sampling facilitate observational economy while addressing substantive issues.

INTRODUCTION AND BACKGROUND

It is very important to recognize that there is no substitute for good data. Statistical thinking is an aid to the interpretation of data. The statistical approach is expected to contribute to the overall insight and perspective of the substantive issue and its resolution in the light of the evidence on hand. What makes the problem of environmental investigations different from studies in physical sciences is that, unlike in the hard sciences, we have a longer span of investigations depending on life stages and their age lengths. Also, the instrumentation changes are necessary in response to the advancing technology. That often puts us in a difficult cycle of no information, new information, and noninformation. When a question is asked of us, we promptly say, "Well, we don't have sufficient information. We need to collect new information." By the time new information is collected, we are ready to say that the information collected is already noninformation. However, noninformation is no information, and thus we go through this cycle involving data that do not produce any information. We somehow need to break this unaffordable cycle.

At the least, we need to be aware of this issue in environmental work. It may be described through the concept of soft data. Not only is there biochemical or environmental variability contributing to large variability usually present in these data sets, but also there are present measurement errors arising from a variety of sources — some known, some unknown, and some unknowable. Thus, the data we have are soft data; and they call not merely for hard analysis, for which we have software packages, but they also need very hard looks if we are expected to make prudent decisions. This is particularly true when we recognize that, while there are questionable statistical routines, there are no routine statistical questions.

Traditional statistical theory and practice have been occupied largely with experiments involving randomization and replication. However, in environmental work, observations most often fall in nonexperimental, nonreplicated, and nonrandom categories. Consequently, problems of model specification and data interpretation acquire special importance and great concern. See, for example, Patil.[1]

In several situations of diverse nature in environmental statistics, experiments involve time and space. In these situations, locally ingenious environmental sampling methods need to be developed, such as encounter sampling, composite sampling, ranked set sampling, etc. See, for example, Patil[1] and Patil et al.[2] for encounter sampling; Boswell et al.,[3-5] Gore et al.,[6,7] and Patil et al.[8] for composite sampling; Patil et al.[9,10] for ranked set sampling, etc. These articles help discuss situations where several of these ingenious/biased sampling and monitoring procedures can be economical together with ingenious ways of drawing useful/unbiased inferences of practical interest.

In this chapter, we discuss some environmental sampling problems in the context of composite sampling and ranked set sampling as practiced in environmental work. With real life examples, we demonstrate that appropriate statistical modeling helps us accomplish satisfactory analysis, permits us to draw reasonable inference, and lets us make decision within an informative and economic setup.

COMPOSITE SAMPLING OF SOILS AND SEDIMENTS

Composite sampling is emerging as a cost-efficient statistical method of environmental sampling. A composite sample may be a physical mix of several sample units (or subunits), or it may be a batch of individual sample units that can be tested as a group. The earliest known application of group testing was the estimation of the transmission rate of plant viruses by insects.[11] The next application of group testing occurred during the World War II, when Dorfman[12] proposed to classify U.S. servicemen as either having or not having syphilis. In a third development, composite sampling has become a standard practice in the sampling of soils, wastewater, effluents, biota, and other bulk materials when the objective is estimation of the mean either with a desired precision or with restrictions on the cost of sampling.

Composite sampling has been most frequently applied in the areas of classifying individual samples, estimating prevalence, and estimating the mean. The usual goal of composite sampling is to obtain the information that would have been obtained from measuring the individual samples, but at a reduced monetary cost, effort, or data variation. Occasionally the goal could be to obtain more efficient procedures, and sometimes it is to obtain information when the individual sample measurements are unavailable.

Here we describe some important methods of composite sampling with examples illustrating their application. For more examples, see Boswell and Patil,[13] and Patil et al.[8]

On-site Surface Soil Sampling for PCB

The Penn State Center for Statistical Ecology and Environmental Statistics has been investigating, among other research areas, the performance of composite sample techniques in environmental problems of characterization and evaluation.

The Pennsylvania Department of Environmental Resources (PA DER) provided this center with data giving the polychlorobiphenyl (PCB) concentration in about 12,000 surface soil samples. These samples were collected from 19 sites in Pennsylvania along the gas pipeline of the Texas Eastern Gas Pipeline Company. The center undertook to assess the applicability of composite sample techniques for problems of estimation and classification. See, for example, Gore et al.[6]

Since sampling and chemical analyses had already been conducted, the study was a retrospective one, in which compositing was "simulated" by averaging the recorded measurements for individual samples. In the absence of measurement error, these averages exactly reproduce the measurements on composite samples that would have been obtained by physical compositing.

The Armagh Site

Location and Features

The Armagh compressor station is located in West Wheatfield Township, Indiana County, about 1.25 mi south of U.S. Route 22. The map in Figure 1 shows the location of the Armagh site in the State of Pennsylvania. The site includes one compressor building along with several other buildings on 79 acres. There are two known liquid pits. The surrounding area contains 64 residences within 1 mi of the station. Some of these houses have private wells; however, a public water supply line was recently installed. None of the private wells has been found to be contaminated as a result of Texas Eastern Gas Pipeline activities. There is one wetland situated within one half mile of the site. Richard Run, which flows to the south of the site, is classified as a cold water fishery. There are no public recreational facilities near the station.

On-site Soils

On-site soils are defined as being within the confines of the station site fencing and are therefore accessible only by Texas Station personnel and authorized site visitors. The areal extent of excavation will be determined by the 10 parts per million (ppm) PCB contour lines which were generated based on the on-site soil characterization data. The cleanup criteria for on-site soils are specified by an average overall PCB concentration of 5 ppm. The objective of the on-site surface soil sampling was to characterize the presence of PCBs at the Armagh site. Sampling locations

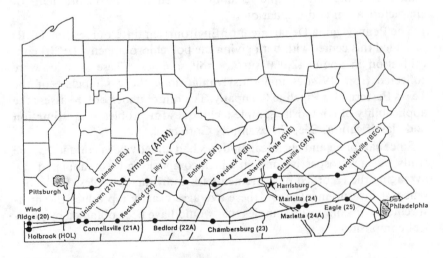

FIGURE 1.   Locations of the Pennsylvania sites.

around potential sources of contamination were selected for sampling in Phase I. As part of Phase II sampling, samples were collected at points around each Phase I sampling location having a total PCB concentration greater than 10 ppm for on-site surface soils.

On-site Surface Soil Sampling

Potential sources of PCB had been identified, and a rectangular grid was laid out around each such source. Four different on-site grids were identified by the alphabetic codes "*A*" through "*D*". Grid points were identified by a two-digit row number and an alphabetic column code. Sampling of the surface soil was done at selected grid points in two distinct phases. The second phase was undertaken to fill in locations not covered during Phase I. Grid "*D*" was not sampled during Phase I, but only during Phase II. Phase II locations were generally farther away from the potential PCB source, and the measured PCB concentrations tended to be lower during this phase.

	Phase I	Phase II
Grid A	78 samples	106 samples
Grid B	16 samples	16 samples
Grid C	36 samples	32 samples
Grid D		74 samples

The distance between consecutive rows as well as between consecutive columns was 25 ft. For the purpose of computerization and facilitating analysis using statistical computer packages, the alphabetic column codes were converted into numerical codes with *A* into 1, *B* into 2, and so on. Row and column codes for grids *B*, *C*, and *D* were shifted to synchronize them with the codes of the grid A. This synchronization enables plotting of all the sampled grid points on the same graph. Schematic plots, which are not to the scale, showing grid points that were sampled in either phase are given in Figure 2.

Soil samples were taken from a 0- to 6-in. depth. After removing vegetation, rocks, and other debris, the sample at each grid point was thoroughly mixed to obtain a homogeneous sample for analysis and quantification. Duplicate and triplicate samples were taken at some locations, but these have not been included in the analysis. The discrepancy between the measurements on primary, duplicate, and triplicate samples can be useful in studying measurement errors, and will be investigated separately in a subsequent report.

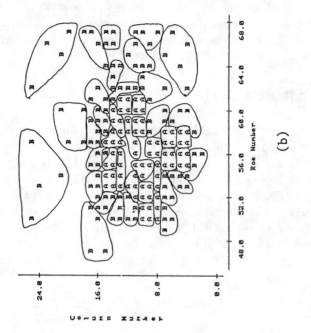

(b)

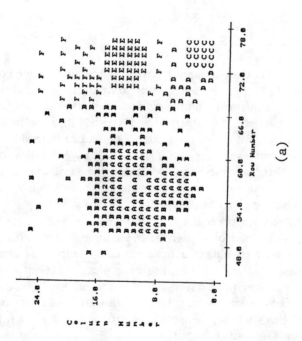

(a)

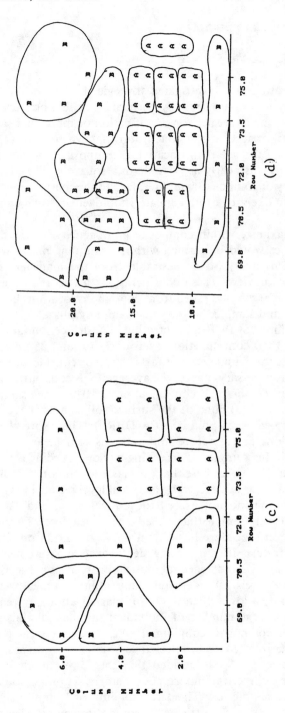

FIGURE 2. Figure 2 (a) Sampling locations on Grids A, B, and C: Phase I and II; A: Grid A, Phase I; B: Grid A, Phase II; C: Grid B, Phase I; D: Grid B, Phase II; E: Grid C, Phase I; and F: Grid C, Phase II. (b) Sampling locations and compositing scheme for Grid A; A: Phase I; B: Phase II. (c) Sampling locations and compositing scheme for Grid B; A: Phase I; B: Phase II. (d) Sampling locations and compositing scheme for Grid C; A: Phase I; B: Phase II.

Simulating Composite Samples

Choice of the Composite Sample Size

Boswell and Patil[14] have investigated strategies for composite sample formation when samples are spatially correlated. After comparing four different compositing strategies for classification of individual samples, Boswell and Patil arrived at the conclusion that when there is spatial dependence among the sampling locations, compositing samples from neighboring points—as nearly as possible in a square region—increases the cost efficiency of composite sampling. Due to the spatial dependence, these samples are likely to exhibit greater homogeneity than randomly selected samples.

In order to maximize within composite homogeneity, it was decided that all composites would be formed within a sampling phase and also within a grid. This may also be desirable from the management and operational point of view. These considerations led to the decision to composite individual samples taken from contiguous locations belonging to the same grid and sampled in the same sampling phase.

The detection limit for PCB concentration in a surface soil sample was 1 ppm. Further, PCB concentration levels of 5, 10, and 25 ppm are of significance; since it is stipulated that for a site to be considered environmentally safe, it is necessary that (1) the average PCB concentration does not exceed 5 ppm, (2) the PCB concentration in 90% of the samples be less than 10 ppm, and (3) none of the surface soil samples have a PCB concentration exceeding 25 ppm (see PA DER[15]). These three levels of PCB concentrations are therefore important as critical levels for the problem of classifying samples as exceeding or not exceeding a specified critical level. It may be noted here that the classification of all individual samples as exceeding or not exceeding a critical level results in (1) identifying all samples that do not exceed the critical level, and (2) actual measurements on every sample that exceeds the critical level. This feature of the classification procedure has a dual implication. On one hand, a low critical level for classification will generate exact measurements on a large number of samples and hence subsequent problems of classification with higher critical levels will not require any additional measurements. On the other hand, a low critical level for classification, by requiring measurement on a large number of individual samples, diminishes the cost efficiency of composite sampling. It then becomes clear that the composite sample size needs to be selected very judiciously. In case of the Armagh site, however, it was decided that the lowest levels of PCB concentration are of practical importance—namely, 5 ppm—should control the composite sample size. The detection limit of 1 ppm in conjunction with this critical level restricts the composite sample size not to exceed five. Further, a composite sample of size five would make it impossible to distinguish between the critical level and the detection limit for composite samples that have a measurement below 1 ppm. This implies that the composite sample size should not exceed four. These considerations led to the decision of a composite sample size of four. In a few cases, where the spatial arrangement made it difficult to identify exactly four nearby sample locations, composite samples of sizes three or five were formed. The estimators, accordingly, are computed as weighted averages rather than simple averages. The derivations of these estimators are developed in Lovision et al.[16]

Estimation of the Mean PCB Concentration

Composite sampling provides an unbiased estimator of the population mean. Further, in the absence of measurement error, the composite sample estimate is numerically equal to the estimate obtained by quantifying each individual sample. Accordingly, the two estimates also have the same squared standard error, σ^2/N, where σ^2 is the population variance and N is the number of physical samples. The advantage of compositing is that the estimate can be obtained with a substantial reduction in the number of quantifications and in the total project cost. In the case of the Armagh site, for example, the compositing described in Table 1 requires only 90 quantifications to estimate the mean PCB concentration with the same precision as the 358 measurements needed for exhaustive quantification.

It is interesting to compare S_X and $\tilde{\sigma}$, the estimates based on individual and composite sample values, respectively. Since compositing is based on

Table 1. Unbiased Estimates of μ and σ

	Measurements					
	Individual Samples			Composite Samples		
	N	$\bar{X}$	S_x	n	$\tilde{\mu}$	$\tilde{\sigma}$
Phase 1						
Grid A	78	363.32	1355.6	20	363.32	1793.72
Grid B	16	64.56	40.0	4	64.56	36.12
Grid C	36	1075.14	2076.5	9	1075.14	2974.74
Phase II						
Grid A	106	81.46	198.0	26	81.46	208.94
Grid B	16	26.01	23.3	4	26.01	36.46
Grid C	32	70.2	79.3	8	70.2	84.1
Grid D	74	36.59	91.1	19	36.59	97.28

proximity of individual samples, composites are expected to be internally homogeneous. Internal homogeneity of composites implies $\tilde{\sigma} > S_X$. From Table 1 it is apparent that not all composites are more homogeneous than individual samples. However, there is only one case (Phase I, grid B) where $\tilde{\sigma} < S_X$. Here both S_X^2 and $\tilde{\sigma}^2$ are unbiased estimates of the population variance σ^2.

A disadvantage of compositing is that one loses the spatial resolution that is provided by quantifying every sample, though the loss is somewhat mitigated by compositing contiguous data. Two methods are available for partial recovery of the spatial information. The first, which is not developed here, employs a parametric model for the spatial variability in which compositing has an estimable effect on the model parameters. It is then possible to estimate the original parameters from the composited data and to predict contaminant levels throughout the region of interest. This method looks on site characterization as the fundamental goal of the analysis. The second method classifies the individual samples according to a specified criterion level. Note that in reality once individual samples have been classified according to a criterion level, actual measurements on all those samples that exceed the criterion level are made during the classification process; and hence classification at a higher criterion level becomes redundant.

Locating Individual Samples with High PCB Concentrations

Taking measurements on composite samples results in a loss of information on measurements of the PCB concentrations in individual samples that constitute the measured composite samples. As a consequence of the loss of this information, it is impossible to use composite sample measurements for locating individual samples with high PCB concentrations. However, under certain circumstances, it is possible to locate individual samples with high PCB concentrations without exhaustive measurements on all the individual samples. In addition to the composite sample measurements, additional measurements on certain individual samples are required to be made in order to locate individual samples with high PCB concentrations.

Casey et al.[17] have considered the problem of detection of extreme pollutant values when the cost of extensive and comprehensive monitoring is prohibitively high. They have demonstrated an efficient method to determine the maximum sample measurement from a finite set of sequential samples without explicitly testing them all. They have assumed that the measurements have high positive autocorrelation. As noted by Casey et al.,[17] there is an advantage if composite samples are internally homogeneous. With internal homogeneity, the sample with the maximum value is more likely to be composited with samples having high values than with samples having low values. This will result in a high measurement on the composite sample that contains the sample with the extreme measurement. We have extended the search procedure described by Casey et al. so that any desired number of extreme values can be searched with a minimum number of measurements on individual samples.

Consider a composite sample of size k. Let $x_1, x_2, \ldots, x_k$ be the PCB concentrations in the k individual samples and let y be the PCB concentration of the composite sample. Further, let $x_{k:k}$ denote the maximum of the PCB concentrations in the k individual samples. That is:

$$x_{k:k} = \max\{x_1, x_2, \ldots, x_k\}$$

It is then easy to observe that:

$$y \leq x_{k:k} \leq k \cdot y$$

This inequality implies that the measurement on every composite sample gives the lower and the upper bounds for the largest measurement among its constituent individual samples. It is then easy to make the following observation.

Consider two composite samples. Let the composite sample sizes be k_1 and k_2, the composite measurements be y_1 and y_2, and the maximum individual concentrations be $x_{k_1:k_1}$ and $x_{k_2:k_2}$ respectively. Then

$$y_1 \leq x_{k_1:k_1} \leq k_1 \cdot y_1$$

$$y_2 \leq x_{k_2:k_2} \leq k_2 \cdot y_2$$

Without loss of generality assume that $y_1 < y_2$. In general this does not imply that $x_{k_1:k_1} \leq x_{k_2:k_2}$. However, if $k_1 \cdot y_1 \leq y_2$, then it can be inferred that $x_{k_1:k_1} \leq x_{k_2:k_2}$, and hence the first composite sample cannot contain the individual sample with the largest PCB concentration. It is thus clear that there is no need to consider the first composite sample when searching for the individual sample with the highest PCB concentration. In this way, we may eliminate a number of composite samples as not containing the individual samples with high PCB concentrations. This elimination process may leave us with a very few composite samples as possibly containing individual samples with large PCB concentrations. These composite samples may then be subjected to exhaustive testing in order to locate the individual samples with high PCB concentrations.

Using the above reasoning, we have the following sweepout method for locating the individual samples with high PCB concentrations.

Identify the composite sample with the largest measurement. Suppose we denote the measurement on this composite sample by y_{max}, and suppose that k_{max} is the size of this composite. Obviously, then, the largest PCB concentration in a constituent individual sample of this composite has the following bounds:

$$y_{max} \leq x_{max} \leq k_{max} \cdot y_{max}$$

Any other composite sample of size k and PCB concentration of y cannot contain an individual sample with a PCB concentration exceeding x_{max} if $k \cdot y \leq y_{max}$, and can be ignored as far as the search for the individual sample with the highest PCB concentration is concerned. Suppose further that all the constituent samples of the composite sample with the PCB concentration of y_{max} are subjected to measurement, so that the measurement on the individual sample with the highest PCB concentration is available. Using the notation introduced above, let x_{max} denote this value. A composite sample of size k^* and PCB concentration y^* cannot have an individual sample with PCB concentration exceeding x_{max} if $k^* \cdot y^* \leq x_{max}$. If however $x_{max} \leq k^* \cdot y^*$, then some retesting of individual samples in the composite of size k^* is required. Depending on operational considerations, it may be possible to curtail the retesting as individual values become available.

To illustrate this method in the case of the Armagh site, we note that the highest PCB concentration in a composite sample was 4897.5 ppm. Since the size of this composite is 4, the highest PCB concentration in an individual sample cannot exceed 19,590 ppm. Exhaustive testing of the constituent samples resulted in the highest PCB concentration in an individual sample, which was 10,000. Note that there was a composite sample with PCB concentration of 3,999.5 ppm, and hence might contain an individual sample with PCB concentration exceeding 10,000 ppm. On measuring every individual sample in this composite, it was indeed found to be the case, as there was an individual sample with PCB concentration of 10,700 ppm. This implies that no composite sample with a PCB concentration of 2,675 ppm or less, can contain an individual sample with PCB concentration exceeding 10,700 ppm. Since there was no composite sample with a measurement exceeding 2675 ppm, the sampling location with the largest PCB concentration has been identified. Note it required only 8 measurements in addition to the 90 composite sample measurements.

Figure 3 shows a scatterplot of individual sample measurements plotted against the simulated composite sample measurements. The two rays from the origin indicate the upper and the lower bounds on the largest individual sample measurement for every composite sample measurement. Thus, corresponding to the composite sample with a measurement of 4,897.5 ppm, the upper bound for an individual sample measurement in this composite is 19,590 ppm; while the lower bound for the same is

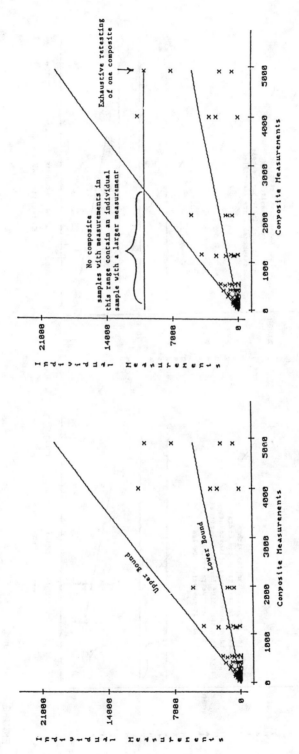

FIGURE 3. The seep-out method for identifying the individual samples with extremely large values.

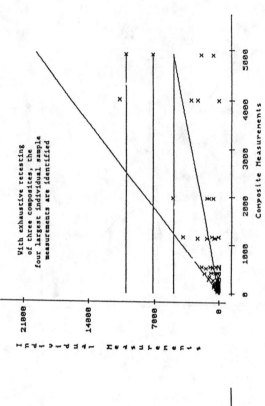

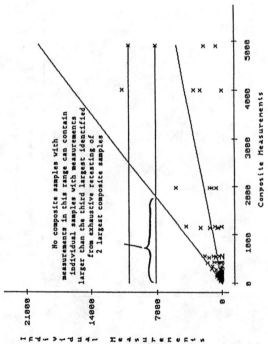

FIGURE 3.   Continued.

4,897.5 ppm, which is the same as the composite sample measurement. Since 4,897.5 ppm was the largest composite sample measurement, individual samples in this composite were measured separately; and an individual sample with a PCB concentration of 10,000 ppm was identified. A horizontal line through the point indicating this individual sample shows that there is only one composite which can possibly contain an individual sample with more than 10,000 ppm of PCB concentration. Exhaustive testing in this composite located an individual sample with a PCB concentration of 10,700 ppm. There is no other composite that can contain an individual sample with a PCB concentration exceeding 10,700 ppm, as is evident from Figure 3b. The exhaustive testing of two composites has thus identified the individual sample with the largest PCB concentration. This search can easily be extended to identify more individual samples with high PCB concentrations. Figure 3b–d show how exhaustive testing of only three composites identifies the individual samples with the four largest PCB concentrations. In other words, with only 12 measurements in addition to the 90 measurements on the simulated composite samples, we were able to identify the four individual samples with the highest PCB concentrations.

Estimation of Trace Metal Storage in Lake St. Clair Postsettlement Sediments Using Composite Samples[18]

During 1985, Canadian and United States agencies and institutions undertook a cooperative study of Lake St. Clair sediments. The objectives of the sediment sampling program were to describe the organic and metal contaminants stored in the sediments and to provide information on the permanence of storage. Sediment cores were collected from 36 locations within the lake (see Figure 4). Core samples were collected by inserting core liners as deeply as possible into the sediments. Penetration into the sands was 3–9 cm, and into the silts and clays it was 6–36 cm. Cores were extruded at 1-cm intervals to a depth of 10 cm and at 1- to 2-cm intervals for sediment depths greater than 10 cm. After noting the sediment texture, the intervals were stored frozen in polyethylene bottles. In the laboratory, frozen samples were weighed and freeze-dried without external heat. After weighing the dried samples to determine water content, the samples were stored in polyethylene bottles. These samples were then gently ground with a mortar and pestle. Composite samples were formed with subsamples of sediment from each section of a core proportional to its contribution to the total mass of sediment in the core. Composite samples were thoroughly mixed and stored in polyethylene bottles.

A 2-g subsample of each composite was weighed into a 250-mL flask, spiked with polonium-209, and extracted into 100 mL of hot (80°C) 10%

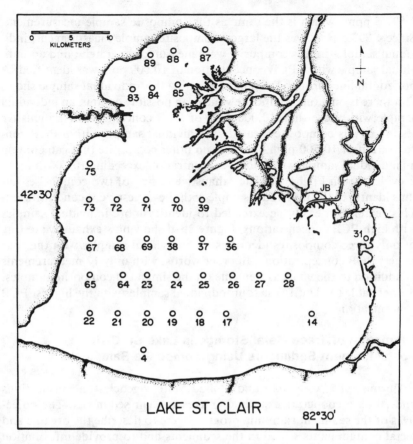

FIGURE 4.   Stations at which cores were recovered during sampling of Lake St. Clair in 1984. (From Rossman, R., *J. Great Lakes Res.*, 14(1), 1988. With permission.)

v/v hydrogen peroxide. Hydrogen peroxide was added each time the sample volume was reduced to 5–10 mL. When the reaction with the hydrogen peroxide subsided, more hydrochloric acid was added to bring the volume to 50 mL. This process was repeated two more times during the 40-hr extraction period. After 40 hr, the solution was allowed to evaporate to a volume of 5–10 mL. This extraction technique dissolved all components of the sediment except silicate minerals. The extract was then separated from the insoluble residue, and the filtered extract was transferred into a 100-mL volumetric flask and brought to volume.

Cadmium, chromium, copper, nickel, lead, and zinc were analyzed by standard lame techniques using an atomic absorption spectrophotometer. Quantification was with standard curves. Bismuth and antimony were analyzed by flameless atomic absorption. Except for cadmium and

Table 2. Mean Coefficients of Variation (mg/kg) for Analysis of Lake St. Clair Sediments

Element	Coefficient of variation	Limit of detection
Bi	8.3	0.054
Cd	18	1.8
Cr	4.5	4.8
Cu	3.4	2.4
Ni	4.3	5.4
Pb	8.0	7.4
Sb	21	0.047
Zn	5.0	4.8

Source: Rossmann.[18]

antimony, the coefficient of variation for each metal was below 10% (Table 2). Detection limits are those obtained for the ranges of concentration found in the samples. For the eight metals considered in the study, a total of 288 analyses were done. Of these, only three results for composite samples were below the detection limit. Thus the composite results were within the certainty of the analysis. Analyses of the U.S. EPA municipal digested sludge and a previously analyzed Lake Michigan sediment sample were used to monitor the quality of the analyses. All results were within the given U.S. EPA range of acceptance or were reasonably close to previous results (Tables 3 and 4).

SAMPLING OF WATER AND OTHER FLUIDS

Composite Sampling of Highway Runoff[19]

Storm water runoff from highways is monitored by manual grab sampling or automatic water quality samplers in conjunction with flow measuring instruments. Discrete runoff samples can be used to characterize the changes in concentration of various pollutants through a storm, but are usually mixed in some way to form a composite sample so that average concentrations can be used to calculate total mass loadings of pollutants. Because runoff characteristics are continuously changing, sampling at discrete points is limited in accuracy. Small storms may pass unsampled, peaks in concentration may occur between samples, or large storms may exceed the container capacity of the sampler. For these reasons, it is desirable to continuously accumulate a composite runoff sample for determination of total pollutant loadings. Wullshleger et al.[20] suggest four methods of combining discrete samples to obtain a composite according to the time they were taken and the flow rate or the volume

Table 3. Results of Analysis of Municipal Digested Sludge Compared to Previous Analyses of the Sludge and the U.S. EPA True Average and Range (mg/kg)

Metal	N	Lake St. Clair study Mean	Standard deviation	Standard mean	Deviation	U.S. EPA reported conc. Mean	Range
Bi	10	23.6	2.26	24.2	2.43	—	—
Cd	10	18.7	1.23	18.2	0.351	20.772	2.49–39.1
Cr	10	198	13.0	195	0.577	204.46	115–294
				217	4.04		
Cu	18	1040	26.0	1040	0.707	1095.3	831–1360
				1020	0.0		
Ni	10	181	5.03	182	1.44	198.31	164–233
				188	0.342		
Pb	10	532	17.6	517	0.0	518.76	305–733
	10	546	10.9	517	12.0		
Sb	9	5.13	0.970	6.99	0.0707	—	—
				5.94	0.156		
Zn	9	1230	35.6	1220	15.4	1320	1190–1450

Source: Rossman.[18]

Table 4. Results of Analysis of a Standard Lake Mud During the Lake St. Clair Study Compared with Previous Analysis of the Standard Lake Mud (mg/kg)

Metal	N	Lake St. Clair study Mean	Standard deviation	Previous studies Mean	Standard deviation
Bi	3	0.372	0.234	0.296	0.0240
Cd	3	5.74	0.0781	—	—
Cr	3	54.7	0.131	56.2	0.17
Cu	3	40.1	0.0961	39.8	1.25
Ni	3	30.1	0.0709	36.3	1.55
Pb	3	68.4	2.33	79.0	1.80
	3	77.7	0.687		
Sb	3	0.451	0.0301	0.550	0.0328
				0.520	0.0329
Zn	3	146	2.33	168	14

Source: Rossman.[18]

Table 5. Highway Runoff Water Quality Comparisons

| | 1–5 Sampling site in Seattle | | | |
| | Composite | | Range of discrete sample conc | Average of national composites (2) |
(1)	Average (2)	Range (3)	(4)	(5)
pH	6.1	5.1–6.9	4.5–7.1	—
Conductivity	87 μmho/cm	30–146	31–409	—
COD	137	75–211	8–914	147
TSS	145	43–320	30–1120	261
VSS	38	12–100	2–696	77
TOC	27	4–47	BDL–83	41
Pb	0.8	0.2–1.5	0.1–5.5	0.96
Zn	0.40	0.2–1.0	0.03–1.9	0.41
Cu	0.03	BDL–0.07	BDL–0.15	0.10
TKN	1.11	0.64–1.96	0.18–3.96	2.99
NO_3–NO_2–N	0.82	0.52–1.65	0.05–2.20	1.14
Total P	0.34	0.20–0.55	0.12–1.08	0.79

Source: Clark et al.[19]
Note: BDL = Below detectable limit. All concentration in mg/L unless stated otherwise.

they represent. Another method is to use a device that continuously removes a fixed fraction of the storm water runoff proportional to the flow rate and automatically accumulates it in a composite sample. Clark et al.[19] developed such a device and took samples from Interstate-5, I-5, in Seattle between February and September 1979. A summary of the analysis of their data is given in Table 5.

A fully automated discrete sampling system was established with a mechanical sampler. A composite sampler was also developed with the following considerations:

1. The composite sampler must produce a representative sample, with the average characteristics of the runoff from an entire storm.
2. The resulting sample was to be used in calculating the entire storm amount, and no other flow measuring device was being used.
3. The sample volume had to be sufficiently small to store.
4. The sampler should sample solids in the storm water and must not be incapacitated by litter and debris.
5. The sampler should need minimal maintenance and should not require electrical power.
6. The cost of the composite sampler should be significantly lower than the conventional discrete sampler.

After it was developed, the composite sampler was tested and compared with the conventional discrete sampler, and the following conclusions were drawn. The composite sampler was capable of accurately
removing a fixed amount of fraction of the total flow in the channel
proportional to flow rate. Operation of the composite sampling system
was simple and required a minimal amount of maintenance. One unit of
the composite sampler cost $900 as opposed to a cost of $6440 for the
conventional discrete sampler. Figures 5–7 show how composite samples
collected with the composite sampler performed in comparison with the
discrete sampler.

Composite Samples Overestimate Waste Loads[21]

Schaffer et al.[21] have reported two case studies that attempt to make a
comparison between grab and composite samples while evaluating wastewater treatment plants. Wastewater treatment plant performance is mon-

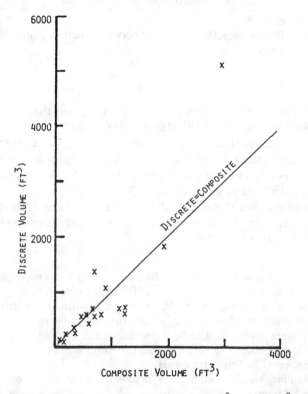

FIGURE 5.   Discrete versus composite runoff volume (1 ft^3 = 0.028 m^3). (From Clark,
                 D. L., et al., Composite sampling of highway runoff, *J. Environ. Eng.*, 107,
                 1981. With permission of the American Society of Civil Engineers.)

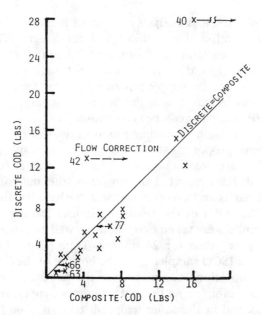

FIGURE 6. Discrete versus composite results for COD (1 lb = 0.453 kg). (From Clark, D. L., et al., Composite sampling of highway runoff, *J. Environ. Eng.*, 107, 1981. With permission of the American Society of Civil Engineers.)

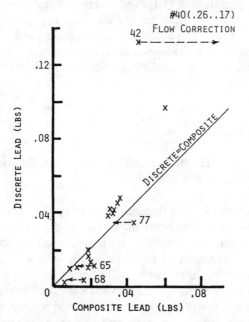

FIGURE 7. Discrete vs composite lead (1 lb = 0.453 kg). (From Clark, D. L., et al., Composite sampling of highway runoff, *J. Environ. Eng.*, 107, 1981. With permission of the American Society of Civil Engineers.)

itored by the collection and analysis of samples from the process stream for physical, chemical, and microbiological constituents. Samples may be broadly classified as "grab" or "composite." A grab sample represents composition of the flow at a given instant in time, irrespective of the flow volume. A composite sample represents an average composition of the flow over time and may or may not be proportional to the flow. Flow proportional (FP) sampling can be accomplished in one of the following ways: fixed time with sample volume proportional to flow (VP) or fixed volume with time proportional to flow (TP). Nonflow proportional (NFP) composites are usually taken as a fixed volume at fixed times.

Schaeffer et al. have reported the results of the analysis of effluent samples at St. Charles and Freeport. At St. Charles, all effluent samples were taken at the outlet of the final chlorination process for clarified water. Grab samples were taken every hour as well as every 24 hr; composite samples were taken for 24 hr/day. Time proportional samples were taken with an ISCO sampler from the bottom of the flume every 15 min; and flow proportional samples, from the center of the flow with a Lakeside Trembler sampler. At Freeport, samples were taken at the same stream locations and in the same temporal pattern. The flow proportional composite samples were collected using a BIF Sanitrol sampler.

Samples from both treatment plants were analyzed for total suspended solids (TSS) and ammonia (NH_3). Flows were monitored continuously at both facilities. Table 6 summarizes the data for Freeport; and Table 7, the data for St. Charles. The tables give the number of observations, the mean, standard deviation, skewness, kurtosis, minimum, and maximum. Table 8 summarizes certain statistical information developed from the analysis of the data.

INDOOR AIR POLLUTION

Sampling Dust from Human Dwellings to Estimate the Prevalence of Indoor Allergens[22]

Asthma, one of the most common respiratory diseases, is characterized by inflammation of the bronchi. In the recent past, an increase in asthma morbidity and mortality have been observed in several countries. This has stimulated interest in both treatment with anti-inflammatory drugs and identification and control of allergens causing the inflammation. Many studies focusing on the quantification of allergen exposure have reported an association between sensitization to dust mite allergens and asthma. *Dermatophagoides* mites and cats produce a variety of allergens, and allergic individuals can exhibit different patterns of sensitivity to the major and minor mite and cat allergens. However, measurement

Table 6. Freeport Effluent Concentrations and Loads

Parameter	Conc (ppm)			Loads (ppm × m³/sec)		
	Max.	Mean	Skew/	Max.	Mean	Skew/
N	Min.	SD	Kurtosis	Min.	SD	Kurtosis
Hourly grabs						
NH₃	18.0	12.8	0.3	4.3	2.5	0.0
167	9.4	1.8	3.1	1.2	0.8	2.0
TSS	1480.0	971.2	0.3	390.0	190.0	0.4
167	704.0	125.3	4.9	86.0	64.0	3.1
Daily time proportioned composites						
NH₃	24.3	11.4	0.2	4.9	2.3	0.2
100	2.8	3.5	4.6	0.4	0.7	5.2
TSS	1158.0	862.3	0.1	316.5	172.1	0.2
99	658.0	102.9	2.8	81.5	40.2	3.9
Daily volume proportioned composites						
NH₃	22.5	11.6	0.0	4.6	2.3	0.1
100	0.6	3.4	4.7	0.1	0.7	4.5
TSS	1222.0	884.8	0.0	314.6	176.6	0.1
99	674.0	116.2	2.9	81.2	41.6	3.5

Source: Schaeffer et al.[21]

Note: The standard deviations (SD) are computed directly from sample
data as s_c; variance correction for compositing, S_w^2, and
autocorrelation τ, are not included.

of a specific allergen, such as the *Dermatophagoides* mites group I allergens (*Fel d I*), can be used to assess allergen exposure. Sufficient data
have been compiled to propose risk asessment guidelines applicable to a
majority of mite- and cat-sensitive individuals. It has been proposed that
greater than 10 μg of total *Dermatophagoides* group I mite allergen
(DER I) per gram of dust should be considered high and representing a
risk of acute attacks of asthma in a majority of mite allergic individuals;
concentrations as low as 2 μg DER I per gram should be considered
moderate and representing a risk for sensitization; and less than 2 μg
DER I per gram poses little risk for a majority of atopic individuals.

The fact that indoor allergens are not uniformly distributed in the dust
of human dwellings makes it difficult to estimate allergen exposure with
a high degree of certainty. That is, part of the error in such quantifications may reside in the selection of the discrete object within a room to
sample. This difficulty may be overcome by collecting samples from
several discrete objects and then physically mixing them before making
measurement. The use of composite samples can also effectively reduce

Table 7. St. Charles Effluent Concentrations and Loads

Parameter	Conc (ppm)			Loads (ppm × m³/sec)		
	Max.	Mean	Skew/	Max.	Mean	Skew/
N	Min.	SD	Kurtosis	Min.	SD	Kurtosis
Hourly grabs						
NH₃	15.5	7.3	0.1	2.9	1.3	0.1
168	2.0	4.0	2.1	0.3	0.7	1.8
TSS–105	122.0	28.0	1.2	23.0	4.9	1.9
191	0.0	24.6	3.8	0.0	4.7	4.4
TSS–180	32.0	8.0	0.9	6.3	1.4	2.0
163	0.0	7.1	3.3	0.0	1.3	5.0
Daily time proportioned composites						
NH₃	17.5	9.6	0.1	3.3	2.2	0.1
50	5.0	2.8	2.8	1.1	0.5	2.8
TSS–105	13.0	39.0	2.2	30.1	8.7	1.9
65						
Daily volume proportioned composites						
NH₃	17.5	12.6	0.0	4.3	2.8	0.8
63	6.0	3.0	2.3	1.7	0.5	4.3
TSS–105	99.0	42.0	0.6	24.1	9.5	0.4
66	7.0	22.5	3.2	1.1	5.1	3.3

Source: Schaeffer et al.[21]

Note: Standard deviations (SD) are computed directly from sample data as s_c; variance correction for compositing, s_w^2 and autocorrelation τ, are not included.

Table 8. Composite Pair Differences—Volume Minus Time Proportioned Concentrations and Loads

Parameter (N)	Conc (ppm)		Loads (ppm × m³/sec)	
	Mean	t	Mean	t
Freeport NH₃ (100)	0.24	1.97	0.05	2.00
Freeport TSS (99)	22.50	3.70	4.47	3.83
St. Charles NH₃ (49)	2.99	9.42	0.65	9.00
St. Charles TSS (65)	2.54	0.90	0.71	1.08

Source: Schaeffer et al.[21]

the number of measurements necessary to provide reliable estimates of indoor allergen levels while minimizing the sample collection effort and analytical test costs. One of the major objectives of this study was to evaluate the use of composite samples as compared to individual samples from discrete objects.

Dust samples, discrete as well as composite, were collected from three specific objects from living rooms and bedrooms of 15 homes using a special filter sampling device connected to a vacuum cleaner. Discrete and composite samples were collected from floor, furniture (upholstery/bed), and window coverings in both rooms of each home. Composite samples were collected in defined sequence by vacuuming the three objects for 5 min each. Discrete samples were collected by vacuuming specified objects for 10 min. The laboratory procedure of dust extraction and allergen quantification is described by Lintner et al.[22] Tables 9 and 10 show the results of comparisons between discrete and composite dust samples. Composite sample values are plotted against the averages of the discrete samples in Figure 8. It is clear from this figure that composite sample values in this example are generally higher than the averages of the corresponding specific sample values. If this were caused by measurement error, then we would expect the variance of the composite samples to be smaller (not larger) than the variance of specific samples, since only one measurement was made instead of three. A comparison between the two variances shows that composite samples have a larger variance than

Table 9. Comparison of Mite Allergen Measurements from Composite and Discrete Dust Samples

Allergen	Comparison	p value
DER I	Bedroom vs living room	0.12
	Living room, composite vs discrete[a]	0.07
	Bedroom, composite vs discrete[b]	0.55
Der p I	Bedroom vs living room	0.01
	Living room, composite vs discrete[a]	0.23
	Bedroom, composite vs discrete[b]	0.53
Der f I	Bedroom vs living room	0.57
	Living room, composite vs discrete[a]	0.09
	Bedroom, composite vs discrete[b]	0.33

Source: Lintner et al.[22]
[a]Contrast for the living room composite: composite = floor/4 + upholstery/2 + window/4.
[b]Contrast for the bedroom composite: composite = floor/3 + bed/3 + window/3.

Table 10.   Comparison of Cat Allergen Measurements from Composite and Discrete Dust Samples (Signs test)

Allergen	Comparison	p value
Fel d I	Bedroom, composite vs discrete Composite = floor/3 + bed/3 + window/3	0.75
Fel d I	Living room, composite vs discrete Composite = floor/3 + bed/3 + window/3	0.55

Source: Lintner et al.[2]

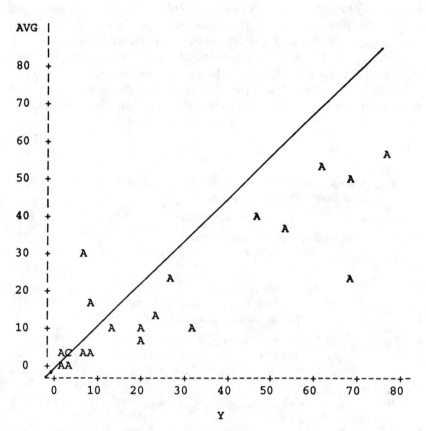

FIGURE 8.   Averages of specific samples plotted against the corresponding composite sample measurements.

specific samples. It remains to be verified that the process of dust extraction causes this discrepancy.

BIOACCUMULATION[23]

The human body or any living organism for that matter when exposed to a polluted environment, accumulates contaminants in its tissue. It is, therefore, very useful to sample tissue of such organisms under investigation, in order to evaluate the amount of accumulation (called bioaccumulation since biological processes cause the accumulation of a particular contaminant in the organism). It is a common observation that the tissue from a single member is not sufficient for making measurement. It is therefore necessary for technological reasons to composite the tissue samples extracted from several organisms so that a measurement is possible.

Compositing tissue samples extracted from several selected organisms represents an attempt to estimate the average concentration. If X_1, X_2, . . . , X_k represent the contaminant concentration of k tissue samples from k individual organisms, then these samples can be pooled to obtain a single composite measurement:

$$Y = \sum_{i=1}^{k} w_i X_i$$

where, for $i = 1, \ldots k$, w_i is the proportion of the contribution from the ith individual to the composite. Rohde[24] showed that the expected value and the variance of Y are given by:

$$E(Y) = \mu, \quad Var(Y) = \sigma^2/k + k\sigma_w^2\sigma^2$$

where μ = population mean
            σ^2 = population variance
            σ_w^2 = variance of the compositing proportions
            k = number of individual samples in each composite

If the w_i's are all equal, $w_i \equiv \frac{1}{k}$, then the numerical value of Y is equal to the average of the k sample values. That is, $Y = \bar{X}$. In this case, by analyzing only one composite sample, an estimate of the mean of k individual samples is obtained. However, due to compositing, the information on the individual sample variability is lost. This is true for a single composite sample. Replicate composite samples can be used in bioaccumulation monitoring programs to obtain a more accurate estimate of the population mean and to increase the precision of this estimate.

The comparison between a single composite and replicate individual samples can be extended to replicate composite samples (see Rohde[24,25]). The mean of n composite sample values $Y_1, Y_2, \ldots, Y_n$ is given by:

$$\bar{Y} = \sum_{j=1}^{n} Y_j/n$$

The expected value and the variance of $\bar{Y}$ are given by:

$$E(\bar{Y}) = \mu, \quad \mathrm{Var}(\bar{Y}) = \sigma^2/nk + k\sigma_w^2\sigma^2$$

In particular, if the composite samples comprise samples of equal mass so that $w_i \equiv \frac{1}{k}$, and hence $\sigma_w^2 = 0$, then:

$$Var(\bar{X}) = \sigma^2/k, \quad Var(\bar{Y}) = \sigma^2/nk$$

where n = the number of replicate samples (individual or composite) used in the estimate of the population variance (σ^2)

k = number of individual samples constituting each compositing sample

In this case it is easy to verify that:

$$\frac{Var(\bar{X})}{Var(\bar{Y})} = n$$

Thus, it can be seen that a collection of replicate composite tissue samples will result in a more efficient estimate of the mean. It should also be noted that for unequal proportions of composite samples, the variance of the composite sample mean increases with σ_w^2 and in extreme cases may even exceed the variance of the individual sample mean. A table of values for σ_w^2 that lead to such an increase is given by Schaeffer and Janardan.[26] Using the Dirichlet model for compositing probabilities, Rohde[25] has shown that:

$$\frac{Var(\bar{X})}{Var(\bar{Y})} = \frac{n + 1}{2}$$

as the increase in the precision that can be achieved at an additional cost of compositing.

Example: The National Human Adipose Tissue Survey

The National Human Adipose Tissue Survey (NHATS) is an annual survey to collect and analyze a sample of adipose tissue specimens from autopsied cadavers and surgical patients. The primary objectives of NHATS include:

- to identify chemicals that are present in the adipose tissue of individuals in the U.S. population
- to estimate the average concentration levels of selected chemicals in adipose tissue of individuals in the U.S. population and in various demographic subpopulations
- to determine if any of the four factors (namely, geographic region, age, race, and sex) affect the average concentration levels of selected chemicals detected in the U.S. population

Every year approximately 800–1200 adipose tissue specimens are collected using a multistage sampling plan. First, the 48 conterminous states are stratified into four geographic areas, which form four strata. Next, a sample of metropolitan statistical areas (MSAs) is selected from every stratum with probabilities proportional to MSA populations. Finally, several cooperators (hospital pathologists or medical examiners) are chosen from every selected MSA and asked to supply a specified quota of tissue specimens. The quota specifies the number of specimens needed in each of the categories defined by the donor's age, race, and sex. The categories are:

- age groups: 0–14 years, 15–44 years, and 45+ years
- race: Caucasian and non-Caucasian
- sex: male and female

The sampling plans were designed to give unbiased and efficient estimates of the average concentration levels of selected chemicals in the entire population and in various subpopulations defined by the demographic variables described above. Levels are characterized by the average or median chemical concentrations; prevalence is the proportion of individuals with chemical concentrations exceeding specified criterion levels.

Results from the Analysis of 1987 NHATS Data

The analysis was performed on data obtained from 48 composite samples formed from 865 adipose tissue specimens from sampled cadavers and surgical patients. Thus, each composite contained an average of 18 specimens. Not all of the chemicals provided sufficient data to perform a meaningful analysis. Two criteria were used to determine which

Table 11. Estimated Average Concentrations (pg/g) with Relative Standard Errors (%) for Selected Dioxins and Furans from FY87 NHATS Composite Samples

	Entire nation	Age group (yr)		
		0–14	15–55	45+
Population percentages	**100**	**23**	**46**	**31**
Compound				
Dioxins				
2,3,7,8-TCDD	5.38	1.98	4.37	9.40
	(6)	(41)	(12)	(4)
1,2,3,7,8-PECDD	10.7	3.30	9.33	18.2
	(4)	(22)	(7)	(4)
1,2,3,4,7,8/	75.1	23.4	70.9	120
1,2,3,6,7,8-HXCDD	(4)	(23)	(6)	(3)
1,2,3,7,8,9-HXCDD	11.7	6.13	10.8	17.1
	(4)	(18)	(7)	(4)
1,2,3,4,6,7,8-HPCDD	110	45.7	99.8	174
	(3)	(11)	(5)	(3)
1,2,3,4,6,7,8,9-OCDD	724	215	692	1150
	(4)	(17)	(7)	(5)
Furans				
2,3,7,8-TCDF	1.88	1.97	1.45	2.45
	(7)	(11)	(15)	(7)
2,3,4,6,7,8-PECDF	9.70	1.87	8.00	18.0
	(8)	(100)	(15)	(8)
1,2,3,6,7,8-HXCDF	5.78	1.80	4.59	10.5
	(13)	(83)	(26)	(13)

Source: Orban et al.[27]

chemicals should be analyzed. First, a chemical must be detected in at least 50% of the composites. Second, a minimum of 30 measurements were considered necessary to achieve sufficient precision of the estimates. Thus, of the 16 chemicals, there were 9 that met both criteria for performing the analyses. For each of the nine chemicals analyzed, Table 11 lists the estimated average concentration in the entire population and in the three age groups.

RANKED SET SAMPLING

Ranked set sampling (RSS) is a method of sampling which is mainly used for estimating a population mean. It utilizes prior information about the characteristic of interest for ranking the randomly selected sampling units from a population before resorting to quantification of some of the units so drawn. It is, in fact, useful when the quantification of a sampling unit is difficult but the randomly drawn units could be ranked by a visual inspection or some other crude method without knowing their exact measurements. Interestingly, it combines the convenience of purposive sampling and the control of simple random sampling (SRS). As the SRS estimator of a population mean, the corresponding RSS estimator provides an unbiased estimator of the population mean; however, the estimator is more efficient than that of the SRS estimator in almost all situations. In the worst circumstances when a ranking is equivalent to a random ordering, its performance reduces to that of SRS. In view of these facts, it has tremendous potential for considerably economical investigation when we need to estimate the mean in environmental problems. In this section, we have made an attempt to illustrate the method and examine its superiority over SRS for estimating the average PCB concentration in the surface soil along the gas pipeline of the Texas Eastern Gas Pipeline Company. We have also tried to explore its application in evaluating the effectiveness of insecticides and pesticides to protect and preserve the environment. For its application in other areas, see Patil et al.[10]

Method

The selection of a ranked set sample involves the drawing of m random samples each of size m from a population. Having drawn the random samples, the units of each sample are ranked by a visual inspection or any other method not involving the exact measurements of the variable of interest. The unit with the smallest rank is quantified from the first sample; the unit having the second smallest rank is measured from the second sample, and so on until the unit with the highest rank is used for the determination of the magnitude of the characteristic of interest from the mth sample. This yields m measurements corresponding to the quantification of m units out of m^2 randomly selected units, each representing a specific rank. The whole procedure is repeated r times which, in turn, gives mr quantified values in such a way that every rank has r quantified values. It is important to note that m^2r units are randomly selected from the population and used for ranking, but only the mr units are utilized for the determination of the magnitude of the characteristic. These measurements constitute a ranked set sample. If N and n denote the population and the sample size respectively, then:

$$N \geq m^2r \quad \text{and} \quad n = mr$$

In order to illustrate the method of drawing an RSS sample let us take $m = 3$ and $r = 2$. The scheme may be diagrammed following Stokes[28] and Muttlak,[29] as shown above.

In this diagram each row denotes a judgment ordered sample and the circled units indicate the units which are to be quantified. It means that 9 units are selected randomly from the population but only 3 units are quantified to get the required ranked set sample in each cycle.

r \ m	1	2	3
1	⊙	·	·
	·	⊙	·
	·	·	⊙
2	⊙	·	·
	·	⊙	·
	·	·	⊙

In general, let $X_{11}, X_{12}, \ldots, X_{1m}; X_{21}, X_{22}, \ldots, X_{2m}; \ldots; x_{m1}, X_{m2}, \ldots, X_{mm}$ be independent random variables all having the same cumulative distribution function $(cdf)F$. Then the ith order statistic from the ith sample is shown by $X_{(i:m)}$. In case the procedure of drawing random samples is repeated r times then the ith order statistic from the ith sample in the jth cycle is denoted by $X_{(i:m)j}$.

The RSS estimator $(\bar{X}_{(m)r})$ of the population mean (μ) is computed by:

$$\bar{X}_{(m)r} = \frac{\sum_{i=1}^{m}\sum_{j=1}^{r} X_{(i:m)j}}{mr}$$

or

$$\bar{X}_{(m)r} = \frac{1}{m}\sum_{i=1}^{m} \bar{X}_{(i:m)}$$

since

$$\bar{X}_{(i:m)} = \frac{1}{r}\sum_{j=1}^{r} X_{(i:m)j}$$

Further as $E\bar{X}_{(i:m)} = \mu_{(i:m)}$ and $\mu = \frac{1}{m}\sum_{i=1}^{m}\mu_{(i:m)}$ we get $E(\bar{X}_{(m)r}) = \mu$, where $\mu_{(i:m)}$ denotes the expected value of the ith order statistic. This suggests that $\bar{X}_{(m)r}$ is an unbiased estimator of the population mean (μ). The expression for the variance of $\bar{X}_{(m)r}$ is given by:

$$Var(\bar{X}_{(m)r}) = \frac{1}{m^2} \sum_{i=1}^{m} \frac{\sigma_{(i:m)}^2}{r}$$

where $\sigma_{(i:m)}^2$ represents the variance of the ith order statistic. Also,

$$Var(\bar{X}_{(m)r}) = \frac{1}{mr}\{\sigma^2 - \frac{1}{m}\sum_{i=1}^{m}(\mu_{(i:m)} - \mu)^2\}$$

where σ^2 denotes the population variance.

Comparison of the RSS Estimator with the SRS Estimator

In order to examine the performance of the two estimators under consideration we compare the precision of the RSS estimator relative to that of the SRS estimator with the same sample size. To compute the SRS estimator of the population mean, we need to draw a random sample of the size mr. For this, one unit is randomly selected from each sample in each cycle. The unit is then quantified. We denote the SRS estimator by $\bar{X}$. It is computed as shown below:

$$\bar{X} = \frac{\sum_{i=1}^{m}\sum_{j=1}^{r}X_{ij}}{mr}$$

Its variance is obtained as follows:

$$Var(\bar{X}) = \frac{\sigma^2}{mr}$$

The relative precision (RP) is defined as mentioned below:

$$RP = \frac{Var(\bar{X})}{Var(\bar{X}_{(m)r})}$$

or,

$$RP = \frac{1}{1 - \frac{1}{m}\sum_{i=1}^{m}\{\frac{\mu_{(i:m)} - \mu}{\sigma}\}^2}$$

Its limits are given by:

$$1 \le RP \le \frac{m+1}{2}$$

Takahasi and Wakimoto[30] gave the rigorous proof of the limits for all continuous distributions with finite variances. It means that almost $(m + 1/2)$ times as many random samples are required to equal the precision of the RSS estimator, provided ranking is perfect. The relative cost (RC) and the relative savings (RS) are computed as shown below:

$$RC = \frac{1}{RP}$$

$$RS = 1 - RC$$

and

$$RS = \frac{1}{m}\Sigma_{i=1}^{m} \left\{ \frac{\mu_{(i:m)} - \mu}{\sigma} \right\}^2$$

It means $RS \geq 0$.

Impact of Imperfect Ranking

As ranking of the randomly drawn units are carried out on the basis of some crude method in the absence of the exact magnitude of the characteristic of interest that a unit possesses, it may not always be feasible to perform the ordering correctly. Even in this situation of imperfect ranking Dell and Clutter[31] have shown that the RSS estimator of the population mean is unbiased and the $RP \geq 1$. However, the magnitude of RP gets reduced in this case. In fact, the higher the magnitude of ranking error the smaller is the magnitude of the RP. As a solution to the impasse David and Levine[32] and Stokes[33] have suggested to use some other variable which helps in ranking relatively more correctly and conveniently than the main variable of interest. The variable is known as a concomitant variable in the literature. It is supposed to be correlated with the variable of interest. If both the variables have the same cumulative distribution function and follow the bivariate normal distribution, then the RP of the RSS estimator with the ranking done on the basis of a concomitant variable, and the SRS estimator is given by:

$$RP = \frac{1}{1 - \frac{\rho^2}{m}\Sigma_{i=1}^{m} \left\{ \frac{\mu_{(i:m)} - \mu}{\sigma} \right\}^2}$$

where ρ denotes the correlation coefficient between the variable of interest and the concomitant variable. It is evident from the expression that the RP depends on the value of ρ^2.

Unequal Allocation of Sample Sizes

With a view to improve the magnitude of RP, McIntyre[34] and Takahasi and Wakimoto[30] have suggested allocating sample size to each group (i.e., rank order) proportional to its standard deviation.

Let r_i denote the number of times (i.e., the size of the sample of the ith group) the units with rank i to be quantified. Then, it is computed as follows:

$$r_i = \frac{n\sigma_{(i:m)}}{\Sigma_{i=1}^{m}\sigma_{(i:m)}}$$

$$i = 1, 2, \ldots, m$$

and

$$r_i + r_2, \ldots, r_m = n, r_i \geq 1$$

If T_i denotes the sum of the measurements for the units having the ith, rank, the RSS estimator ($\overline{X}_{(m)u}$) of the population mean is given by:

$$\overline{X}_{(m)u} = \frac{1}{m}\Sigma^m_{i=1}\frac{T_i}{r_i}, \; E(\overline{X}_{(m)u}) = \mu$$

and

$$Var(\overline{X}_{(m)u}) = \frac{1}{m^2}\Sigma^m_{i=1}\frac{\sigma^2_{(i:m)}}{r_i}$$

In order to estimate $Var(\overline{X}_{(m)u})$, r_i should be greater than or equal to two. The RP of the RSS estimator relative to the SRS estimator is defined as follows:

$$RP = \frac{\sigma^2/n}{\frac{1}{m^2}\Sigma^m_{i=1}\frac{\sigma^2_{(i:m)}}{r_i}}$$

and

$$0 \leq RP \leq m$$

See Takahasi and Wakimoto.[30] However, it is important to note that the RP under the equal allocation is less than or equal to that of the unequal allocation provided the allocation is carried out proportional to the standard deviation of each group. Further, if $r_1 = r_2 = \ldots = r_m$ then the RSS design is said to be balanced; otherwise it is unbalanced.

Illustration

With the aim to illustrate the effectiveness of RSS relative to SRS in estimating the level of concentration of polychlorinated biphenyls (PCB) at the Armagh site along the gas pipeline of the Texas Eastern Gas Company, we examine the schemes mentioned below:

1. balanced allocation of samples using all possible combinations of each set size
2. balanced allocation of samples for a specific sample
3. unbalanced allocation of samples

For this purpose we use the measurements of the contaminant obtained by using grids A and C each with phase I and II together. Table 12

Table 12. Number of Observations, Mean, SD, CV, Coefficient of Skewness, and Kurtosis of PCB Values in Grids A and C

Characteristics	Grid	
	A	C
Number of observation	184	68
Mean	200.9	600.2
SD	902.9	1585
CV	4.49	2.64
Coefficient of skewness	9.27	4.48
Coefficient of kurtosis	99.69	20.88

gives the number of observations, mean, standard deviation, coefficient of variation, coefficients of skewness, and kurtosis of PCB values for grids A and C separately. For applying RSS protocol we have considered set sizes two, three, and four. Using all possible combinations of the PCB values for each set size, the magnitude of the relative savings (RS) due to the RSS estimator relative to that of the SRS estimator has been computed using balanced allocation of samples. The findings are mentioned in Table 13. We observe that the value of RS increases with the set size, but its amount is higher for grid C than grid A because the values of PCB under the former are more homogeneous and symmetric than those of the latter. Also, the values of $\bar{X}_{(m)r}$, RP, and RS have been computed for a specific sample. These results are mentioned in Table 14. It is evident that the results suffer from the sampling fluctuation. We find from Table 15 that the values of RS are quite substantial due to unequal allocation of samples. It is obvious that RSS with this allocation performs much better than with equal allocation. Though it is difficult to determine the proportion of samples in advance in the absence of the standard deviation at each rank order for an unknown population, one could take help of other surveys of similar nature conducted earlier or conduct a preliminary survey based on a small sample size.

Table 13. Relative Saving (RS) Considering All Possible Combinations of Each Set Size Under Perfect Ranking with Balanced Allocation

Set size (m)	Grid	
	A	C
	RS	RS
2	4	9
3	7	16
4	10	22

Table 14. The Values of $\bar{X}_{(m)r}$, RP, and RS Under Perfect Ranking with Balanced Allocation in the Case of a Specific Sample

Set size	Grid					
	A			C		
(m)	$\bar{X}_{(m)r}$	RP	RS	$\bar{X}_{(m)r}$	RP	RS
2	155.035	1.02	2	711.3	1.14	12
3	286.187	1.19	16	223.6	1.01	1
4	166.817	1.07	7	635.9	1.56	36

RSS may also be used for forming relatively more homogeneous composite samples compared to those based on random groupings. With m samples of size m we form m composite samples by physically mixing the units of the same rank before resorting to quantification. Thus, we get mr composite samples out of m^2r units drawn from the population. The standard deviation of these measurements is expected to have smaller variance than the same number of measurements obtained after quantifying the composite samples obtained by random groupings of the units. The results are summarized in Table 16 and 17 for grids A and C, respectively.

In view of the findings we may conclude that it has enormous capability for application in evaluating the impact of chemical treatments, for example, on vegetation, much more economically. Its dependence on

Table 15. The Values of $\bar{X}_{(m)u}$, RP, and RS Under Perfect Ranking with Unbalanced Allocation of Samples

Set size (m)	Grid A				Grid C			
	Proportion of samples (exact no.)	$\bar{X}_{(m)u}$	RP	RS	Proportion of samples (exact no.)	$\bar{X}_{(m)u}$	RP	RS
2	1:10 (8, 84)	205.9	1.724	42	1:10 (3, 31)	535.2	2.041	51
2	1:15 (6, 86)	203.1	1.818	45	1:15 (2, 32)	520.4	2.174	54
3	1:4:20 (2, 10, 48)	203.6	2.174	54	1:1.7:1.5 (5, 8, 8)	560.1	1.471	32
3	1:4:25 (2, 8, 50)	201.1	2.326	57	1:2:7 (2, 4, 15)	615.2	1.923	48
4	1:3:5:16 (8, 5, 9, 28)	247.1	1.695	41	1:2:3:4 (2, 3, 5, 6)	576.6	2.083	52
4	1:3:9:27 (2, 2, 10, 30)	226.1	1.316	24	1:1:3:5 (2, 2, 4, 8)	802.4	1.449	31

Table 16.   Sample Size, Mean, and Standard Deviation for Individual Samples, Composite Samples, and Composites of Ranked Samples for Grid A

Set size	Item	Sample size	Mean	SD
2	Individual samples	184	200.72	902.9
	Composite samples	92	200.72	627.9
	Composites of ranked samples	92	200.72	618.4
3	Individual samples	180	183.8	870.7
	Composite samples	60	183.8	490.6
	Composites of ranked samples	60	183.8	470.4
4	Individual samples	176	187.8	880.2
	Composite samples	44	187.8	509.8
	Composites of ranked samples	44	187.8	321.5

Table 17.   Sample Size, Mean, and Standard Deviation (SD) for Individual Samples, Composite Samples, and Composites of Ranked Samples for Grid C

Set size	Item	Sample size	Mean	SD
2	Individual samples	68	601	1585
	Composite samples	34	601	1067
	Composites of ranked samples	34	601	982.8
3	Individual samples	63	599	1630
	Composite samples	21	599	865
	Composites of ranked samples	21	599	663.3
4	Individual samples	64	590	1618
	Composite samples	16	590	761.0
	Composites of ranked samples	16	590	952.7

prior information of the characteristic of interest for ranking can be tackled effectively to a great extent by utilizing the experience and the expertise of the field personnel. For applications in other areas, see Patil et al.[10]

ACKNOWLEDGMENT

Adapted from the Keynote Address by G. P. Patil to the Symposium on Environmental Statistics of the American Chemical Society, 1992. It has been prepared with partial support from the Statistical Analysis and Computing Branch, Environmental Statistics and Information Division, Office of Policy Planning, and Evaluation, U.S. Environmental Protection Agency, Washington, D. C. under a Cooperative Agreement Number CR-815273. The contents have not been subjected to Agency review and, therefore, do not necessarily reflect the views of the Agency and no official endorsement should be inferred.

REFERENCES

1. G. P. Patil (1991). Encountered data, statistical ecology, environmental statistics, and weighted distribution methods. *Environmetrics*, 2(4), 377–423.
2. G. P. Patil, C. Taillie, and S. Talwalker (1992). Encounter sampling and modeling in ecological and environmental studies using weighted distribution methods. Technical Report 92-0402, Center for Statistical Ecology and Environmental Statistics, Department of Statistics, Pennsylvania State University, University Park, PA.
3. M. T. Boswell, S. D. Gore, and G. P. Patil (1990). Efficiency of various composite sample retesting schemes to classify samples with presence/absence measurements. Technical Report 90-0901, Center for Statistical Ecology and Environmental Statistics, Department of Statistics, Pennsylvania State University, University Park, PA.
4. M. T. Boswell, S. D. Gore, G. D. Johnson, and G. P. Patil (1992). Composite sampling protocols for site characterization and evaluation of cleanup attainment. Technical Report 92-0401, Center for Statistical Ecology and Environmental Statistics, Department of Statistics, Pennsylvania State University, University Park, PA.
5. M. T. Boswell, S. D. Gore, G. Lovison, and G. P. Patil (1992). Annotated bibliography of composite sampling. Technical Report 92-0802, Center for Statistical Ecology and Environmental Statistics, Department of Statistics, Pennsylvania State University, University Park, PA.

6. S. D. Gore, G. P. Patil, and C. Taillie (1992). Studies on the applications of composite sample techniques in hazardous waste site characterization and evaluation: II. Onsite surface soil sampling for PCB at the Armagh site. Technical Report No. 92-0305, Center for Statistical Ecology and Environmental Statistics, Department of Statistics, Pennsylvania State University, University Park, PA.

7. S. D. Gore, G. P. Patil, and C. Taillie (1992). Manual on design and analysis with composite samples. Technical Report No. 92-1002, Center for Statistical Ecology and Environmental Statistics, Department of Statistics, Pennsylvania State University, University Park, PA.

8. G. P. Patil, S. D. Gore, and C. Taillie (1992). Monograph on composite sampling — a novel method to accomplish observational economy in environmental studies. Technical Report No. 92-1001, Center for Statistical Ecology and Environmental Statistics, Department of Statistics, Pennsylvania State University, University Park, PA.

9. G. P. Patil, A. K. Sinha, and C. Taillie (1992). Ranked set sampling annotated bibliography. Technical Report 91-1201, Center for Statistical Ecology and Environmental Statistics, Department of Statistics, Pennsylvania State University, University Park, PA.

10. G. P. Patil, A. K. Sinha, and C. Taillie (1992). Ranked set sampling. Technical Report 91-1202, Center for Statistical Ecology and Environmental Statistics, Department of Statistics, Pennsylvania State University, University Park, PA.

11. M. A. Watson (1936). Factors affecting the amount of infection obtained by aphis transmission of the virus Hy. III. *Philos. Trans. R. Soc. London, Ser. B.*, 226, 457-489.

12. R. Dorfman (1943). The detection of defective members of large populations. *Ann. Math. Stat.*, 14, 436-440.

13. M. T. Boswell and G. P. Patil (1987). A perspective of composite sampling. *Commun. Statist. — Theory Meth.*, 16(10), 3069-3093.

14. M. T. Boswell and G. P. Patil (1990). Composite sample designs for characterizing continuous sample measures relative to a criterion. Technical Report No. 90-1001, Center for Statistical Ecology and Environmental Statistics, Department of Statistics, Pennsylvania State University, University Park, PA.

15. PA DER (1991). Case summary: Texas Eastern Gas Pipeline Company. Division of Special Investigations, Bureau of Waste Management, PA Department of Environmental Resources, Harrisburg, PA.

16. G. Lovison, S. D. Gore, and G. P. Patil (1992). Design and analysis of composite sampling procedures: A review. Technical Report 92-1007, Center for Statistical Ecology and Environmental Statistics, Department of Statistics, Pennsylvania State University, University Park, PA.

17. D. Casey, P. N. Nemetz, and D. Uyeno (1985). Efficient search procedures for extreme pollutant values. *Environ. Monitoring Assessment*, 5, 165-176.

18. R. Rossman (1988). Estimation of trace metal storage in Lake St. Clair post-settlement sediments using composite samples, *J. Great Lakes Res.*, 14(1), 66-75.

19. D. L. Clark, R. Asplund, J. Ferguson, and B. W. Mar (1981). Composite sampling of highway runoff. *J. Environ. Eng.*, 107, 1067–1081.
20. R. E. Wullshleger, A. E. Zanoi, and C. A. Hanson (1976). Methodology for the study of urban storm generated pollution and control. Envirex, Inc., Contract 68-03-0335, for Environmental Protection Agency.
21. D. J. Schaeffer, H. W. Kerster, D. R. Bauer, K. Rees, and S. McCormick (1983). Composite samples overestimate waste loads. *Water Pollut. Control Fed. J.*, 55(11), 1387–1392.
22. T. J. Lintner, C. L. Maki, K. A. Brame, and M. T. Boswell (1992). Sampling dust from human dwellings to estimate indoor allergens. Technical Report 92-0805, Center for Statistical Ecology and Environmental Statistics, Department of Statistics, Pennsylvania State University, University Park, PA.
23. Tetra Tech (1987). Bioaccumulation monitoring guidance: strategies for sample replication and compositing. U.S. EPA, Office of Marine and Estuarine Protection, Washington, DC.
24. C.A. Rohde (1976). Composite sampling. *Biometrics*, 32, 273–282.
25. C. A. Rohde (1979). Batch, bulk and composite sampling. "*Sampling Biological Populations*," R. M. Cormack, G. P. Patil, and D. S. Robson, Eds., International Co-operative Publishing House, Fairland, MD, 365–377.
26. D. J. Schaeffer and K. G. Janardan (1978). Theoretical comparison of grab and composite sampling programs. *Biometrical J.*, 20, 215–227.
27. J. D. Orban, R. Lordo, and J. Schwemberger (1990). Statistical methods for analyzing composite sample data applied to EPA's human monitoring program. MS.
28. S. L. Stokes (1986). Ranked set sampling. *Encyclopedia of Statistical Sciences* S. Kotz et al. (Eds.) John Wiley & Sons, New York, 585–588.
29. H. A. Muttlak (1988). Some aspects of ranked set sampling with size biased probability of selection. PhD. Thesis, University of Wyoming, Laramie, WY.
30. K. Takahasi and K. Wakimoto (1968). On the unbiased estimates of the population mean based on the sample stratified means of ordering. *J. Ann. Inst. Stat. Math.*, 20, 1–31.
31. T. R. Dell and J. L. Clutter (1972). Ranked set sampling theory with order statistics background. *Biometrics*, 28, 545–553.
32. H. A. David and D. N. Levine (1972). Ranked set sampling in the presence of judgement error. *Biometrics*, 28, 553–555.
33. S. L. Stokes (1977). Ranked set sampling with concomitant variables. *Commun. Stat. Theory Methods*, A6, 1207–1211.
34. G. A. McIntyre (1952). A method of unbiased selective sampling, using ranked sets. *Aust. J. Agric. Res.*, 3, 385–390.

CHAPTER 4

Predictive Models of Fish Response to Acidification: Using Bayesian Inference to Combine Laboratory and Field Measurements

William J. Warren-Hicks and Robert L. Wolpert

SUMMARY

Bayesian inference techniques are used to develop a statistical model predicting brook trout (*Salvelinus fontinalis*, Mitchell) population response to acidification. A statistical framework is created for combining two independent types of data: laboratory bioassay data (fish survival/death) and field survey observations (presence/absence) of fish response to acid-base chemical conditions. The Bayesian framework combines these two types of data using a uniform noninformative prior distribution. Combining the bioassay and field data in the model development process overcomes the effects of multicollinearity that occur in models developed from the field data only. In addition, Bayesian methods including graphic displays of the data provide valuable tools for solving issues associated with model development from observational data sets.

MOTIVATION

The National Acid Precipitation Assessment program (NAPAP) recently concluded over 10 years of research into the effects of acidification on the environment. In part, the NAPAP assessment was concerned with the regional effects of acidification on potentially susceptible surface waters. Models predicting fish population response to acidification, as the one presented in this study, were used to estimate regional population changes under a variety of acidic deposition scenarios. The regional estimates of fish response under differing deposition scenarios were used to assess the magnitude of the acidification effects on fish populations.

The primary sources of data available for exploring the relationship between acidification and fish response to acidification are field surveys of potentially impacted lakes and laboratory bioassay experiments. The field survey data provide information on fish population response under actual field conditions. Because the number of confounding factors affecting fish response in the field is large, identifying the effects of acid-base surface water chemistry on fish survival can be difficult. The laboratory data are generated under strictly controlled conditions, and as such are assumed to provide an accurate representation of the relationship between fish response and concentrations of specific water chemicals. Data available from each of these sources indicate a strong association between fish survival and low pH and calcium (Ca) concentrations, and elevated concentrations of inorganic aluminum (Al).

A major problem encountered in the development of predictive models of fish response to acidification is the presence of strong multicollinearity among the chemicals acting as predictor variables.[1,2] For example, field measurements of Al and Ca tend to be strongly correlated with pH. Changes in the relationship between the measured chemical concentrations over time can affect the applicability of the predictive model. For example, most atmospheric deposition comes in contact with terrestrial components of the ecosystem prior to reaching a lake. The terrestrial components can assimilate acidic deposition by (1) dissolution of carbonate-containing materials; (2) cation exchange reactions with minerals releasing Ca^{2+}, Mg^2, Na^+, or K^+ in exchange for H^+, or (3) chemical reactions releasing Al^{3+}, Fe^2, or Mn^{2+}. In an area receiving deposition of strong acids, an increase in the export of cations and metals and an increase in acid anions can be expected.[3] On examination of a cross section of lakes in this region, those lakes with high pH will be expected to have higher levels of Ca. If over time the amount of acidic loading decreases, then in a particular lake an increase in pH without a corresponding change in Ca can occur.

These changes in the relationship between pH and Ca concentrations will result in a change in the collinearity structure of the data over time. If the model is applied to a data set exhibiting the same collinearity structure as the data used to develop the model, then the model can be expected to perform adequately.[4] However, if the collinearity structure among the water chemistry variables changes over space or time, then the model may not be appropriate. One method for reducing the effects of collinearity on model prediction uncertainty is to incorporate additional information. In this study, Bayesian methods will be used to combine laboratory bioassay data with field monitoring data to reduce the effect of multicollinearity on model predictions.

THE DATA

The field data set was drawn from a survey of some 1469 Adirondack lakes conducted between 1984 and 1987 by the Adirondack Lake Survey Corporation (ALSC), as reported in Baker et al.[2] It consists of data from 313 lakes: 177 from the 1984–1985 survey, and 136 from the 1986–1987 survey. Although the ALSC reported measurements of many variables from each lake, only four were considered in the present study: the presence or absence of brook trout and the three water chemistry variables pH, inorganic Al, and Ca. The 1984–1985 field survey data are used for model calibration, while the 1986–1987 survey is used for model verification and testing.

The bioassay data consist of survival observations of an early brook trout life stage (swim-up fry). Similar studies from three different investigators were included.[5-7] Each investigator manipulated pH and Al concentrations (and, in one case, Ca concentration) in repeated bioassays, placing a number (usually 25) of fish in flow-through exposure chambers and recording daily observations of the number of survivors. Observations were terminated and the experiment brought to a close, after a predetermined period ranging from 10–21 days. Variations in experimental protocol among investigators were minor and appear to be inconsequential.[8] The final bioassay data set consists of 163 experiments, including 4096 individual fish responses.

BAYESIAN INFERENCE

The major difference between Bayesians and non-Bayesians is the concept of probability employed.[4] For the non-Bayesian, probability is regarded as representing the frequency with which an event would occur in repeated trials. For the Bayesian, probability is regarded as representing a degree of reasonable belief based on existing information. This information takes on two forms: sample information and prior information. Probability statements about the parameters of interest are made based only on these two sets of information. A major advantage of the Bayesian framework is that it provides a method for statistically combining these sources of data.

Sample information enters the Bayesian framework through the likelihood function. For a Bayesian, the likelihood is regarded as a function of the model parameters, θ, for a given set of data. Only the sample data available for the particular problem of interest are used in the analysis. Prior information represents that information available prior to taking the sample. The prior information represents a direct probability statement about θ, which can be generated from both subjective and empirical evidence. Prior data can be generated from historical studies, output from a previous Bayesian analysis, interviews with scientific experts, the investigator's own beliefs, or a special class of priors called noninformative or diffuse prior information. A noninformative prior contains no information about θ (i.e., does not favor any possible value of θ over any other).

Bayes theorem for continuous random variables is written as:

$$p(\theta|y) = \frac{L(y|\theta)\pi(\theta)}{\int L(y|\theta)\pi(\theta)d\theta}$$

$$= \frac{p(\theta,y)}{p(y)}$$

where θ = a vector of parameters
y = a vector of responses
$L(y|\theta)$ = likelihood function of the measured responses
$\pi(\theta)$ = prior distribution
$p(y)$ = predictive distribution of y
$p(\theta|y)$ = the posterior probability distribution function of θ
$p(\theta,y)$ = joint distribution of θ and y

The posterior distribution contains all current information about the parameters of interest. It can be thought of as a revised description of the parameters, the prior belief having been updated by the sample data y. Information about θ, for example, the mean and standard deviation, can be estimated directly from the posterior distribution. In this study, the means of the posterior distributions are used for the coefficients in the logistic regression model; the standard deviations represent the uncertainty in the coefficients. The mean of the posterior distribution is frequently chosen in Bayesian approaches because it can be shown to be an optimal estimate under common error assumptions.[9] Unlike classical statistics, however, regions under the posterior distribution are direct probability statements about the uncertainty in the parameters. These areas, sometimes labeled highest posterior density (HPD) regions, are the probability regions in which the parameter occurs (for example, 95% change of occurrence). Direct probability statements about uncertainty essentially require Bayesian analysis, and the thrust of classical statisticians has been to find alternate ways of indicating accuracy.[9] The posterior distribution of a single parameter (marginal distribution) can be plotted. Plotting the marginal distributions of parameters in different models against each other on a single graph is an excellent method of model comparison. If additional data are collected after calculating the posterior distribution, the posterior becomes the prior distribution and the posterior is updated. Therefore, Bayes' theorem provides an easy way to combine sources of data.

MODELS AND LIKELIHOOD FUNCTIONS

Three distinct models will be developed in this section. The first will be a four-dimensional model calibrated from field observations of brook trout presence/absence. The second is a four-dimensional model calibrated from laboratory observations of brook trout survival. The third is a six-dimensional hierarchical model, consisting of exchangeable field and laboratory submodels that reflect features of both the field and laboratory settings. Further details on these and other models are described in Wolpert and Warren-Hicks.[10]

The models take the following form: conditional on the observed vector x_i of explanatory variables and on an uncertain parameter vector θ, the indicator variables s_i of fish presence in the ith lake (for i in a set I^F indexing the lakes) and the survivor counts s_i of fish in the ith laboratory bioassay experiment (for i in a set I^L indexing the bioassays) are all taken to be independent binomial $Bi(n_i, p_i)$ random variables. In each model considered, n_i is known and a specified functional form for the field presence probabilities $p_i = p^F(x_i, \theta)$, for $i \in I^F$; and laboratory survival probabilities $p_i = p^L(x_i, \theta)$, for $i \in I^L$ is assumed. Thus, in every case the log likelihood function can be written in the form:

$$l(\theta) = c + \sum_{i \in I^F \cup I^L} \left[s_i \log(p_i) + (n_i - s_i) \log(1 - p_i) \right]$$

$$p_i = \begin{cases} p^F(x_i, \theta) \text{ for } i \in I^F \\ p^L(x_i, \theta) \text{ for } i \in I^L \end{cases}$$

only the dimension of θ and the form of the probability functions $p^F(x_i, \theta)$ and $p^L(x_i, \theta)$ change from one model to another. The explanatory variables in all cases consist of an intercept constant $x_{i0} = 1$ and three recorded water chemistry indicators on a logarithmic scale: $x_{i1} = \text{pH}$, $x_{i2} = \log_{10}[\text{Al}]$, and $x_{i3} = \log_{10}[\text{Ca}]$.

A Logistic Regression Model for Field Observations

The most commonly recommended statistical procedure for studying the relationship of binary data (in this case, indicators s_i of the presence of brook trout in a lake) to one or more explanatory variables is binary regression, especially multiple logistic regression (e.g., Cox,[11] McCullagh and Nelder,[12] Bishop et al.[13]). In this procedure, the success probabilities p_i depend on a linear function of the vector x_i of explanatory variables $[x_{ij}]$, and so can be written in the form:

$$p_i = G\left(\Sigma_j x_{ij}\beta_j^F\right)$$

for some parameter vector β^F and specified cumulative distribution function (CDF) $G(x)$ (often the normal $\Phi(x)$ or the logistic $\Psi(x) = e^x/(1 + e^x)$). In the following notation, the sum is written as the ith component $x\beta_i^F$ of the matrix product $x\beta^F$. Experimental evidence concerning β^F is reflected in the log likelihood function, which in the logistic case is simply:

$$l\left(\beta^F\right) = \sum_{I \in I^F}\left[s_i x\beta_i^F - n_i \log\left(1 + e^{x\beta_i^F}\right)\right]$$

Routine numerical methods suffice to find the point where $l\left(\beta^F\right)$ attains its maximum (the *maximum likelihood estimator* [MLE] β^F).

A Logistic Regression Model for Laboratory Bioassay Experiments

A logistic regression model similar to that above can be used to describe the number s_i of survivors from among n_i swim-up fry in a controlled laboratory bioassay experiment (indexed by $i \in I^L$) with explanatory variable vector x_i. The likelihood function for the laboratory parameter vector β^L is simply:

$$l\left(\beta^L\right) = \sum_{I \in I^L}\left[s_i x\beta_i^L - n_i \log\left(1 + e^{x\beta_i^L}\right)\right]$$

An Exchange Hierarchical Bayesian Model

From a naive perspective, the two independent logistic regression models (field and laboratory) can be combined to form a single eight-dimensional model with log likelihood function:

$$l\left(\beta^F,\beta^L\right) = l\left(\beta^F\right) + l\left(\beta^L\right)$$

or

$$l(\beta^F,\beta^L) = \sum_{i \in I^F}\left[s_i x\beta_i^F - n_i \log\left(1 + e^{x\beta_i^F}\right)\right]$$

$$+ \sum_{i \in I^L}\left[s_i x\beta_i^L - n_i \log\left(1 + e^{x\beta_i^L}\right)\right]$$

However, what is missing is some way to link the field and laboratory bioassay evidence together, i.e., to relate the logistic regression parameter vectors β^F governing the field presence and β^L for the bioassay survival. Without such a link the laboratory data are completely independent of the field data, and cannot help resolve the multicollinearity.

In the Bayesian approach, the link is provided by a joint prior distribution for the two parameters vectors, $\pi^B(d\beta^F\,d\beta^L)$. Such a Bayesian approach seems especially appropriate for this application since it is only a *prior belief* that the laboratory data offer evidence about the field conditions that leads us to combine field and laboratory data. This argument is a major reason for attacking this problem from the Bayesian perspective.

The hierarchical model developed here employs a joint prior — $\pi^B(d\beta^F\ d\beta^L)$ — expressing the prior belief that while the survival rates of swim-up fry in laboratory bioassay experiments might be quite different from the fractions of lakes that support brook trout, still the "relative potency" of pH, inorganic aluminum, and calcium ought to be the same in the two settings. If a small increase of 0.01 in pH has the same effect as a 40% decrease in the inorganic aluminum concentration or a 3.9% increase in calcium concentration in the laboratory, for example, the same ought to hold in the field.

One way to describe a model embodying this assumption of equal relative potency is to introduce hyperparameters α_A and α_C for the *potencies* (relative to pH) of aluminum and calcium, respectively, so that the presence and survival probabilities depend on the three chemistry variables only through the quantity (which we think of as the "threat") $T_i = -pH_i + \alpha_A \log_{10}[Al_i] + \alpha_C \log_{10}[Ca_i]$, through the relations:

$$\log \left(\frac{p_i^F}{1 - p_i^F} \right) = \beta_0^F - \beta_1^F T_i \qquad \log \left(\frac{p_i^L}{1 - p_i^L} \right) = \beta_0^L - \beta_1^L T_i$$

Thus the log likelihood is:

$$l(\alpha_A, \alpha_C, \beta_0^F, \beta_1^F, \beta_0^L, \beta_1^L) = \sum_{i \in I^F} [s_i (\beta_0^F - \beta_1^F T_i) - n_i \log (1 + e^{\beta_0^F - \beta_1^F T_i})]$$

$$+ \sum_{i \in I^L} [s_i (\beta_0^L - \beta_1^L T_i) - n_i \log (1 + e^{\beta_0^L - \beta_1^L T_i})]$$

The field and laboratory submodels remain logistic, but they are no longer independent. A Bayesian analysis in which the laboratory data and field observations are related *only* through the relative potencies can be expressed in hierarchical form with a (hyper)prior distribution $\pi^\alpha (d\alpha_A d\alpha_C)$ and conditional distributions $\pi^{F|\alpha} (d\beta_0^F d\beta_1^F | \alpha_A, \alpha_C)$ and $\pi^{L|\alpha} (d\beta_0^L\ d\beta_1^L | \alpha_A \alpha_C)$. If these are identical, the resulting model will be *exchangeable* and will treat laboratory data and field observations in an entirely symmetric fashion.

Since the concern in this application is *not* symmetric, an equivalent but more convenient choice of parameters is used: select $\theta = (\beta_0^F, \beta_1^F, \beta_2^F, \beta_3^F, \delta, \sigma)$ and treat $\beta^L = \sigma\beta^F + \delta$ as a derived quantity, leading to the log likelihood expression $l(\beta^F, \delta, \sigma) = l(\beta^F) + l(\sigma\beta^F + \delta)$, or:

$$l(\beta_0^F, \beta_1^F, \beta_2^F, \beta_3^F, \delta, \sigma) = \sum_{i \in I^F} [s_i x \beta_i^F - n_i \log (1 + e^{x\beta_i^F})]$$

$$+ \sum_{i \in I^L} [s_i (\sigma x \beta_i^F + \delta) - n_i \log (1 + e^{\sigma x \beta_i^F + \delta})]$$

Estimation Techniques

Calculating maximum likelihood estimates and Bayesian posterior means of the model parameters is a challenge whenever nonlinear models are evaluated. Maximum likelihood estimates for the parameters of each model were found using an interactive modified Newton-Raphson routine developed by Stallard.[14] Calculating the Bayesian posterior means and predictive means requires evaluating the ratio of two 6-dimensional integrals. Tensor product quadrature rules are hopelessly inefficient for such a task, but Monte Carlo methods with appropriate variance-reduction techniques (described in Wolpert[15]) permit simultaneous calculation of the relevant means, variances, and covariances to a precision of less than 1% in a few minutes on a desk top Unix workstation.

Noninformative Prior Information

The likelihood functions for the models were combined with a uniform noninformative prior in the Bayesian estimation scheme. Opinions differ on what (if anything) would constitute a noninformative prior distribution. After several sensitivity tests with various prior distributions, the resulting parameter estimates were shown to be quite similar. Therefore, the uniform distribution was chosen, principally for ease of implementation. In addition, it is possible to regard the laboratory data as generating an informative prior distribution for the field data; details of this formulation are presented in Wolpert and Warren-Hicks.[10]

DISCUSSION

The effects of multicollinearity can be examined using both classical and Bayesian tools. Of particular interest are the large standard deviations of the field-based model parameters presented in Tables 1 and 2 (maximum likelihood and Bayesian estimates, respectively). The large variances indicate that the model is badly misspecified, probably due to the effect of multicollinearity. As further evidence of multicollinearity in the field data, note the predictive distributions of the field-based model for two sets of predictor variables; one set near the locus of multicollinearity [x = 1, 4.9, 1.3, 1.7] and the other far away [x = 1, 5.5, -8,8, -0.3] (Figure 1). Note that the model predictions far away from the

Table 1. Maximum Likelihood Parameter Estimates

Model:	Field	Laboratory	Hierarchical
β_0	-10.04 ± 2.60	-24.31 ± 0.93	-8.53 ± 1.82
β_{pH}	1.72 ± 0.60	5.59 ± 0.20	1.77 ± 0.34
β_{Al}	-0.10 ± 0.30	-0.88 ± 0.06	-0.28 ± 0.06
β_{Ca}	1.01 ± 1.51	0.68 ± 0.13	0.22 ± 0.06
μ			-0.85 ± 0.42
$\log \sigma$			1.15 ± 0.19

Note: Each entry is the maximum likelihood estimate (MLE) plus or minus one standard error.

multicollinearity are poorly determined, indicating that the model is not appropriate for application to data sets with collinearity structures different from the calibration data set. In addition, a frequently used classical statistics measure of multicollinearity, the condition number, is 39.9 for the standard error covariance matrix derived from the field observations (indicating extreme multicollinearity). Another manifestation of multicollinearity in the field-based model is illustrated in the likelihood contour plot of Figure 2, showing the profile likelihood for the pH and aluminum logistic regression coefficients in the neighborhood of the maximum likelihood estimate. Note how poorly the logistic regression parameters are determined by the field observations, and how well they are determined by the laboratory data. The experimenter's ability to control the explanatory variables in the laboratory experiment substantially reduces the multicollinearity which plagues the field observations, as reflected in the lower condition number (13.74) for the standard error covariance matrix derived from the laboratory measurements.

Several statistical tools can be used to judge the influence of incorporating bioassay data into the predictive model. For example, Figure 3 presents overlaid plots of the posterior marginal densities for the field pH coefficient from the field model, the laboratory pH coefficient from

Table 2. Posterior Parameter Means

Model:	Field	Laboratory	Hierarchical
β_0	-10.54 ± 2.64	-24.38 ± 0.93	-9.23 ± 1.06
β_{pH}	1.76 ± 0.62	5.61 ± 0.20	1.90 ± 0.21
β_{Al}	-0.12 ± 0.30	-0.89 ± 0.06	-0.30 ± 0.04
β_{Ca}	1.18 ± 1.56	0.69 ± 0.13	0.23 ± 0.05
μ			-2.81 ± 0.61
$\log \sigma$			1.08 ± 0.11

Note: Each entry is the posterior means plus or minus one posterior standard deviation.

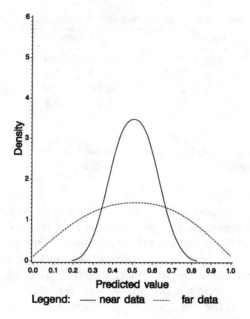

FIGURE 1. Predictive densities: variable sets near and far from the collinear surface.

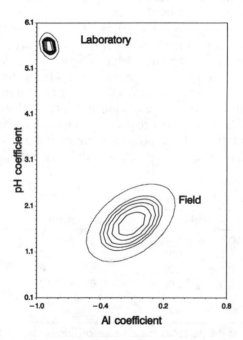

FIGURE 2. Profile likelihood contour plot: field and laboratory model coefficients.

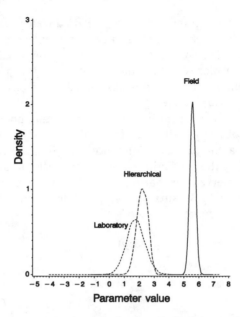

FIGURE 3. Marginal posterior densities for the pH coefficient in three models.

the laboratory model, and the pH coefficient from the hierarchical model. Note that the distribution of the hierarchical coefficient is centered at about the same value as the field-based parameter, but the spread is much narrower (i.e., the uncertainty is much less) reflecting the influence of the laboratory data. Similar features were exhibited by plots (not shown) for the inorganic Al and Ca coefficients. This finding increases the confidence that the hierarchical model is reasonably specified. In all cases, the hierarchical coefficients are considered superior to the field-only model coefficients because they incorporate both laboratory and field information.

The parameters appear better determined in the hierarchical model than in the field-based model (Table 2). The posterior standard deviations display a reduction in parameter uncertainty for the hierarchical model compared to the field-based model. In the combined data model, the laboratory information serves to help resolve the individual effects of pH, inorganic Al, and Ca on fish response to acidification. These effects are masked in the field-only model by the multicollinearity exhibited by the field data set.

These models are intended to help investigators predict the effects on fish of various possible changes in the levels of acidification. As such, they should be judged on the basis of their simplicity; on their fit to the field data; and, in particular, on their ability to predict the presence or absence of brook trout populations for specified water chemistry concentrations that may differ from those in the present data set. While the hierarchical model has two more parameters, it was created in the hope that it would offer more accurate predictions, and less prediction uncertainty when applied to future data sets that differ from the present one by exhibiting less (or different) correlation among the predictor variables. Unfortunately, a definitive assessment of the model performance in meeting this objective will have to await the availability of suitable data sets. While a comparison of the model predictions in a future data set is not possible, the predictions can be evaluated in a portion of the present data set reserved for that purpose: the 136 observations from the second (1986–1987) ALSC survey. Two possible criteria for assessing the fit of the models include:

1. Deviance: $= -2[l(x\beta;y) - l(\hat{\theta};y)]$

2. Brier score: $= 1 - \sum_{i=1}^{N} (\text{predicted}_i - \text{observed}_i)^2/N$

For the field-based model, the deviance[16] and Brier score[17] values are 0.93 and 67.02, respectively. Similar values for the hierarchical model are 0.93 and 72.37. By these criteria, the hierarchical model is comparable to the field-only model, despite the influence of the laboratory bioassay data. Since both ALSC surveys suffer from the same collinearity, this does not offer an opportunity to confirm whether the hierarchical model fulfills its intended purpose, but it does show that the model fits existing data sets as well as the field-only model. This information, combined with the evidence that the hierarchical model parameters are much better specified than the individual models, indicates that the hierarchical model predictions can be expected to perform in a superior fashion to the field-only model. This inference is especially true when the hierarchical model is applied to observations exhibiting a different correlation structure than those used to calibrate the model.

CONCLUSION

The problems created by multicollinearity in a field data set were addressed using a hierarchical Bayesian model for combining laboratory and field observations. The hierarchical model parameters were shown to be better specified than their field-only model equivalents, while model predictions of the presence and absence of brook trout in a reserved, independent data set were shown to be comparable to the simpler model. In addition, the predictive densities and marginal posterior distributions of model parameters that arise in the Bayesian analysis were shown to lend insight into the system under study.

Bayesian methods provide a flexible and coherent framework for combining disparate information in the face of uncertainty about both the model and data. Data sets in the natural, social, and medical sciences often feature a collinearity structure that evolves over time due to changes in the environment, changes in the social, economic, and demographic features of a population; and changes in medical technology, respectively. The methods presented here offer a way to resolve uncertainties arising in prediction problems with multicollinear data.

REFERENCES

1. Reckhow, K. H., R. W. Black, T. B. Stockton, Jr., J. D. Vogt, and J. G. Wood. 1987. Empirical models of fish response to lake acidification. *Can. J. Fish. Aquat. Sci.* 44:1432–1442.
2. Baker, J. P., D. P. Bernard, S. W. Christensen, M. J. Sale, J. Freda, K. Heltcher, L. Rowe, P. Scanlon, P. Stokes, G. Suter, and W. Warren-Hicks. 1990. Biological Effects of Changes in Surface Water Acid-Base Chemistry. State-of-Science/Technology Report 13. National Acid Precipitation Assessment Program, Washington, DC.
3. Dillon, P. J., N. D. Yan, and H. H. Harvey. 1984. Acidic deposition: effects on aquatic ecosystems. *CRC Crit. Rev. Environ. Control* 13:167–194.
4. Kennedy, P. 1985. *A Guide to Econometrics.* MIT Press, Cambridge, MA.
5. Baker, J. P. 1981. Aluminum toxicity to fish as related to acid precipitation and Adirondack surface water quality. Ph.D. Thesis, Cornell University, Ithaca, NY.
6. Ingersoll, C. G. 1986. The effects of pH, aluminum, and calcium on survival and growth of brook trout (*Salvelinus fontinalis*) early life stages. Ph.D. Dissertation, University of Wyoming, Laramie.
7. Holtze, K. E. and N. J. Hutchinson. 1989. Lethality of low pH and Al to early life stages of six fish species inhabiting PreCambrian Shield waters in Ontario. *Can. J. Fish. Aquat. Sci.* 46:1188–1202.
8. Warren-Hicks, W. J. 1990. Predicting brook trout population response to acidification: Using Bayesian inference to combine laboratory and field data. Ph.D. Dissertation, Duke University, Durham, NC.
9. Berger, J. O. 1985. *Statistical Decision Theory and Bayesian Analysis.* Springer-Verlag, New York.
10. Wolpert, R. L. and W. J. Warren-Hicks. 1992. Bayesian Hierarchical Logistic Models for Combining Field and Laboratory Data. In Bayesian Statistics 4, J. M. Bernardo, J. O. Berger, A. P. Dawid, and A. F. M Smith Eds., Oxford University Press, Oxford, pp. 525–546.
11. Cox, D. R. 1970. *Analysis of Binary Data.* Chapman & Hall, London, UK.
12. McCullagh, P. and J. A. Nelder. 1989. *Generalized Linear Models.* Chapman & Hall, London, UK.

13. Bishop, Y. M., S. E. Fienberg, and P. W. Holland. 1975. *Discrete Multivariate Analysis: Theory and Practice*. MIT Press, Cambridge, MA.
14. Manton, K. G. and E. Stallard. 1988. *Chronic Disease Modeling*. London Griffin, UK.
15. Wolpert, R. L. 1991. Monte Carlo integration in Bayesian statistical analysis. *J. Contemporary Math.* 116:101–115 [published by the American Mathematical Society, ISBN 0-8218-5122-5].
16. Pregibon, D. 1981. Logistic regression diagnostics. *Ann. Statistics* 9:705–724.
17. Harrell, F. E. and K. L. Lee. 1985. A Comparison of the Discrimination of discriminant Analysis and Logistic Regression under Multivariate Normality, In: P. K. Sen, Ed. *Biostatistics: Statistics in Biomedical, Public Health, and Environmental Sciences*. Worth, Amsterdam, Holland, pp. 167–179.

CHAPTER 5

A New Approach for Accommodation of Below Detection Limit Data in Trend Analysis of Water Quality

Neerchal K. Nagaraj and Susan L. Brunenmeister

ABSTRACT

Water quality data are often collected in monitoring programs to serve as a basis for the estimation of trends. A problem arises in trend estimation when data series contain observations reported as below detection limit (BDL). Several published methods that deal with BDL observations are generally oriented toward obtaining the "best value" to substitute for the censored values. We suggest a weighted regression approach, as a modification of the usual substitution approach, which incorporates the fact that a detection limit is a measure of precision of the data. When the definition of a detection limit associated with a datum is unknown, a conservative lower bound (with a theoretical justification) for the precision of the observation is given. We illustrate this approach using water quality monitoring data collected by the Chesapeake Bay program. The results show that weighted regression estimates of linear trends are much less sensitive to the method of substitution for BDL values than the trend estimates obtained by unweighted regression. By lessening the bias of simple substitution methods, the proposed weighting approach may obviate the need for "best value" substitution methods which are more cumbersome to implement. The results also indicate that use of weighted regression in multiply-censored data series eliminates the need to apply the highest detection limit to all data in the series (when there are points below the highest limit in the data series) in order to avoid trends due to changing detection limits. This is an important result since the latter practice which is generally used in (unweighted) trend estimation results in the loss of the advantage of improved detection limits.

0–87371–936–0/94/$0.00 + $.50

INTRODUCTION

Water quality data are often collected in monitoring programs to serve as a basis for the estimation of trends. Such time series of observations may be censored at one detection limit, or at several. Changes in detection limits during the course of a monitoring program often occur as a result of intentional improvements to the level of detection or by switches in the chemical analysis methods that are used. Varying detection limits in a data series can also occur as a result of quality assurance procedures that specify periodic redeterminations of the detection limit. Although data can be censored at both lower and upper limits, below detection limit (BDL) observations have been the common problem we have encountered in water quality data from the Chesapeake Bay program. In general, seasonally low BDL nutrient concentrations may be typical of estuarine and marine monitoring situations.

Several published methods that deal with BDL observations are generally oriented toward obtaining the "best value" to substitute for the censored values (cf. References 1–9). In the second section, we give a brief description of these substitution methods. A more detailed review (cf. Reference 4) suggests that overall best performance is obtained by what are termed "robust methods" of substitution, and Monte Carlo studies (cf. References 2 and 6) support the claim. However, these substitution methods are not easily adapted to correlated data. This is a significant drawback, since time series of environmental data are usually correlated.

The American Chemical Society Committee on Environmental Improvement (ACSCEI) defined the limit of detection (LOD) as "the lowest concentration level that can be determined to be statistically different from a blank" (cf. Reference 10). They recommended that the LOD or detection limit be calculated as three times the standard deviation of blank determinations. A detection limit may refer to an instrument detection limit (IDL) based on signal to noise ratios or a method detection limit (MDL) based on replicate determinations of a blank or field blanks (cf. Reference 11). Once the detection limit is determined, it is used as a censoring threshold for subsequent measurements, and the concentration levels observed below that limit are reported as BDL.

A detection limit, defined as a multiple of the standard deviation of a number of measurements, thus contains information about the precision of observations above and below the detection limit. We present, in the third section, a new substitution approach that incorporates this fact. In some cases an analyst may be faced with data containing observations censored at one or more detection limits for which the method of calculation is not known. For this case, we derive a conservative estimate of the variance associated with a censoring threshold.

SUBSTITUTION METHODS

The consensus of authors we consulted concerning the current state of methodology for dealing with BDL data is that "bad data is better than no data" (cf. References 1–9). That is, if possible, one should obtain the actual values for the observations reported as BDL. However, this is usually not possible; indeed, the ACSCEI recommends reporting below detection limits as "not detected" with the associated limit of detection appended as a separate variable (cf. Reference 10). Since values for BDL observations are needed to perform many statistical procedures, estimation methods for BDL data have been studied.

The most commonly used method is the method of simple substitution. These methods consist of substituting a single value for an unknown BDL observation. Following the substitution, the usual statistical analysis is carried out as if the fill-in data is real data. The fill-in values may be one of the following: zero, the detection limit, or half the detection limit. This simple method generally leads to biased estimates. The performance of these simple substitutions in Monte Carlo studies were generally poor (cf. References 6, 2). However, these methods are definitely the simplest to implement.

A second class of methods are based on distributional assumptions. Under the assumption that the observations follow a certain distribution (such as lognormal) the parameters are estimated based on the observed concentrations above the detection limit and the percentage of BDL data. Either maximum likelihood or probability plotting methods can be used. The maximum likelihood method is more involved to implement than the plotting method. The plotting method is also complex when more than one detection limit is involved. These methods are described in References 2 and 5. In Tobit regression, maximum likelihood estimates of model parameters are obtained under a distributional assumption (cf. Reference 8).

A combination of the substitution method and the distributional assumption is recommended by Helsel. Under the distributional assumption, the first few (depending on how many BDL observations are in the data set) percentiles are estimated. These values are substituted for the BDL observations and the target analysis is conducted as usual. This method is computationally involved, especially when there are multiple detection limits; however, this approach consistently produced small estimation errors (cf. Reference 4).

The methods described in the literature so far concern only uncorrelated data; extensions of these methods to correlated data and their performances have not been explored. These methods, even in cases in which it is possible to extend them to account for autocorrelation in the data, will be computationally intensive.

WEIGHTING METHOD

The treatment of BDL data suggested in this report is novel. It takes advantage of one common feature of the definition of a detection limit: measurement of uncertainty. Thus we note that a detection limit, in addition to being a threshold for data censoring, also contains information regarding variability in the data. We, therefore, propose to incorporate this information in the analysis via weighting. We distinguish two situations: one in which the detection limit calculation method is known, and one in which it is unknown.

Because detection limits are measures of uncertainty, it is clear that the varying detection limits in the data are an indication of heteroscedasticity. We postulate that the standard deviation s_t of the tth observation is given by

$$s^2_t \text{ proportional to } r^2_t$$

where r_t is a measure of the uncertainty in the tth observation. The quantity r_t could simply be one third of the detection limit (if the ACSCEI definition applies) or some other number which reflects a knowledge of the uncertainty in the data.

In the case where the detection limit calculation is unknown, a conservative estimate of the variance can be arrived at as follows. When a datum is reported as below detection, its true value could lie anywhere between zero and the detection limit. An estimate for the datum is given by the midpoint of this interval. The uncertainty in this estimate depends on the exact distribution of the data, which is unknown. However, an upper bound exists for the variance of an arbitrary distribution of data in this interval. Technically, this can be stated as follows:

RESULT 1. *If X is any random variable taking values in an interval* $[a, b]$, *then the variance of X can not exceed* $(b - a)^2/4$. (See Appendix 1 for the proof.)

Thus, in such a case, a conservative estimate of the variance of the data below the detection limit is the square of the detection limit/4. This estimate of variance does not apply to data above detection, and in this case other estimates of uncertainty might be explored. However, the convention of obtaining detection limits for trace level concentration is based on the belief that one is more certain about the measurements that are large. Thus, the operant detection limit is a conservative estimate of the uncertainty in the above detection limit datum as well.

If we conclude that the detection limit or the above conservative estimate appropriately represents the changing variance of multiply-censored data series, then the best estimation procedure for trend estimation is weighted least squares where observations are weighted according to the uncertainty associated with them (cf. Reference 12). Thus the observations are weighted by the inverse of the estimates of their standard deviations. In our approach, the BDL data are replaced with one half of the operant detection limit and observations are weighted by $w_t = (1/r_t)$, (cf. Reference 12, Equation 2.2, p. 11).

A step by step description of our procedure is as follows :

1. Replace BDL data with one half of the corresponding detection limit:

$$x_t = (d_t)/2$$

where d_t is the detection limit corresponding to the tth observation.
2. Select a weighting factor w_t for each x_t, where $w_t = 1/r_t$ and r_t is $d_t/2$.
3. The trend model in our example is:

$$x_t = b_o + b_1 t + e_t$$

where e_t is the zero mean noise term in the regression. We also included dummy variables to account for the month effects.
4. Apply weight w_t for the tth observation. That is, our model becomes:

$$w_t x_t = b_o w_t + b_1 w_t t + w_t e_t$$

Thus, the weighting factor w_t accounts for the heteroscedasticity among the observations due to the varying degree of precision in the measurements as reflected by varying detection limit.

In general, heteroscedasticity in the data can occur due to many factors, including those due to sampling variability. It is important to recognize the need for a composite measure of uncertainty due to all contributing factors. However, construction of such a model requires estimates of the individual sources of variability and an understanding of their interrelationships, which may be difficult to obtain.

AN EXAMPLE

Since the inception of the Chesapeake Bay Water Quality Monitoring program, detection limits for orthophosphate concentrations sampled at mainstem monitoring stations successively declined due to modifications in the water chemistry analysis methods (Figure 1). In this example, we apply our method to a time series of orthophosphate concentrations measured from June 1984 through February 1991 at a monitoring station in Maryland waters near the Maryland/Virginia border.

In the field sampling procedure, grab samples were taken at different depths in the water column at the station in order to characterize the

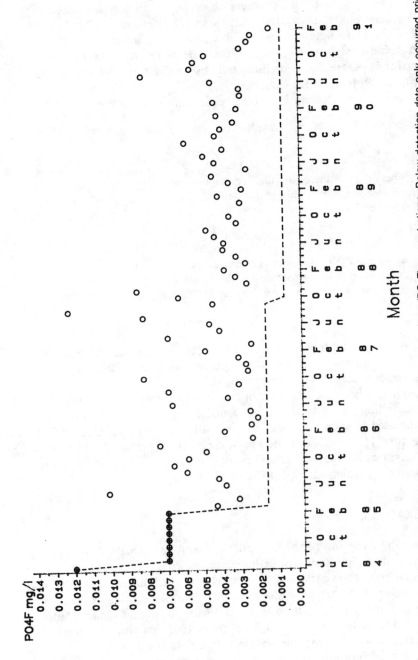

FIGURE 1.   Sample time series of observed mean monthly ortho-phosphate (PO₄F) concentrations. Below detection data only occurred prior to March 1985 in this data set and are shown set to the detection limit. The prevailing detection limit in each period is shown by the dashed line.

depth profile of nutrient concentrations. These samples were taken at the surface, bottom, and two midwater depths which depend on the density structure of the water column. We selected samples measured above the pycnocline (AP) because this series appeared to contain a downward trend and inspection of the data revealed no further significant autocorrelation in the data.

During the winter, the station was sampled once a month; in other months, it was sampled twice a month. To obtain a monthly series of observations, we averaged the within-month concentrations to obtain monthly values (see Figure 1). In this data set, within-month concentrations were either BDL or above detection at the same detection limit, therefore obviating the problem of estimating a monthly average from BDL and above BDL data. In this plot, BDL data were set equal to the detection limit; the only BDL data in the series occurred prior to March 1985, and all these values were BDL.

The monthly values and the range of uncertainty, r_t, associated with each monthly value in this sample data set is shown in Figure 2. In this plot, BDL data were set equal to one half the detection limit. We calculated r_t as one third the prevailing detection limit (see Figure 1), since detection limits of orthophosphate were determined as three times the standard deviation of duplicate samples.

To investigate the level of censoring on our estimates, we created four additional data sets based on this original sample data. This was done by applying an artificial detection limit (ADL) to the data from March 1985 onward. We selected four ADLs to correspond to approximate additional censoring levels of 5% (ADL = 0.00240), 15% (ADL = 0.00265), 30% (ADL = 0.00315), and 50% (ADL = 0.00420) of the data. Thus, each of these ADLs, applied individually, replaced the real detection limits of 0.0016 and 0.0006 that occurred from March 1985 onward. These artificial data sets retained the real detection limits of 0.012 and 0.007 prior to March 1985. These contrived detection limit series are shown in Figure 3 along with observed mean monthly concentrations (BDL data prior to March 1985 are shown set to one half the detection limit).

To examine the effect of the substitution choice for BDL data, we constructed three categories of sample data sets by setting BDL observations equal to 0, to one half the prevailing detection limit, or to the prevailing detection limit.

To illustrate the effects of applying the weighting procedure, we modeled each sample series twice: (1) weighting each observation according to the prevailing detection limit (as shown in Figures 1 and 3) and (2) applying no weights (which computationally means applying the same weight) to the observations. Our rationale for weighting all observations, not only BDL data, in a time period by the prevailing detection limit is

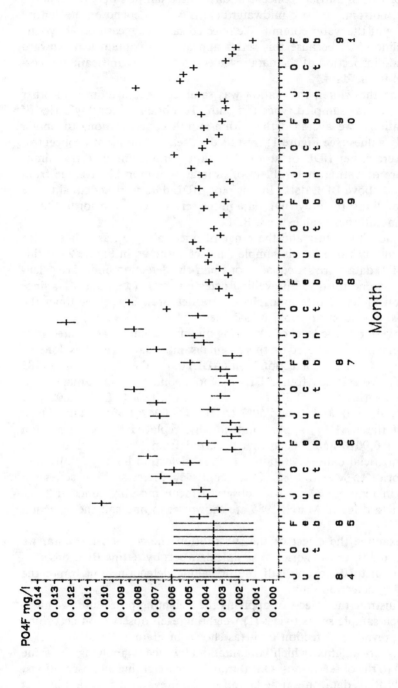

FIGURE 2. Sample time series of observed mean monthly ortho-phosphate (PO_4F) concentrations. Below detection data prior to March 1985 are shown set to ½ the prevailing detection limit. The error bars around each observation are equal to ⅓ the prevailing detection limit.

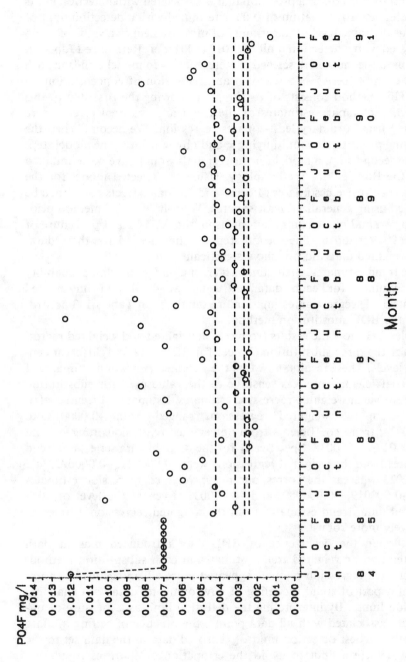

FIGURE 3. Sample time series of observed mean monthly ortho-phosphate (PO_4F) concentrations. Below detection data prior to March 1985 are shown set to the prevailing detection limit. The actual detection limits prior to March 1985 are shown by a single dashed line. Additional censoring levels (ADL's) corresponding to 0.00240, 0.00265, 0.00315 and 0.0042 are shown.

based on the idea of data precision that is associated with detection limits (cf. Reference 10): a datum of 0.03 determined with a detection limit of 0.03 has the same level of uncertainty of measurement as a value of 0.015 associated with a detection limit of 0.015 (100%; cf. Reference 13, p. 82).

We used the method described in Reference 14 to model trends in each of the example series as a seasonal autoregression of concentration on time. This method consists of two steps. Following the first step of this method, preliminary estimates of the trend and seasonal effects were obtained using ordinary least-squares regression. We accomplished the weighting procedure by multiplying x_t and t by w_t prior to the model step. For the second step, a model for the residuals would have been obtained using the Box-Jenkins methodology. Using the selected model for the error terms, revised estimates of trend and seasonal effects could then be obtained using generalized least squares. We followed this method prior to step two and made computations using the AUTOREG procedure of SAS/ETS Version 6. The second step was not needed for these data, since weighted data did not show significant autocorrelation.

The trend estimates, their standard errors, and associated probability level obtained for each data set using weighted and unweighted (weights = 1) regression estimation are tabulated in Tables 1–3 according to the BDL substitution method.

Tables 1–3 show the results from the unweighted and weighted regressions for the three substitution choices for BDL data and different censoring levels. These results show that the weighted regression estimates of trend coefficients were less sensitive to the value used for substitution than were the unweighted regression estimates. Estimates of trend coefficients from the weighted regressions ranged from −0.000013 to −0.000024; trend coefficients from unweighted regressions ranged from −0.000002 to 0.000038. Specifically for the original data series, the trend estimates from the weighted regressions were −0.000021, −0.000022, and −0.000023 whereas the corresponding unweighted regression estimates were −0.000019, −0.000009, and −0.000038. Even at the level of 50% censored data, trend estimates from the weighted regression were comparatively more robust.

Although the occurrence of BDL data introduced bias in both weighted and unweighted trend estimates in each substitution method, the weighted regression method appeared more advantageous in an important aspect of trend modeling of environmental data with changing detection limits. By interpreting the detection limit as a measure of uncertainty associated with all data points, the practice of setting all data below the highest detection limit of censored data in the data set to the highest detection limit to avoid the estimation of spurious trends becomes unnecessary. This is illustrated by the stability of weighted trend estimates over censoring levels within each substitution method. Using

Table 1. Trend Estimates from Weighted and Nonweighted Regressions with Different Levels of Censoring

Censoring level		Weighted	Unweighted
Original	b	−0.000023	−0.000038
	s.e.(b)	0.000009	0.000009
	p	0.0243	0.0001
+5%	b	−0.000023	−0.000036
	s.e.(b)	0.000009	0.000009
	p	0.0208	0.0002
+15%	b	−0.000024	−0.000036
	s.e.(b)	0.000009	0.000009
	p	0.0117	0.0002
+30%	b	−0.000023	−0.000036
	s.e.(b)	0.000009	0.000009
	p	0.0135	0.0002
+50%	b	−0.000022	−0.000030
	s.e.(b)	0.000008	0.000008
	p	0.0068	0.0002

Note: Below detection observations set to the detection limit. Estimated slope (b), standard error of the slope (s.e. [b]) and associated *p* value (*p*) are indicated.

weighted regression therefore avoids the loss in improved data quality imposed by this practice.

Use of weighted regression also improved other data aspects. When the data series contained a high percentage of BDL data, a significant amount of autocorrelation occurred in the model residuals. This autocorrelation was not nearly as prevalent in residuals from our weighted regressions as in the residuals from the unweighted regression.

CONCLUSIONS

In this chapter we proposed a weighting scheme which interprets the detection limit as a measure of uncertainty. The approach seems reasonable since detection limits are usually calculated as multiples of the standard deviation of blank determinations. For instances in which detection limit calculations are unknown, we present a weighting function that can be used which does not rely on distributional assumptions. A mathematical result shows that it is the most conservative estimate of uncertainty, over all possible distributions.

Table 2. Trend Estimates from Weighted and Unweighted Regressions with Different Levels of Censoring

Censoring level		Weighted	Unweighted
Original	b	−0.000022	−0.000009
	s.e.(b)	0.000010	0.000008
	p	0.0317	0.3149
+5%	b	−0.000020	−0.000010
	s.e.(b)	0.000010	0.000009
	p	0.0552	0.2988
+15%	b	−0.000018	−0.000009
	s.e.(b)	0.000011	0.000009
	p	0.0952	0.3560
+30%	b	−0.000019	−0.000011
	s.e.(b)	0.000011	0.000010
	p	0.0936	0.2628
+50%	b	−0.000019	−0.000014
	s.e.(b)	0.000011	0.000010
	p	0.0974	0.1696

Note: Below detection observations set to one half of the detection limit. Estimated slope (b), standard error of the slope (s.e. [b]) and associated p value (p) are indicated.

The advantage of using weighted regression rather than unweighted regression in describing trends in our sample data set was threefold. First, weighted trend estimates were much less sensitive to the level of censoring in the data than were unweighted trend estimates. Second, weighted trend estimates were much less sensitive to the choice of data substitution for BDL data (i.e., zero, one half the detection limit, or the detection limit) than were unweighted trend estimates. Finally, the effects of autocorrelation were less in the weighted regression models than in the unweighted regression models.

The results also showed that the use of weighted regression eliminates the need to censor data to the maximum detection limit when dealing with multiply-censored data series to avoid spurious trends. Thus, trend estimation with weighted regression preserves the effect of improved detection limits in data sets. The performance of the weighted regression estimation procedure could be further evaluated using Monte Carlo methods. Such a study would provide a good topic for further research. In such a study, the questions we addressed in our specific example can be studied in a general way. Questions that we have attempted to address in a general way could be more extensively examined.

Table 3. Trend Estimates from Weighted and Nonweighted Regressions with Different Levels of Censoring

Censoring level		Weighted	Unweighted
Original	b	−0.000021	−0.000019
	s.e.(b)	0.000010	0.000011
	p	0.0433	0.0912
+5%	b	−0.000017	−0.000017
	s.e.(b)	0.000012	0.000012
	p	0.1481	0.1603
+15%	b	−0.000013	−0.000018
	s.e.(b)	0.000013	0.000013
	p	0.3169	0.1582
+30%	b	−0.000015	−0.000012
	s.e.(b)	0.000015	0.000014
	p	0.3067	0.3945
+50%	b	−0.000016	−0.000002
	s.e.(b)	0.000016	0.000015
	p	0.3319	0.9026

Note: Below detection observations set to zero. Estimated slope (b), standard error of the slope (s.e. [b]) and associated *p* value (*p*) are indicated.

ACKNOWLEDGMENTS

We thank Peter Bergstrom, Karen Jones, and Lacy Williams of Computer Science Corporation for helpful comments on the manuscript. This work was supported on EPA Contract No. 68–W00043 to Computer Science Corporation.

REFERENCES

1. Gilliom, R. J., Hirsch, R. M., and Gilroy, E. J., "Effect of Censoring Trace-Level Water Quality data on Trend-Detection Capability," *Environ. Sci. Technol.*, 18:530–535 (1990).
2. Gilliom, R. J. and Helsel, D. R., "Estimation of Distributional Parameters for Censored Trace Level Water Quality Data. 1. Estimation techniques," *Water Res. Res.*, 22(2):135–146 (1986).
3. Gleit, A., "Estimation for Small Normal Sets with Detection Limits," *Environ. Sci. Technol.*, 19(12):1201–1206 (1985).
4. Helsel, D. R., "Less than Obvious: Statistical Treatment of Data Below the Detection Limit," *Environ., Sci. Technol.*, 24(12):1766–1744 (1990).

5. Helsel, D. R. and Cohn, T. A., "Estimation of Descriptive Statistics for Multiply Censored Water Quality Data," *Water Res. Res.*, 24(12):1997-2004 (1988).
6. Helsel, D. R. and Gilliom, R. J., "Estimation of Distributional Parameters for Censored Trace Level Water Quality Data. 2. Verification and Applications," *Water Res. Res.*, 22(2):147-155 (1986).
7. Porter, P. S., Ward, R. C, and Bell, H. F., "The Detection Limit," *Environ. Sci. Technol.*, 22(8):856-861 (1988).
8. Robinson, P. M., "On the asymptotic properties of estimators of models containing limited dependent variables," *Econometrica*, 50(1):27-41 (1982).
9. Travis, C. C. and Land, M. L., "Estimating the Mean of Data Sets with Nondetectable Values," *Environ. Sci. Technol.*, 24(7):961-962 (1990).
10. ACS Committee on Environmental Improvement, "Guidelines for Data Acquisition and Data Quality Evaluation in Environmental Chemistry," *Anal. Chem.*, 52:2242-2249 (1980).
11. ACS Committee on Environmental Improvement, "Principles of Environmental Analysis," *Anal. Chem.*, 55:2210-2218 (1980).
12. Carroll, R. J. and Ruppert, D., *Transformation and Weighting in Regression*, (New York, NY: Chapman & Hall, 1988).
13. Taylor, J. K., *Quality Assurance of Chemical Measurements*, (Chelsea, MI: Lewis Publishers, Inc, 1987).
14. Nagaraj, N. K. and Brunenmeister, S. L., "Application of Seemingly Unrelated Regression Estimation (SURE) to Characterizing Trends in Total Phosphorus in the Upper Chesapeake Bay, October 1984 to September 1989," in *Proceedings of a Conference: New Perspectives in the Chesapeake system: A research and Management Partnership*, (Solomons, MD: Chesapeake Bay consortium, 1991) CRC Publication No. 137:355-369.

APPENDIX 1

In this appendix we give the proof of Result 1.

RESULT 1. *Let* X *be a random variable taking values in the interval* [a, b]. *Then variance of* X *does not exceed* $(b - a)^2/4$.

PROOF. *Let* $Y = (X - a)/(b - a)$. *Since* $Var(X) = (b - a)^2 Var(Y)$, *it is enough to show that* $0 \leq Var(Y) \leq 1/4$.

Since Y takes values in the interval $[0, 1]$, Y^2 also takes values in the interval $[0, 1]$, $Y^2 \leq Y$ and we have:

$$0 \leq E(Y^2) \leq E(Y) \leq 1$$

Therefore

$$Var(Y) = E(Y^2) - [E(Y)]^2$$

$$\leq E(Y) - [E(Y)]^2$$

$$\leq E(Y) [1 - E(Y)].$$

Now the conclusion follows noting that maximum of the function $p(1 - p)$, where p ranges in the interval $[0, 1]$, is $1/4$.

SECTION II

Spatial Statistics

N. Phillip Ross

INTRODUCTION

Chemostatistics and geostatistics are areas involving the analysis of statistical data and information that describe where chemicals and contaminants come from and go to and what effect they have on the environment. This branch of statistics involves every stage of statistical analysis; namely, design; data collection; efficient data storage and retrieval; exploratory data analysis; statistical model building; statistical inference; diagnostic checking; and finally reporting of substantive conclusions in forms that are readily, easily, and clearly understandable. Environmental statistics can be thought of as involving two areas; temporal and spatial statistics. This section describes the importance of and problems involved in spatial statistics in a general chapter (Chapter 6) and in two chapters describing examples involving designing and optimizing an air pollution monitoring network (Chapter 7) and the analysis of air quality and emissions trends (Chapter 8).

Several of the problem areas discussed in these chapters include: analyzing multivariate data, kriging and other methods for imputing values in regions not directly monitored, and handling artificially censored data such as those below the detection level. In the effort to develop a more complete picture it is observed that there may be a larger penalty for overprediciton than for underprediction. These and other pitfalls involved in the effort to determine the overall picture are discussed.

Much attention has been given to the determination and analysis of temporal data, but it may be that spatial statistics are more important. For example, people living in scarcely populated areas feel that they have as much right to know the pollution levels as those who live in large urban areas. Although some people are interested in the overall picture of pollution and contamination, most want to know what the situation is for their specific locality.

Spatial chemostatistics are thus a part of environmental statistics that require attention for the unique problems involved in the overall contribution to understanding and describing the state of the environment.

CHAPTER 6

Spatial Chemostatistics

Noel Cressie

ABSTRACT

Chemostatistics is concerned with statistics for measured chemical data. Sometimes knowledge of where such data come from can be very useful in predicting values of a process at locations where no data were taken. Examples include resource evaluation, atmospheric-deposition monitoring, and hazardous waste site characterizations. Results from chemical assays give a highly multivariate picture of the underlying process. However, the picture is incomplete because of sparseness of the samples in space, censoring due to limits of detection, measurement error, and so forth. This chapter will discuss these issues.

INTRODUCTION

Chemostatistics is concerned with statistics for measured chemical data. It embraces every stage of a statistical analysis, namely, design, data collection, efficient data storage and retrieval, exploratory data analysis, statistical model building, statistical inference, diagnostic checking, and finally reporting of substantive conclusions. Ideally, in the light of the conclusions, the sequence of stages can start again in order to answer a new scientific question.

Some chemists may feel that their subject is an exact science with little need for statistics, a subject for which the very nature is the study of uncertainty. Statistics attempts to model order in "disorder," such as quantification of equation error or measurement error. Even when the disorder is discovered to have a perfectly rational explanation at one scale, there is often a smaller (or larger) scale where the data do not fit the theory exactly; and the need arises to investigate the new uncertainty.

0–87371–936–0/94/$0.00 + $.50
© 1994 by Lewis Publishers

There are at least two areas of chemistry research where (access to) good statistical training will be essential in the future. First, chemists are more than ever using instruments such as chromatographs, spectrometers, etc. that give rise to enormous amounts of data, usually multivariate. Second, the interface of chemistry with medical research, biological research, etc. has led to the study of complicated systems; experiments need to be well planned and good predictive models may be stochastic.[43]

This chapter is about statistics—in particular, spatial statistics—for environmental problems. For example, consider the collection and analysis of wet deposition data (i.e., deposition of anions and cations: H^+, NH_4, Na, Ca, SO_4, NO_3, Cl, etc. which have been dissolved in precipitation) in North America. Daily data collection over a number of years at, say, 100 locations yields a massive data set. However, most of it is temporal so that spatially the data are still rather sparse. Nevertheless, spatial prediction is just as important as temporal prediction, since people living in those cities and rural districts without monitoring stations have an equal right to know how little or how much their water or their air is polluted. The studies done on this large scale are usually observational, although occasionally one has the opportunity to "design" an experiment. Vong et al.[41] describe the changes in rainwater pH and sulfate associated with the closure of a copper smelter.

Chemostatistics has already generated statistical applications that encompass experimental design and optimization, calibration, multivariate data analysis and pattern recognition, sampling strategies, and signal processing. In this chapter, I shall emphasize the spatial, multivariate, and truncation aspects of environmental data.

The next section presents the geostatistical approach with particular attention given to univariate spatial prediction (kriging). The multivariate aspects are brought out in the section entitled "Multivariate Geostatistics." Finally, in the last section, problems associated with limits of detection are discussed.

GEOSTATISTICS

Environmental data are often spatial over regions where the spatial index is continuous. Not surprisingly, then, geostatistics is having an increasingly important role to play in environmental modeling efforts. For a statistician interested in spatial statistics, this leads to a whole new set of exciting problems. As a consequence, standard geostatistics for mining applications has to be augmented with new methods. For example, Journel[22,24] has written informative expositions of the use of geostatistics in environmental problems. Istok and Cooper[20] demonstrated how to predict groundwater contaminant concentrations using geostatistics and Myers[32] implemented it to assess the movement of a multipollutant plume. Geostatistical studies of acid deposition have been carried out by various authors.[2,3,9,12,28]

The ensuing presentation of geostatistics follows a statistical modeling approach; more can be found in Part I of the book by Cressie.[8] There is another approach, relying on probabilities obtained from the sampling scheme, that Journel[23] and Isaaks and Srivastava[19] sometimes prefer. However, its limitations are sorely felt when one wishes to go beyond marginal- and bivariate-distribution (or moment) estimation, to kriging. Provided good diagnostics are used to verify the goodness of fit of a spatial statistical model, the modeling approach advocated in this chapter is extremely powerful.

First, a measure of the (second-order) spatial dependence exhibited by the spatial data is needed. A model-based parameter (which is a function) known as the variogram is defined here; its estimate provides such a measure. Statisticians are used to dealing with the autocovariance function. However, the class of processes with a variogram contains the class of processes with an autocovariance function, and thus kriging can be carried out on a wider class of processes than the one traditionally used in statistics.

Let $\{Z(s): s \in D \subset \mathbb{R}^d\}$ be a real-valued stochastic process defined on a domain D of the d-dimensional space $\mathbb{R}^d$ ($d = 1, 2, 3, \ldots$), and suppose that differences of variables lagged $\mathbf{h}$-apart vary in a way that depends only on $\mathbf{h}$. Specifically, suppose:

$$\text{var}(Z(\mathbf{s} + \mathbf{h}) - Z(\mathbf{s})) = 2\gamma(\mathbf{h})$$

$$\text{for all } \mathbf{s}, \mathbf{s} + \mathbf{h} \in D; \tag{1}$$

typically the spatial index s is two- or three-dimensional (i.e., $d = 2$ or 3). The quantity $2\gamma(\cdot)$, which is a function only of the *difference* between the spatial locations s and $s + \mathbf{h}$, has been called the *variogram* by Matheron.[29]

When $2\gamma(\mathbf{h})$ can be written as $2\gamma^0(||\mathbf{h}||)$, for $\mathbf{h} \in \mathbb{R}^d$, the variogram is said to be *isotropic*; otherwise it is said to be *anisotropic*, in which case the process $Z(\cdot)$ is also referred to as anisotropic. Anisotropies are caused by the underlying physical process evolving differentially in space. Sometimes the anisotropy can be corrected by an invertible linear transformation of the lag vector $\mathbf{h}$. That is, *geometric* anisotropy is described by:

$$2\gamma(\mathbf{h}) = 2\gamma^0(||A\mathbf{h}||)$$

$$\mathbf{h} \in \mathbb{R}^d$$

where A is an invertible $d \times d$ matrix and $2\gamma^0$ is a function of only one variable.

Replacing Equation 1 with the stronger assumption:

$$\text{cov}(Z(s + h), Z(s)) = C(h)$$

$$\text{for all } s, s + h \in D \tag{2}$$

and specifying the mean function to be constant; that is:

$$E(Z(s)) = \mu$$

$$\text{for all } s \in D \tag{3}$$

defines the class of *second-order* (or wide-sense) *stationary* processes in D, with (auto) covariance function $C(\cdot)$. Time series analysts often assume Equation 2 and work with the quantity $\rho(\cdot) \equiv C(\cdot)/C(0)$. Conditions in Equations 1 and 2 define the class of *intrinsically stationary* processes. Assuming only Equation 2

$$\gamma(h) = C(0) - C(h); \tag{4}$$

that is, the semivariogram can be related very simply to the covariance function.

For the purposes of this section, assume that the variogram is known; in practice, variogram parameters are estimated from the spatial data. Suppose it is desired to predict $Z(s_0)$ at some unsampled spatial location s_0 using a linear function of the data $\mathbf{Z} \equiv (Z(s_1), \ldots, Z(s_n))'$:

$$\hat{Z}(s_0) = \sum_{i=1}^{n} \lambda_i Z(s_i) \tag{5}$$

It is sensible to look for coefficients $\{\lambda_i: i = 1, \ldots, n\}$ for which Equation 5 is uniformly unbiased and which minimize the mean-squared prediction error, $E(Z(s_0) - \hat{Z}(s_0))^2$.

The uniform unbiasedness condition imposed on Equation 5 is $E(\hat{Z}(s_0)) = \mu = E(Z(s_0))$, for all $\mu \in \mathbb{R}$, which is equivalent to:

$$\sum_{i=1}^{n} \lambda_i = 1 \tag{6}$$

If the process is second-order stationary and Equation 6 is assumed:

$$E(Z(s_0) - \sum_{i=1}^{n} \lambda_i Z(s_i))^2 = C(0) - 2\sum_{i=1}^{n} \lambda_i C(s_i - s_0) +$$

$$\sum_{i=1}^{n}\sum_{j=1}^{n} \lambda_i\lambda_j C(s_i - s_j) \qquad (7)$$

If the process is intrinsically stationary (a weaker assumption) and Equation 6 is assumed:

$$E(Z(s_0) - \sum_{i=1}^{n} \lambda_i Z(s_i))^2 = 2\sum_{i=1}^{n} \lambda_i \gamma(s_i - s_0) -$$

$$\sum_{i=1}^{n}\sum_{j=1}^{n} \lambda_i\lambda_j \gamma(s_i - s_j) \qquad (8)$$

Using differential calculus and the method of Lagrange multipliers, optimal coefficients $\lambda \equiv (\lambda_1, \ldots, \lambda_n)'$ can be found that minimize Equation 8 subject to Equation 6; they are:

$$\lambda = \Gamma^{-1}\left[\gamma + \frac{(1 - 1'\Gamma^{-1}\gamma)1}{1'\Gamma^{-1}1}\right] \qquad (9)$$

and the minimized value of Equation 8 (kriging variance) is:

$$\sigma_k^2(s_0) = \gamma'\Gamma^{-1}\gamma - \frac{(1 - 1'\Gamma^{-1}\gamma)^2}{1'\Gamma^{-1}1} \qquad (10)$$

In Equations 9 and 10, $\gamma \equiv (\gamma(s_1 - s_0), \ldots, \gamma(s_n - s_0))'$, $1 \equiv (1, \ldots, 1)'$, and Γ is the $n \times n$ symmetric matrix with $(i, j)^{th}$ element $\gamma(s_i - s_j)$, which is assumed to be invertible.

The kriging predictor given by Equations 5 and 9 is appropriate if the process $Z(\cdot)$ contains no measurement error. If measurement error is present, then a "noiseless version" of $Z(\cdot)$ should be predicted.[7]

Thus far, kriging has been derived under the assumption of a constant mean. More realistically, assume:

$$Z(s) = \mu(s) + \delta(s)$$

$$s \in D \qquad (11)$$

where $E(Z(s)) = \mu(s)$, for $s \in D$, and $\delta(\cdot)$ is a zero mean, intrinsically stationary stochastic process with var $(\delta(s + h) - \delta(s)) = $ var $(Z(s + h) -$

$Z(s)) = 2\gamma(\mathbf{h})$, $\mathbf{h} \in \mathbb{R}^d$. In Equation 11, the *large-scale variation* $\mu(\cdot)$ and the *small-scale variation* $\delta(\cdot)$ are modeled as deterministic and stochastic processes, respectively, but with no unique way of identifying either of them individually. *Universal kriging* predictors can be derived for the case $\mu(s) = \mathbf{x}(s)'\beta$, a linear combination of variables that could include trend surface terms or other explanatory variables thought to influence the behavior of the large-scale variation (e.g., Section 3.4 of the book by Cressie[8]).

MULTIVARIATE GEOSTATISTICS

Hazardous waste site characterization and remediation problems are typically multivariate and spatial in nature. However, databases are typically aspatial and statistical methods are often both aspatial and univariate. Multivariate exploratory spatial data analyses[6,17] are an important (and time-consuming) part of building the statistical models subsequently used in characterization/remediation. This section assumes that a valid multivariate spatial statistical model has already been established.

Consider a vector-valued spatial process:

$$\{\mathbf{Z}(s): s \in D\}$$

where $\mathbf{Z}(s) \in \mathbb{R}^k$ *and* $D \subset \mathbb{R}^d$; *usually* $d = 1, 2,$ or 3. Write $\mathbf{Z}(s) \equiv (Z_1(s), \ldots, Z_k(s))'$. Assume

$$\mathbf{Z}(s) = \mu(s) + \delta(s) \qquad (12)$$

where $\mu(\cdot)$ is a $k \times 1$ mean vector composed of fixed effects, and $\delta(\cdot)$ is a $k \times 1$ random vector with zero mean.

Let the data consist of $\{\mathbf{Z}(s_1), \ldots, \mathbf{Z}(s_n)\}$ at spatial locations $\{s_1, \ldots, s_n\}$ in D. Define $\mathbf{Z}_j \equiv (Z_j(s_1), \ldots, Z_j(s_n))'$ and $\delta_j \equiv (\delta_j(s_1), \ldots, \delta_j(s_n))'$. Also, let $\mu_j(s) = \mathbf{x}_j(s)'\beta_j$

where $\quad \beta_j = p_j \times 1$ vector of parameters

$\quad\quad\quad \mathbf{x}_j(s) = p_j \times 1$ vector of "explanatory" variables for $Z_j(s)$

Finally, let X_j be the $(n \times p_j)$ matrix whose ℓ^{th} row is $\mathbf{x}_j(s_\ell)'$; $j = 1, \ldots, k$. Then $\mathbf{Z} \equiv (\mathbf{Z}_1', \ldots, \mathbf{Z}_k')'$ satisfies:

$$\mathbf{Z} = X\beta + \delta \qquad (13)$$

where $X \equiv \operatorname{diag}(X_1, \ldots, X_k)$, $\beta \equiv (\beta_1', \ldots, \beta_k')'$, and $\delta \equiv (\delta_1', \ldots, \delta_k')'$. Also:

$$\mathbf{Z(s)} = X(\mathbf{s})\beta + \delta(\mathbf{s}) \tag{14}$$

where $X(\mathbf{s}) \equiv \mathrm{diag}(\mathbf{x}_1(\mathbf{s})', \ldots, \mathbf{x}_k(\mathbf{s})')$. Then $E(\delta) = \mathbf{0}$, $E(\delta(\mathbf{s})) = \mathbf{0}$ and write $\Sigma_{j\ell} \equiv \mathrm{cov}(\delta_j, \delta_\ell)$, $c_{j\ell}(\mathbf{s}) \equiv \mathrm{cov}(\delta_j, \delta_\ell(\mathbf{s}))$, $c_j(\mathbf{s}) \equiv \mathrm{cov}(\delta, \delta_j(\mathbf{s}))$, $\Sigma(\mathbf{s}) \equiv \mathrm{var}(\delta(\mathbf{s}))$, and $C_{j\ell}(\mathbf{s},\mathbf{u}) \equiv \mathrm{cov}(\delta_j(\mathbf{s}), \delta_\ell(\mathbf{u}))$. Thus, $\Sigma \equiv \mathrm{var}(\delta)$ has $(j,\ell)^{\mathrm{th}}$ block equal to $\Sigma_{j\ell}$, $c(\mathbf{s}) \equiv \mathrm{cov}(\delta, \delta(\mathbf{s}))$ has $(j,\ell)^{\mathrm{th}}$ block equal to $c_{j\ell}(\mathbf{s})$ (or has ℓ^{th} column equal to $c_\ell(\mathbf{s})$), and $\Sigma(\mathbf{s})$ has $(j,\ell)^{\mathrm{th}}$ element equal to $C_{j\ell}(\mathbf{s},\mathbf{s})$.

Suppose that each component process of $Z(\cdot)$ possesses a variogram:

$$2\gamma_{jj}(\mathbf{h}) \equiv \mathrm{var}(Z_j(\mathbf{s} + \mathbf{h}) - Z_j(\mathbf{s}))$$

$$\mathbf{h} \in \mathbb{R}^d, \, j = 1, \ldots, k$$

There are two ways to generalize this notion to account for cross dependence between $Z_j(\cdot)$ and $Z_\ell(\cdot)$. The most natural one for multivariate spatial prediction (cokriging) is seen below to be:

$$2\gamma_{j\ell}(\mathbf{h}) \equiv \mathrm{var}(Z_j(\mathbf{s} + \mathbf{h}) - Z_\ell(\mathbf{s}))$$

$$\mathbf{h} \in \mathbb{R}^d \tag{15}$$

which, apart from a mean correction, is the quantity proposed by Clark et al.[5]

The other generalization:

$$2\nu_{j\ell}(\mathbf{h}) \equiv \mathrm{cov}(Z_j(\mathbf{s} + \mathbf{h}) - Z_j(\mathbf{s}), Z_\ell(\mathbf{s} + \mathbf{h}) - Z_\ell(\mathbf{s}))$$

$$\mathbf{h} \in \mathbb{R}^d \tag{16}$$

can only be used for cokriging under the special condition that the matrix covariance function (Equation 18) below is symmetric.[25,30] However, it has been the generalization traditionally recommended.[30,31,33,42] Relationships between the $\{\gamma_{j\ell}(\cdot)\}$, $\{\nu_{j\ell}(\cdot)\}$, and cross-covariances $C_{j\ell}(\cdot,\cdot)$ are given by Myers[33] and Ver Hoef and Cressie.[40]

Let

$$E(\mathbf{Z(s)}) = \mu$$

$$\mathbf{s} \in D \tag{17}$$

$$\mathrm{cov}(\mathbf{Z(s)},\mathbf{Z(u)}) = C(\mathbf{s},\mathbf{u})$$

$$\mathbf{s}, \mathbf{u} \in D \tag{18}$$

where $\quad \mu \equiv (\mu_1, \ldots, \mu_k)'$
and $C(\mathbf{s},\mathbf{u}) = k \times k$ matrix (not necessarily symmetric)

The cokriging predictor of $Z_1(s_0)$ is a linear combination of all the available data values of all the k variables:

$$\hat{Z}_1(s_0) = \sum_{i=1}^{n} \sum_{j=1}^{k} \lambda_{ji} Z_j(s_i) \qquad (19)$$

Notice that Equation 19 assumes that all components of $\mathbf{Z}(s_i)$ are available for each i. Should this not be the case, a straightforward modification of Equation 19 through Equation 23 is possible.[25,31]

Asking for a predictor that is uniformly unbiased; that is, $E(\hat{Z}_1(s_0)) = \mu_1$, for all μ, yields the necessary and sufficient condition:

$$\sum_{i=1}^{n} \lambda_{1i} = 1, \quad \sum_{i=1}^{n} \lambda_{ji} = 0 \qquad (20)$$

$$j = 2, \ldots, k$$

Therefore, the best linear unbiased predictor (i.e., the cokriging predictor) is obtained by minimizing:

$$E(Z_1(s_0) - \sum_{i=1}^{n} \sum_{j=1}^{k} \lambda_{ji} Z_j(s_i))^2 \qquad (21)$$

subject to the constraints (Equation 20). In terms of covariances and cross-covariances,[30] the cokriging equations are easily seen to be:

$$\sum_{i=1}^{n} \sum_{j=1}^{k} \lambda_{ji} C_{j\ell}(s_i, s_{i'}) - m_\ell = C_{1\ell}(s_0, s_{i'}) \qquad (22)$$

$$i' = 1, \ldots, n, \; \ell = 1, \ldots, k$$

combined with the linear equations in Equation 20. These linear equations result in Lagrange multipliers $m_1, \ldots, m_k$. The minimum mean-squared prediction error, or (co)kriging variance, is:

$$\sigma_1^2(s_0) = C_{11}(s_0, s_0) - \sum_{i=1}^{n} \sum_{j=1}^{k} \lambda_{ji} C_{1j}(s_0, s_i) + m_1 \qquad (23)$$

The covariance-matrix function $C(s, u)$ usually has to be estimated from the data. The assumption $C(s, u) = C^*(s - u)$ allows estimation of $C^*(\cdot)$ from $\mathbf{Z}(s_1), \ldots, \mathbf{Z}(s_n)$. It is also possible to formulate cokriging in terms of variograms and cross-variograms. Substituting $\{-\gamma_{j\ell}(\cdot)\}$ for

$\{C_{j\ell}^*(\cdot)\}$ in Equations 22 and 23 yields the appropriate cokriging equations.

In contrast, substitution of $\{-\nu_{j\ell}(\cdot)\}$ for $\{C_{j\ell}^*(\cdot)\}$ in Equations 22 and 23 is recommended in much of the geostatistical literature; however, it is only appropriate should $C_{j\ell}^*(\cdot)$ be a *symmetric* matrix. To provide models for $\{\nu_{j\ell}(\cdot)\}$, but to have a restrictive condition on $\{C_{j\ell}^*(\cdot)\}$ that usually cannot be checked, is self-defeating. After all, the rationale behind working with cross-variograms is to finesse the need for cross-covariances. Further, if the condition can be checked and does not hold, the cokriging equations in terms of $\{\nu_{j\ell}(\cdot)\}$ are *wrong*; an example is given by Ver Hoef and Cressie.[40]

On substituting $\gamma_{j\ell}(\mathbf{s},\mathbf{u}) \equiv (1/2)\mathrm{var}(Z_j(\mathbf{s}) - Z_\ell(\mathbf{u}))$ for $C_{j\ell}(\mathbf{s},\mathbf{u})$ in the notation defined in the introduction of this chapter and consequently replacing Σ with Γ, $\Sigma_{j\ell}$ with $\Gamma_{j\ell}$, $\mathbf{c}_{j\ell}(\mathbf{s})$ with $\gamma_{j\ell}(\mathbf{s})$, $\mathbf{c}_j(\mathbf{s})$ with $\gamma_j(\mathbf{s})$, $c(\mathbf{s})$ with $\gamma(\mathbf{s})$, and $\Sigma(\mathbf{s})$ with $\Gamma(\mathbf{s})$, the cokriging predictor of $Z(\mathbf{s}_0)$ is given by Ver Hoef and Cressie[40] in its most general form as:

$$\hat{\mathbf{Z}}(\mathbf{s}_0) \equiv \gamma(\mathbf{s}_0)'\Gamma^{-1}\mathbf{Z} + \{X(\mathbf{s}_0) - \gamma(\mathbf{s}_0)'\Gamma^{-1}X\}(X'\Gamma^{-1}X)^{-1}X'\Gamma^{-1}\mathbf{Z} \quad (24)$$

provided that the condition in Equation 20 and $(k-1)$ analogous conditions on the other sets of coefficients pertaining to $\hat{Z}_2(\mathbf{s}_0), \ldots, \hat{Z}_k(\mathbf{s}_0)$ hold. They also show that the mean-squared-prediction-error matrix of Equation 24 is:

$$\mathbf{M} \equiv E(\mathbf{Z}(\mathbf{s}_0) - \hat{\mathbf{Z}}(\mathbf{s}_0))(\mathbf{Z}(\mathbf{s}_0) - \hat{\mathbf{Z}}(\mathbf{s}_0))'$$

$$= -\Gamma(\mathbf{s}_0) + \gamma(\mathbf{s}_0)'\Gamma^{-1}\gamma(\mathbf{s}_0)$$

$$- \{X'\Gamma^{-1}\gamma(\mathbf{s}_0) - X(\mathbf{s}_0)'\}'(X'\Gamma^{-1}X)^{-1}\{X'\Gamma^{-1}\gamma(\mathbf{s}_0) - X(\mathbf{s}_0)'\} \quad (25)$$

and has the property that $E(\mathbf{Z}(\mathbf{s}_0) - B'\mathbf{Z})(\mathbf{Z}(\mathbf{s}_0) - B'\mathbf{Z})' - \mathbf{M}$ is nonnegative definite for any $nk \times k$ matrix B. In that sense, the cokriging predictor (Equation 24) is optimal.

Cressie and Helterbrand[10] develop a *spatial* principal components analysis that allows multivariate data reduction. Special cases result in Switzer and Green's[37] maximal autocorrelation functions (MAFs), which are based on $\{2\nu_{j\ell}(\cdot)\}$. However, in its full generality, knowledge of cross-variograms $\{2\gamma_{j\ell}(\cdot)\}$ are needed. Spatial principal components can also be applied to multivariate cross-validation, allowing one to diagnose lack of model fit and detect spatial outliers in a multivariate setting.

LIMITS OF DETECTION

Chemical analyses transform canisters of contaminated soil and vials of polluted water into data on a log sheet or a magnetic disk. The data recorded are pollutant concentrations, although at times no concentration is reported, the datum being declared a "nondetect" (ND). In the analysis by the chemist, stringent quality-control protocols have to be met before a measurement is released. These protocols derive from a desire not to declare the presence of a pollutant (and an associated measurement) when in fact none is present.

As a consequence, a "limit of detection" (LOD) is often given, below which any measurement obtained is declared ND. For example, the 1988 Love Canal study[4] defined the LOD as the concentration C_0 for which the power of an $\alpha = 0.05$ test of H_0: true concentration in the field sample is zero, vs H_1: true concentration in the field sample is greater than zero, is $\pi = 0.95$.

In the discussion that follows, I shall briefly critique this measurement strategy and then go on to investigate how spatial modeling might proceed with such censored data. There seems to be considerable disarray within the chemistry profession regarding terms (and their meanings) associated with detection (see, for example, p. 326 of the volume edited by Currie[11] and the article by Lambert et al.[27]). At first, it seems straightforward: The American Chemical Society Committee on Environmental Improvement (ACSCEI) in 1980 defined the LOD as "the lowest concentration level that can be determined to be statistically different from a blank."[1] While the general idea behind this statement is laudable, its implementation requires some statistical imagination. Witness the Love Canal criterion stated above: there is a cutoff value M_0 above which would occur only $\alpha = 5\%$ of measurements on blank samples. However, for samples from a pollutant with true concentration C_0, $\pi = 95\%$ of the measurements would be above M_0. Notice the arbitrariness of the choice of α and π; and for the purposes of implementation of this criterion, notice the necessity of a probability distribution for measurement given true concentration.

It is apparent to every modern-day statistician that this artificial censoring of data that fall below an LOD determined by Neyman-Pearson hypothesis testing[34] destroys information and must be reexamined. In a desire to release only the highest quality data (if not released, an ND is entered), the chemist has emphasized a measurement-by-measurement perspective and lost sight of the larger goals of the study. Exploratory data analysis[39] followed by methods such as logistic regression[27] and kriging with different scales of variation[8] attempt to filter out the measurement error and make inference on true concentrations. However, the task is made much more difficult if measurements below the LOD are replaced by NDs.

Even the practice of declaring as an ND any measurement below the LOD confuses two quantities, the measurement random variable M and the true concentration C. For the Love Canal study, it is guaranteed that:

$$Pr(M \geq M_0|C = 0) = 0.05 \tag{26}$$

$$Pr(M \geq M_0|C = C_0) = 0.95 \tag{27}$$

where M_0 is determined from Equation 26, and the LOD C_0 is determined from Equation 27. However, what is not controlled is:

$$Pr(M \geq C_0|C = 0) \tag{28}$$

Would it be acceptable that 15% of blank samples register as detects above the LOD C_0? There are many other methods that have been used to determine the LOD. Although the ACSCEI recommended in 1980 that it be calculated as 3 times the standard deviation of blank samples, that multiplicative factor has in different studies ranged between 2 and 10 (and beyond) (see p. 16 of the overview chapter written by Currie[11]). In the worst case, the method of determination of the LOD is not stated at all. On occasions, the LOD may change in the course of the study, as new laboratories or instruments are used.

Further, because the measurement process of gas chromatography/mass spectroscopy has an extra "identification" stage that is meant to confirm the presence of the target compound (in the absence of confirmation, an ND is declared), it is possible that a measurement is bigger than the LOD but an ND is still entered.[27] In any environmental study, the statistician must know how the LODs are defined and why the ND entries have been declared so.

Finally, to complete the tragedy of errors, the data analyst—whose task it is to make sense out of the data—often substitutes for NDs the value zero, LOD, one half LOD, or a prediction of some sort in order to apply continuous distribution theory. (Nonparametric methods based on ranks could be used, but are invalid if measurements associated with NDs could be greater than the LOD.) Thus, what the measurement process takes away, the statistical analysis process tries to re-create.

The current practice of artificially censoring pollutant concentration measurements has resulted from an overzealous use of classical mathematical statistics. Imagine a situation where *all* measurements are reported. The environmental process can now be modeled at both high and low concentrations, although it is recognized to be convolved with measurement error (noise). First, by "borrowing strength" from all parts of the process observed (be they high or low concentrations), the noise can be filtered out, even in those parts of low concentration. For example, if the data are regional, they often exhibit spatial continuity and the geostatistical methods (such as kriging) of the "Geostatistics" and "Multivariate Geostatistics" sections can be used. Hence, it may be possible to determine whether a regional concentration is below an action level even though that level is within measurement error of zero. (It is virtually impossible to do this from NDs if the LOD is above the action level.) Second, it is clear that resources should be put into determining $\Pr(M \le m|C = c)$; $c \ge 0$ (from, e.g., blank and spiked samples). Third, one no longer has to worry about the *global* implications of a measurement-by-measurement misclassification of NDs.

The practice of artificially censoring concentration measurements should stop. The general consensus of authors writing on this topic is that the original measurements are better than the censored data.[14-16,18,27,35,38] Old habits are hard to lose; data sets of censored measurements will be around for a while. How can an environmental process with spatial continuity be modeled from detects and NDs?

Assume that an ND is recorded if and only if the measurement $M<$LOD. To make this discussion simple, assume that the underlying process of measured concentrations is a stationary lognormal random field (e.g., see Section 3.2.2 of the book by Cressie[8]). Then the observed process is a truncated field, where each potential datum is obtained by truncating the lognormal process at LOD. Thus, the observed process is $Y(s) = \exp(Z(s))$, where s is a spatial location in a domain D:

$$Z(s) = \begin{cases} W(s), & \text{if } W(s) \ge \ell \equiv \log(\text{LOD}) \\ \ell, & \text{otherwise} \end{cases} \tag{29}$$

and $\{W(s): s \in D\}$ is a stationary Gaussian (or normal) process. That is, the process $W(\cdot)$ satisfies Equations 2 and 3, and all joint finite-dimensional distributions are normal.

Because $\{Y(s): s \in D\}$ and $\{Z(s): s \in D\}$ are in a one-to-one relation, I can equivalently consider inference on $W(s_0)$ (the Gaussian scale) at a known spatial location s_0, based on data $Z(s_i) = \log Y(s_i)$; $i = 1, \ldots, m$, and $Z(s_{m+1}) = \ldots = Z(s_n) = \ell$. Write

$$U \equiv (W(s_1), \ldots, W(s_m))' = (Z(s_1), \ldots, Z(s_m))' \tag{30}$$

$$V \equiv (W(s_{m+1}), \ldots, W(s_n))'$$

although it is only known that $V < \ell 1$. Stein[36] shows how to calculate:

$$F(w; u,\ell) \equiv \Pr(W(s_0) \le w|U = u, V < \ell 1)$$
$$= \frac{\int_{(-\infty,\ell]^{n-m}} \Phi((w - a(u) - b'v)/\tau))p(v|u)dv}{\int_{(-\infty,\ell]^{n-m}} p(v|u)dv} \tag{31}$$

where conditional on $U = u$ and $V = v$, $W(s_0)$ is normally distributed with mean $a(u) + b'v$ and variance τ^2; $\Phi(\cdot)$ is the cumulative distribution function of a standard normal random variable; and, for suitable parameters **c**, *B*, and *T*:

$$p(\mathbf{v}|\mathbf{u}) = (2\pi)^{-(n-m)/2}|T|^{1/2} \exp\{-(1/2)(\mathbf{v} - \mathbf{c} - B\mathbf{u})'T^{-1}(\mathbf{v} - \mathbf{c} - B\mathbf{u})\} \quad (32)$$

is the conditional (normal) density of **V** given **U** = **u**.

With knowledge of the conditional distribution (Equation 31), the optimal predictor (i.e., the predictor that minimizes mean-squared prediction error) of $\exp(W(\mathbf{s}_0))$ is:

$$E(\exp(W(\mathbf{s}_0))|\mathbf{U} = \mathbf{u}, \mathbf{V} < \ell\mathbf{1}) = \int_{-\infty}^{\infty} \exp(-w)F(dw; \mathbf{u}, \ell) \quad (33)$$

Should a different criterion of optimality be chosen, a different optimal predictor of $\exp(W(\mathbf{s}_0))$ would result, but it will always depend on $F(w; \mathbf{u},\ell)$ given by Equation 31 (e.g., see pp. 107 and 108 of the book by Cressie[8]). Stein[36] goes on to show how Monte Carlo techniques can be used to approximate Equation 31 and uses the results to assess the accuracy of indicator kriging and indicator cokriging.[21,24,26] These latter methods, which yield estimators of $F(w; \mathbf{u},\ell)$, are on occasions substantially different from the optimal estimator (Equation 31).

Finally, if models for the uncensored process $W(\cdot)$ are given in terms of conditional distributions $f(w(\mathbf{s}_0)|w(\mathbf{s}_i); i = 1, \ldots, n)$ and $f(w(\mathbf{s}_j)|w(\mathbf{s}_0), w(\mathbf{s}_i); i \neq j = 1, \ldots, n)$, Gelfand et al.[13] show how the Gibbs sampler (a Monte Carlo technique for estimating joint and marginal distributions) can be used to generate $f(w(\mathbf{s}_0)|\mathbf{U} = \mathbf{u}, \mathbf{V} < \ell\mathbf{1})$, the density of the distribution Equation 31. However, it is rare to find such Markov random field models in environmental applications, where one is typically interested in $W(\mathbf{s}_{0,1})$, $W(\mathbf{s}_{0,2})$, . . . , given (truncated) $Z(\mathbf{s}_1)$, . . . , $Z(\mathbf{s}_n)$. Now, in order to use the Gibbs sampler, *all* conditional distributions involving the random variables $W(\mathbf{s}_{0,1})$, $W(\mathbf{s}_{0,2})$, . . . , $W(\mathbf{s}_1)$, . . ., $W(\mathbf{s}_n)$ would have to be specified. This is not an easy modeling task, even for a Gaussian random field. (Further discussion is on pp. 364 and 365 of the book by Cressie[8].)

ACKNOWLEDGMENTS

Funding for this research came from the National Science Foundation; the National Security Agency, and the Office of Technology Development, Office of Environmental Restoration and Waste Management, U.S. Department of Energy, through the Ames Laboratory. Ames Laboratory is operated by Iowa State University for the U.S. Department of Energy under Contract No. W-7405-ENG-82.

REFERENCES

1. American Chemical Society Committee on Environmental Improvement (ACSCEI) (1980). Guidelines for data acquisition and data quality evaluation in environmental chemistry. *Anal. Chem.*, 52, 2242–2249.
2. Bilonick, R. A. (1985). The space-time distribution of sulfate deposition in the northeastern United States. *Atmos. Environ.*, 19, 1829–1845.
3. Bilonick, R. A. (1988). Monthly hydrogen ion deposition maps for the northeastern U.S. from July 1982 to September 1984. *Atmos. Environ.*, 22, 1909–1924.
4. CH2M-Hill (1988). Love Canal Emergency Declaration Area Habitability Study, Final Report. CH2M-Hill, Reston, VA.
5. Clark, I., Basinger, K. L., and Harper, W. V. (1989). MUCK: A novel approach to co-kriging. In *Proceedings of the Conference on Geostatistical, Sensitivity, and Uncertainty Methods for Groundwater Flow and Radionuclide Transport Modeling*, B. E. Buxton (Ed.), Battelle Press, Columbus, OH, 473–493.
6. Cressie, N. (1984). Towards resistant geostatistics. In *Geostatistics for Natural Resources Characterization*, Part 1, G. Verly, M. David, A. G. Journel, and A. Marechal (Eds.), Reidel, Dordrecht, 21–44.
7. Cressie, N. (1988). Spatial prediction and ordinary kriging. *Math. Geol.*, 20, 405–421.
8. Cressie, N. (1991). *Statistics for Spatial Data*. John Wiley & Sons, New York.
9. Cressie, N., Gotway, C. A., and Grondona, M. O. (1990). Spatial prediction from networks. *Chemometrics Intelligent Lab. Syst.*, 7, 251–271.
10. Cressie, N. and Helterbrand, J. D. (1993). Multivariate spatial statistics in a GIS. *Geogr. Syst.*
11. Currie, L. A. (Ed.) (1988). *Detection in Analytical Chemistry: Importance, Theory, and Practice*. American Chemical Society, Washington, DC.
12. Eynon, B. P. and Switzer, P. (1983). The variability of rainfall acidity. *Can. J. Stat.*, 11, 11–24.
13. Gelfand, A. E., Smith, A. F. M., and Lee, T. M. (1992). Bayesian analysis of constrained parameter and truncated data problems using Gibbs sampling. *J. Am. Stat. Assoc.*, 87, 523–532.
14. Gilbert, R. O. (1987). *Statistical Methods for Environmental Pollution Monitoring*. Van Nostrand Reinhold, New York.
15. Gilliom, R. J., Hirsch, R. M., and Gilroy, E. J. (1990). Effect of censoring trace-level water-quality data on trend-detection capability. *Environ. Sci. Technol.*, 18, 530–535.
16. Gleit, A. (1985). Estimation for small normal sets with detection limits. *Environ. Sci. Technol.*, 19, 1201–1206.
17. Haslett, J., Bradley, R., Craig, P., Unwin, A., and Wills, G. (1991). Dynamic graphics for exploring spatial data with application to locating global and local anomalies. *Am. Statistician*, 45, 234–242.
18. Helsel, D. R. (1990). Less than obvious: statistical treatment of data below the detection limit. *Environ. Sci. Technol.*, 24, 1766–1774.

19. Isaaks, E. H. and Srivastava, R. M. (1989). *An Introduction to Applied Geostatistics*. Oxford University Press, Oxford.

20. Istok, J. D. and Cooper, R. M. (1988). Geostatistics applied to groundwater pollution. III: Global estimates. *J. Environ. Eng.*, 114, 915–928.

21. Journel, A. G. (1983). Nonparametric estimation of spatial distributions. *J. Int. Assoc. Math. Geol.*, 15, 445–468.

22. Journel, A. G. (1984). New ways of assessing spatial distribution of pollutants. In *Environmental Sampling for Hazardous Wastes*, G. Schweitzer (Ed.), American Chemical Society, Washington, DC, 109–118.

23. Journel, A. G. (1985). The deterministic side of geostatistics. *J. Int. Assoc. Math. Geol.*, 17, 1–14.

24. Journel, A. G. (1988). Nonparametric geostatistics for risk and additional sampling assessment. In *Principles of Environmental Sampling*, L. H. Keith (Ed.), American Chemical Society, Washington, DC, 45–72.

25. Journel, A. G. and Huijbregts, C. (1978). *Mining Geostatistics*. Academic Press, London.

26. Lajaunie, C. (1990). Comparing some approximate methods for building local confidence intervals for predicting regionalized variables. *Math. Geol.*, 22, 123–144.

27. Lambert, D., Peterson, B., and Terpenning, I. (1991). Nondetects, detection limits, and the probability of detection. *J. Am. Stat. Assoc.*, 86, 266–277.

28. Le, D. N. and Petkau, A. J. (1988). The variability of rainfall acidity revisited. *Can. J. Stat.*, 16, 15–38.

29. Matheron, G. (1963). Principles of geostatistics. *Econ. Geol.*, 58, 1246–1266.

30. Myers, D. E. (1982). Matrix formulation of co-kriging. *J. Int. Assoc. Math. Geol.*, 14, 249–257.

31. Myers, D. E. (1984). Cokriging—new developments. In *Geostatistics for Natural Resources Characterization*, Part 1, G. Verly, M. David, A. Journel, and A. Marechal (Eds.), Reidel, Dordrecht, 295–305.

32. Myers, D. E. (1989). Borden field data and multivariate geostatistics. In *Hydraulic Engineering*, M. A. Ports (Ed.), American Society of Civil Engineering, New York, 795–800.

33. Myers, D. E. (1991). Pseudo-cross variograms, positive-definiteness, and cokriging. *Math. Geol.*, 23, 805–816.

34. Neyman, J. and Pearson, E. S. (1933). On the problem of the most efficient tests of statistical hypotheses. *Philos. Trans. R. Soc. A*, 231, 140–185.

35. Porter, P. S., Ward, R. C., and Bell, H. F. (1988). The detection limit. *Environ. Sci. Technol.*, 22, 856–861.

36. Stein, M. L. (1992). Prediction and inference for truncated spatial data. *J. Computational Graphical Stat.*, 1, 91–110.

37. Switzer, P. and Green, A. A. (1984). Min/max autocorrelation factors for multivariate spatial imagery. Technical Report No. 6, Department of Statistics, Stanford University, Stanford, CA.

38. Travis, C. C. and Land, M. L. (1990). Estimating the mean of data sets with nondetectable values. *Environ. Sci. Technol.*, 24, 961–962.

39. Tukey, J. W. (1977). *Exploratory Data Analysis*. Addison Wesley, Reading, MA.

40. Ver Hoef, J. M. and Cressie, N. (1993). Multivariable spatial prediction. *Math. Geol.*, 25, 219–240.

41. Vong, R. J., Moseholm, L., Covert, D. S., Sampson, P. D., O'Loughlin, J. F., Stevenson, M. N., Charlson, R. J., Zoller, W. H., and Larson, T. V. (1988). Changes in rainwater acidity associated with closure of a copper smelter. *J. Geophys. Res. D*, 93, 7169–7179.

42. Wackernagel, H. (1988). Geostatistical techniques for interpreting multivariate spatial information. In *Quantitative Analysis of Mineral and Energy Resources*, C. F. Chung, A. G. Fabbri, and R. Sinding-Larsen (Eds.), Reidel, Dordrecht, 393–409.

43. Wald, S., Sjostrom, M., and Hellberg, S. (1987). Chemometrics: multivariate analysis and design. *Bull. Int. Stat. Inst.*, 52, Book 4, 477–495.

CHAPTER 7

Design of the Clean Air Act Deposition Monitoring Network

David M. Holland, Ralph Baumgardner, Tim Haas, and Gary Oehlert

INTRODUCTION

The Clean Air Act Amendments (CAAA) of 1990 will have far-reaching effects on the monitoring of the deposition of atmospherically transported chemicals. The amendments incorporate significant new regulatory requirements for sulfur and nitrogen oxide emission reductions that are designed to reduce risk to public health and protect the natural environment. To determine the effectiveness of these emission reductions in reducing chemical exposure and deposition loadings to aquatic and terrestrial ecosystems, the CAAA calls for the "establishment of a national network to monitor, collect, and compile data, with quantification of certainty in the status and trends of air emissions, deposition, . . ."

To meet the challenge posed by the CAAA, the Environmental Protection Agency (EPA) initiated the development of the Clean Air Act Status and Trends Monitoring Network (CASTNET) which is structured to provide an integrated, interagency approach to large-scale rural monitoring and assessment. Its participants include EPA, other federal agencies, state agencies, and Canadian environmental agencies.

The design of CASTNET is structured to meet the information needs of policymakers, scientists, state and local agencies, and planners who demand high-quality deposition data. Ecological modeling investigations that focus on the complex relationships between ecosystem health and pollutant exposure require precise predictions of deposition loadings. Senior EPA managers involved in the decision-making process require information on temporal trends to answer the fundamental question: are the CAAA improving the nation's air quality? In addition, there is an urgent need to collect deposition data at sensitive aquatic and terrestrial ecosystems that are not currently monitored. Acidic deposition is known to cause adverse effects in lakes, streams, and forests—the extent and magnitude of which are not well characterized. Existing networks are severely deficient in the monitoring of wet deposition in high elevation areas subject to high precipitation, snowfall, and cloud coverage in the eastern and western United States, as well as precipitation in coastal areas. Predictions of deposition for these remote areas are inaccurate, and the correction of this problem is a high priority of the CASTNET monitoring program.

OBJECTIVES OF CASTNET NETWORK DESIGN

A key feature of CASTNET is that the data will be used to assess a variety of environmental problems. CASTNET must be well designed to provide adequate spatial coverage, reasonable temporal trend resolution, and measurements of deposition in critical ecosystems. The network should provide data to precisely characterize the spatial distribution of deposition in the United States through the interpolation of data among monitoring sites. Uncertainties for these predictions will depend on the type of model, natural variability, measurement precision, and network design (primarily through site density and location). In addition, CASTNET should evaluate the effect on sulfate deposition of the legislated emission reductions, a process that requires assessing the sensitivity of existing networks in detecting and quantifying trend in deposition over the CASTNET regional areas.[1] This type of analysis must distinguish short-term and long-term trends caused by meteorologic processes from the trend caused by reduced emissions.

This chapter addresses the current state of wet sulfate deposition monitoring in the conterminous United States. A statistical evaluation of data generated from active deposition networks has been completed to determine the current effectiveness of these networks in meeting the CASTNET design goals. These analyses have focused on determining:

- spatial patterns of rural deposition across the United States
- trend in sulfate concentration from data observed to date at existing monitoring sites and regional areas
- capabilities of existing monitoring networks to detect and quantify future sulfate reduction scenarios
- critical aquatic and terrestrial ecosystems where deposition monitoring is needed

Modeling errors surrounding the spatial interpolation and estimation of trend were used to provide guidance on the need to augment existing monitoring with additional sites to reduce these uncertainties. The design process was complemented by an evaluation of how many additional sites were needed in sensitive ecosystems to comply with the legislated requirement to increase monitoring in these areas.

The "redesign" of existing monitoring is complicated by the multi-objective nature of CASTNET. The seemingly well-defined goals of quantifying a downward trend in deposition or optimizing the spatial prediction mask a huge array of plausible objectives. For example, what form of trend is required, linear or some other parametric form? Are we looking for trends in average deposition or peak values? A number of different ions are subsumed under the term deposition, each with potentially different spatial-temporal characteristics. The design approach summarized in this chapter focuses on annual sulfate values. Ongoing investigations are underway to provide insight toward the design of other primary air and deposition variables.

SPATIAL INTERPOLATION

A spatial interpolation technique commonly referred to as kriging has been used to predict the spatial distribution of annual sulfate deposition. Kriging is a stochastic spatial prediction method that has been used extensively in the field of geostatistics.[2] A brief description of the main features of kriging is given based on discussion in Cressie.[3]

Consider a stochastic spatial process:

$$\{Z(x): x \in D\} \tag{1}$$

where $D \subset \mathbf{R}^2$. The goal of spatial prediction using kriging is to predict $Z(x_0)$ from data, $Z(x_1), \ldots, Z(x_n)$. The second-order spatial dependence of the process is characterized by an unknown function called the semivariogram. A common unbiased empirical semivariogram estimator[4] is:

$$\gamma(h) \equiv \frac{1}{2N_h} \sum_{N_h} [Z(x_i) - Z(x_i + h)]^2 \tag{2}$$

where N_h is the number of paired sites with lag (or separation) distance h. Precise estimation of this function is largely determined by the spatial configuration of sites. This function form assumes an isotropic (independent of direction) covariance structure. Commonly, kriging is applied to a local neighborhood of monitoring sites while using a regionally defined semivariogram.

The literature indicates a large number of kriging applications to environmental data. Most of these studies use conventional geostatistical techniques such as ordinary kriging which require that the underlying process be modeled as a homogeneous spatial random field. However, this assumption may not be valid for describing the sulfate deposition process that is heavily influenced by meteorology (primarily precipitation) and a diverse variety of chemical sources. The combined effect of these phenomena on deposition tends to produce a spatially nonhomogeneous sulfate deposition process across the conterminous United States.

Haas[5,6] employed a moving window, regression, residual form of kriging (MWRRK) procedure to overcome nonstationarity problems. The first-order (mean) nonstationarity is modeled by a spatial regression and the second-order (covariance) nonstationarity is accounted for by a moving window of sites. This procedure allows estimation of a process that exhibits an unknown spatial trend and a spatially heterogeneous covariance structure. MWRRK is actually an adaptation of an estimator developed by Neuman and Jacobson[7] to allow best linear unbiased estimation of a stochastic process having an unknown trend function and a spatially correlated error covariance structure.

A brief description of MWRRK follows. At each prediction location, the smallest circular window of monitoring sites to support fitting a spherical model[3] to the empirical semivariogram (Equation 2) is defined. The window size is found iteratively by starting with a radius equal to about 300 mi. If fewer than 30 sites exist within the window, the window radius is enlarged by increments of 50 mi until the required number of sites are included in the window. To model the potentially large variation in the spatial trend component of deposition, a three-stage generalized least-squares procedure is used to fit a quadratic regression of deposition on the site location coordinates. The error covariance matrix for the second and third stages is estimated from the semivariogram of residuals from the previous regression. If the percentage of modeled variation in deposition at any regression stage is less than 20%, the spatial trend model is abandoned and ordinary kriging using the spherical model is applied to the original data for that window. Otherwise, the residuals from the third regression are considered to be stationary and ordinary kriging is applied to the regression residuals to predict the residual value at the window center.

Predicted deposition at the window center is then found by adding the converged regression of local trend to the predicted residual. Thus, MWRRK performs either an iterative, general linear model regression followed by residual kriging or ordinary kriging within each window. Since MWRRK estimates a semivariogram for each prediction window, there is less potential for variogram inflation due to second order nonstationarity problems.

Using annual (1990) deposition from the National Atmospheric Deposition program/National Trends Network (NADP/NTN), MWRRK was used to predict the spatial distribution of sulfate deposition and associated interpolation error. Uncertainty is expressed as relative error (RE) which is an approximate measure of normalized error variance.[8] It is formed by dividing the square root of the kriging variance by the estimate at the point of prediction. The kriging predictions are visualized by illustrating deposition over a tiled pattern of hexagons across the conterminous United States superimposed over the location of the monitoring sites that are indicated as points (see Figure 1). Deposition at each hexagon centroid is computed as the average of deposition at the six hexagon vertices. During 1990, 179 monitoring sites complied with Level II data quality criteria were detailed by Olsen et al.[9] These criteria for site selection were developed by the Unified Deposition Data Base Committee and depend on measures of site representativeness and data completeness for the period of interest.

Figure 1 illustrates typically high sulfate deposition levels in the midwest and northeast of approximately 2.5–3.5 g/m^2 with levels tapering off to 0.5–1 g/m^2 with movement toward the west and south. All of the predictions over the hexagon grid shown in Figure 1 were obtained using the three-stage regression procedure. Overall, deposition can be estimated with a RE of about 15–30% in most of the eastern United States (see Figure 2). These errors are considerably higher in the west, but these results need to be viewed in conjunction with the generally lower deposition levels found in the west.

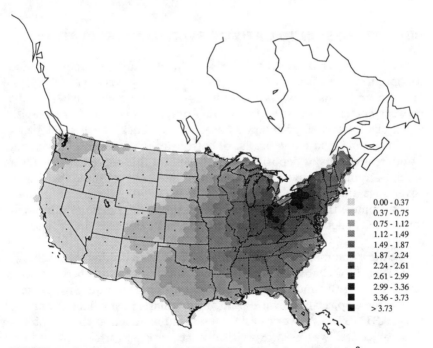

FIGURE 1. Spatial pattern of 1990 NADP/NTN sulfate deposition (g/m^2).

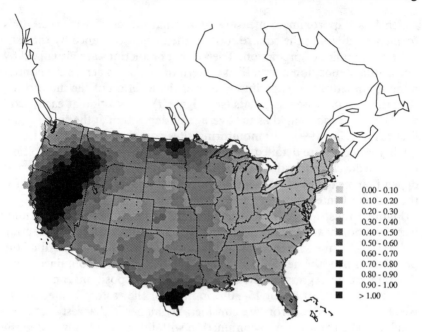

FIGURE 2. Spatial pattern of relative error (percent per 100) of 1990 NADP/NTN sulfate deposition.

NETWORK DESIGN TO OPTIMIZE SPATIAL INTERPOLATION

Several network design criteria exist in the literature that could be used to optimize the design of long-term monitoring networks. Trujillo-Ventura and Ellis[10] discuss multiple objective functions that describe network spatial coverage and the ability of a network to detect standard violations for several pollutants. Caselton and Zidek[11] have developed a design approach based on the use of entropy to reduce uncertainty about future realizations of deposition. To implement this approach, one must specify a number of potential monitoring sites and estimate the spatial covariance among these sites. The optimal design is the set of stations that maximize the expected information about the potential nonmonitored sites. A detailed discussion of this technique is given in Caselton et al.[12]

The most common design criteria associated with kriging methodology are: (1) ensure reliable ability to predict $Z(x_o)$ at locations between sites by minimizing the uncertainty of prediction and (2) optimize the estimation of the empirical semivariogram,[13] particularly at small lag distances. For CASTNET, a network design optimization function that combines these objectives was used. Specifically, network performance[14] was defined as the weighted sum of the mean relative error and a simulated standard deviation of the relative error at one point near the center of the CASTNET geographic region. Two different weightings of the function components (50/50 and 90/10) were used to gain some insight toward the sensitivity of site location with the level of weighting.

The mean relative error was calculated by numerically integrating the relative error surface and dividing that volume by the area of integration. The distribution of the RE at one location near the center of each region can be estimated through simulation. The simulation algorithm uses the $\hat{\gamma}(h,\Theta)$ in the center circular window to generate 40 multivariate normal vectors, each of length equal to the number of old plus new sites. For each vector, a spherical model is fit and used to estimate a new relative error for each vector. This approach does not attempt to minimize the uncertainty of the sample variogram directly, but instead minimizes the effect of imprecise variogram estimation on the relative error. The effect on the interpolation error due to estimating the semivariogram remains an open research issue and one that future CASTNET research will address.

Numerical optimization procedures were used to find the optimal location of 10 new sites by minimizing the design function for each CASTNET region. A potentially more informative measure of network performance would include the regional average simulated relative error standard deviation instead of the average variability at one point within each region. However, performing the simulation at each grid point required to integrate the relative error surface to compute the mean RE is beyond the capabilities of current high-performance workstations. Thus, it was decided to compute the RE standard deviation at one point near the center of each region. Since each site location is defined by two independent variables, the optimization problem of locating 10 sites is a 20-dimensional, nonlinear programming problem. Random starting locations for the new sites were used because optimization over the entire conterminous United States suggested that this starting method could lead to reductions in the weighted optimization function.

Augmenting existing monitoring with 10 sites per region did not significantly reduce average relative error in any of the regions for either weighting of the design function. The effect of adding new sites is detailed in Table 1 for an equal weighting of the optimization function components. Results for the Appalachian Mountains, the Mississippi Delta, and southern California are not given due to bias problems in interpolation caused by current nonrepresentative monitoring in these areas.

Due to the low spatial correlation of the residuals derived from the spatial regression, particularly in the west, it may be necessary to double or triple the number of sites before significant reductions in the mean relative error are realized. The eastern regions tend to exhibit a higher

Table 1. Optimization Results of Adding 10 Sites per Region Based on an Equal Weighting of Average Relative Error and the Variance of Relative Error

Region	Existing sites		Converged
Northeast	1.66[a]	18.9[b]	18.8[b]
		4.9[c]	2.1[c]
Mid-Atlantic	1.91	18.6	18.5
		3.5	1.9
South	1.42	28.9	28.9
		5.1	3.1
Midwest	2.36	23.3	23.3
		2.0	1.7
Plains	0.79	45.9	46.1
		14.3	6.5
Rockies	0.28	42.1	41.9
		11.3	5.5
Inter-Mountain	0.20	92.1	91.6
		484.2	139.6
Northwest	0.52	27.4	27.3
		8.8	4.2

[a]Average sulfate deposition (g/m^2).
[b]Average relative error (%).
[c]Standard deviation of relative error at regional centroid.

degree of micro- and small-scale (up to 100 km) spatial correlation relative to the west, but the variance of the residual process is still large at larger spatial scales. For most of the regions, the variability of the relative error at the center of the region is reduced substantially by 40–70% (see Table 1). The addition of 10 new sites per region tends to reduce the variability of 40 realizations of the semivariogram obtained through simulation which reduces the variance of the simulated estimates of RE.

NETWORK MODEL TO ESTIMATE TREND

A spatial-temporal model and smoothing technique developed by Oehlert[15] is used to estimate the trend of annual precipitation weighted sulfate concentration with quantifiable uncertainty. Sulfate concentration was investigated because the legislated SO_2 emission reductions would likely affect concentrations of sulfate in precipitation. The model was applied to data obtained from the acid deposition system[16] that complies with completeness requirements described by Sisterson et al.[17] Trend results are obtained for 94 stations which had no missing years during the 5-year period, 1982–1986. This particular time period was chosen as a compromise between spatial coverage in the east and temporal length. It should be noted that trends over short time periods are heavily influenced by meteorology, and the results may be misleading. As more data become available through the Clean Air Act emission reduction period, the trend results will be more informative across the longer length of monitoring.

This model is used to estimate changes in the mean level of annual sulfate concentration over a period of several years. The time trend is modeled as a simple linear trend or other parametric shape, although the methodology does not require this formulation. The model structure incorporates three types of noise which appear in wet deposition data: (1) sampling and analytical errors, (2) effects due to short-term meteorology lasting up to a few months that produce autocorrelation in the data, and (3) effects of long-term meteorology that can influence large-scale spatial patterns in deposition. Estimating the trend in a tiled pattern of rectangles is the primary objective of this analysis. For this analysis, the size of each rectangle was arbitrarily chosen to be about 1400 km². This discretized method of modeling and displaying trend complements our ability to display spatial patterns of deposition.

An overview of the motivation and structure underlying the model follows. Existing networks have s monitoring stations each with y years of data. Let $Y_i = (Y_{i1}, \ldots, Y_{iy})$ be a vector containing the annual sulfate concentration at site i, let $Y' = (Y'_1, \ldots, Y'_s)$ be a vector of concentrations for all the stations, and let $j(i)$ indicate the rectangle in which station i occurs. Then the model characterizes the structure of each deposition series, Y_i, as:

$$Y_i = \alpha_{j(i)} + t\beta_{j(i)} + L + N_i + \delta_i \qquad (3)$$

where $\alpha_{j(i)}$ = expected concentration for the rectangle
t = time vector of length y, centered to have mean zero
$\beta_{j(i)}$ = expected concentration slope for the rectangle
L = long-term noise series common to all stations
N_i = short-term station specific noise series
δ_i = station-specific effect accounting for measurement errors and bias that may be induced by elevation and proximity to point sources

A log transformation of annual precipitation weighted concentration is used to help stabilize the interannual variance of sulfate concentration.

The novel component of this model is the inclusion of the continental spatial scale component (L) and the difficulty in modeling this component. While the latter subject is open to debate, there is general agreement on the existence of wide scale long-term phenomena in precipitation data. Thus wet deposition, and to a lesser extent, sulfate concentration in precipitation should inherit at least part of this large-scale spatial variation through the dependence on precipitation volume.

Ordinary least squares (OLS) is used to produce estimates of the mean concentration (α) and slope (β) for each station. To account for the correlated error structure across time and space, we use the regression residuals to estimate the spatial and temporal covariance structure for input toward estimating the variance of site and regional trend estimates. To estimate the trend for each rectangle, we assume that neighboring rectangles have similar values and apply a discrete smoothness prior distribution to the slope to model this relationship. Further details of the modeling approach are given in Oehlert.[15]

The estimated tiled rectangular pattern of trend and associated uncertainty for the logarithm of annual sulfate concentration is shown in Figures 3 and 4 for the period 1982–1986. The estimates of trend and standard errors can be roughly interpreted as percent change per 100 due to use of log transformed data. The six divisions in the figures represent 10, 15, 25, 25, 15, and 10% of the total number of rectangles (3042). The trend in sulfate is increasing over the south, northeast, and central Appalachians. The midwest has a decreasing trend in sulfate for this time period. The highest uncertainties (Figure 4) occur in nonmonitored areas for this time period, particularly along coastal areas.

NETWORK ABILITY TO DETECT AND QUANTIFY TRENDS

This analysis addresses the fundamental policy question of whether existing deposition monitoring will be capable of revealing changes in sulfate concentration in wet deposition due to legislated emission reductions. Knowledge of the capability of current monitoring to reliably estimate projected CAAA Phase I and II changes in deposition is needed before an assessment can be made of the need to augment existing networks with new sites. This section describes the application of Oehlert's model to estimate the statistical power (probability of correctly inferring a trend) of existing monitoring to reliably detect and quantify the magnitude of a projected change in sulfate.

The regional acid deposition model[18] (RADM) was used to project a future deposition trend scenario across the CAAA emission reduction period (1994–2003) for the mid-Atlantic, southern, midwest, and northeast CASTNET regions. The reduction for each region occurs in seven

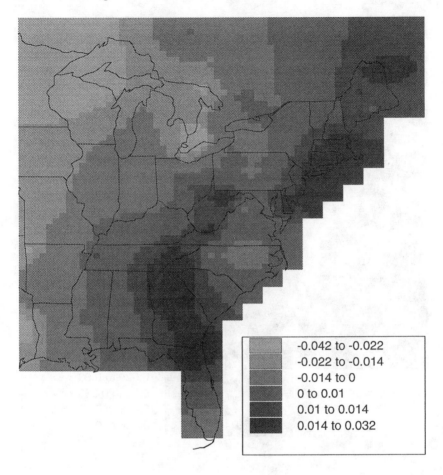

	-0.042 to -0.022
	-0.022 to -0.014
	-0.014 to 0
	0 to 0.01
	0.01 to 0.014
	0.014 to 0.032

FIGURE 3. Estimated rectangle values of trend (ln(mg/L)/year) in log sulfate concentration for the period 1982–1986.

unequal steps at the years: 1994, 1995, 1996, 2000, 2001, 2002, and 2003. The total sulfate decrease for these regions is approximately 30% by the year 2003. Incremental decreases for these years were determined by spreading the modeled CAAA Phase I and II decreases across the reduction period. This scenario is motivated by the belief that not all of the emission controls will occur in the mandated years of 1995 and 2000, but will in fact occur near the mandated years. It was impossible to investigate the entire range of reduction scenarios that may be of interest, but we believe that this projection is sufficiently plausible to give some indication of potential policy conclusions regarding the ability of existing networks to distinguish changes due to emission reductions.

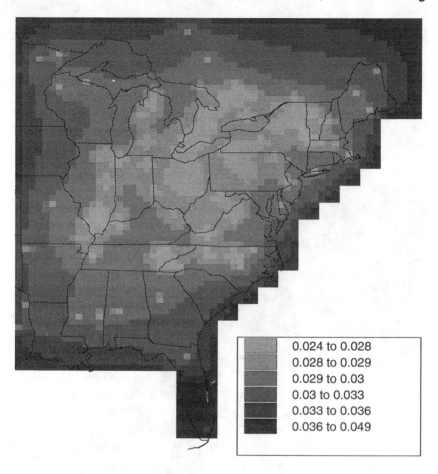

FIGURE 4. Estimated standard errors of rectangle trend values for log sulfate concentration (ln(mg/L)/year) over 1982–1986.

Oehlert's model can be utilized for network design by computing the variance of trend in the eastern regions where RADM projections exist. The magnitude of uncertainty for each estimate of regional trend depends on locations of the stations, amount of smoothing, and covariance matrix of the station trend estimates. The latter depends on estimated parameters required to model the short- and long-term noise patterns of the data. Given all of these modeled inputs, alternative configurations of new sites can be evaluated by varying the locations of some number of new sites and seeing how the variance of regional trend changes. Therefore, the model can be used to optimally locate new stations, locate redundant stations, and estimate the statistical power of seeing future changes in deposition.

For network design purposes, the data were adjusted for the effects of precipitation and linear time trend before computation of the variance of a station trend which is the essential part of this statistical power analysis. Precipitation-adjusted annual averages were formed by adjusting the monthly data for each station by a linear effect for precipitation volume and forming annual values by taking volume weighted means of the adjusted monthly values. This adjustment for precipitation produces a substantial reduction in the spatial covariance compared to the unadjusted data covariance. Then these annual volume-weighted, precipitation-adjusted values are detrended across years. Residuals from this linear adjustment for time are used to compute the variance of a station trend. Our basis to detrend over time is based on the economic factors that changed over 1982–1986 and the expected changes in deposition that these factors caused. If, in fact, emissions remained relatively constant over this period, then we have underestimated variability and consequently overestimated power of trend detection and quantification.

Reliable detection of a decreasing trend in concentration of sulfate in precipitation should occur by the end of the Phase I period (1996) in the eastern United States. The statistical power or probability of correctly detecting (i.e., concluding that sulfate concentration levels are monotonically decreasing) the RADM decrease is quite good, with all regions having a power exceeding 0.9 by the end of 1996. By 1997, the power is essentially 1.0. To address the more difficult question of quantifying the estimated decrease, we have investigated how likely is the estimated decrease to be within ±20% of the assumed true RADM decrease? Quantification according to this standard is much more difficult than trend detection. For the eastern regions, we should have a 90% chance of being within 20% of the true RADM projection by the end of 2001.

The potential statistical gain (i.e., reduction in variability in the regional trend) was investigated by adding 24 sites to rectangles within the border and coastal regions in the United States where relatively high variance of the trend exists. Regional variances were computed for 1997, 2000, and 2003 after adding these sites. The addition of sites does reduce the variance of regional trends, but not to the extent necessary to produce useful improvements in detectability or quantifiability. Future investigations will examine the effect of adding 50 sites in the eastern United States.

In summary, the use of this model indicates that with existing monitoring, we have an excellent chance of detecting downward decreases in sulfate concentration and deposition by 1996, when it is projected that sulfate will have decreased about 18% from pre-Phase I levels. However, precise quantification of the projected decrease will not occur until the year 2001 at which time an approximate 22% decrease will have occurred in the major eastern regions.

ECOSYSTEM DESIGN CONSIDERATIONS

The CAAAs require evaluation of the short- and long-term effects of acidic deposition on aquatic and terrestrial ecosystems with emphasis on high elevation forests and surface waters. In addition to the broad regional spatial scale of monitoring that the spatial and temporal analyses have addressed, there is a need to increase the level of monitoring information in these biologically sensitive ecosystem areas. Current knowledge of trend and deposition loadings in these areas is nearly nonexistent due to a lack of data. This sparsity of data severely restricts the ability of ecological models to predict the adverse impact of deposition on ecosystems that are the most sensitive to pollutants in precipitation.

The majority of NADP/NTN wet deposition sites in the east are at elevations less than 800 m, with only three sites at higher elevations. At high elevations, cloud and fog deposition can result in deposition amounts several times higher relative to that of precipitation alone.[19] Inadequate information exists on the levels of these types of deposition. Monitoring in remote areas will require a methods development effort to overcome current monitoring deficiencies at these sites. Further, relatively few existing sites are located near coastal or estuarine areas. It is essential to increase our knowledge of the level of nitrogen inputs in these sensitive areas. Although many of the wet deposition sites in the west are in sensitive ecosystems, the coverage across mountain areas is not uniform.

SITING RECOMMENDATIONS AND CONCLUSIONS

The regional spatial patterns of sulfate deposition and time trends of regional sulfate concentration means were investigated using data collected by the major wet deposition networks in North America. Statistical analyses were conducted to determine whether new monitoring sites would lead to significant reductions in the prediction uncertainties of spatial interpolation and the variance of estimates of regional trend. The structure of these models allows different configurations of network siting to be considered to provide sensible guidance in placing future network sites. For the large multistate regional spatial scales pertinent to the design of CASTNET, the improvement in reducing average interpolation error and in reducing the length of monitoring time required to reliably estimate a trend was insignificant. Adding new sites does produce a greater degree of confidence in the predicted pattern of relative error obtained through kriging. Ongoing investigations will continue this research for biologically significant ecosystems that cover smaller spatial scales where additional monitoring is expected to produce reductions in the average RE and variance of the trend.

Given the critical need to establish deposition monitoring in areas sensitive to the impact of deposition that are currently not monitored, it was decided that qualitative decisions should be paramount in placing new sites based on the representation of high elevation and coastal areas with varied geographic and meteorologic conditions. The CASTNET design calls for the addition of eight wet and/or cloud deposition sites in the Appalachian highlands above 800 m (see Figure 5). Five new sites will be located along the eastern coastline to provide critical measurements of nitrogren deposition loadings in estuarine/coastal ecosystems. Two of these sites would monitor for fog deposition. In the west, three additional sites are recommended for the Rockies and three sites in the Cascades in the Pacific Northwest. The addition of these new sites will provide the following information gains:

1. estimation of deposition loadings and trend at critical high elevation and coastal ecosystems where these data do not currently exist
2. improved ability to predict deposition gradients in the Appalachian Mountains
3. allow the estimation of regional trend in small geographic regions where current siting is minimal or does not exist

FIGURE 5. Proposed wet deposition/cloud/fog monitoring sites.

REFERENCES

1. Clean Air Status and Trends Network," Environmental Protection Agency, Atmospheric Research Exposure and Assessment Laboratory (February 1992).
2. Journel, A. G. and C. J. Huijbregts. *Mining Geostatistics* (New York, NY: Academic Press, 1978).
3. Cressie, N. *Statistics for Spatial Data* (New York, NY: John Wiley & Sons, 1991).
4. Matheron, G. "Principles of Geostatistics," *Econ. Geol.* 58, 1246–1266 (1963).
5. Haas, T. C. "Lognormal and Moving Window Methods of Estimating Acid Deposition," *J. Am. Stat. Assoc.* 85:950–963 (1990).
6. Haas, T. C. "Kriging and Automated Variogram Modeling Within A Moving Window," *Atmos. Environ.* 24A:1759–1769 (1990).
7. Neuman, S. P. and E. A. Jacobson. "Analysis of Nonintrinsic Spatial Variability by Residual Kriging with Application to Regional Groundwater Levels," *Math. Geol.* 16:499–521 (1984).
8. Venketram, A. "On the Use of Kriging in the Spatial Analysis of Acid Precipitation Data," *Atmos. Environ.* 22:1963–1975 (1988).
9. Olsen, A. R., E. C. Voldner, D. S. Bigelow, W. H. Chan, T. L. Clark, M. A. Lusis, P. K. Misra, and R. J. Vet. "Unified Wet Deposition Data Summaries for North America: Data Summary Procedures and Results for 1980–1986," *Atmos. Envir.* 24A:661–672 (1990).
10. Trujillo-Ventura, A. and J. H. Ellis. "Multiobjective Air Pollution Monitoring Network Design," *Atmos. Environ.* 25A:469–479 (1991).
11. Caselton, W. F. and J. V. Zidek. "Optimal Network Monitoring Design," *Stat. Probab. Lett.* 2:223–227 (1984).
12. Caselton, W. F., L. Kan, and J. V. Zidek. "Quality Data Network that Minimized Entropy," *Statistics in the Environment and Earth Sciences* (London: Griffin, 1992).
13. Zimmerman, D. L. and K. E. Homer. "A Network Design Criterion for Estimating Selected Attributes of the Semivariogram," (Iowa City, IA: Univ. of Iowa Technical Report #195, 1991).
14. Haas, T. C. "Redesigning Continental-Scale Monitoring Networks," Accepted for publication by *Atmos. Environ.*
15. Oehlert, G. W. "Regional Trends in Sulfate Wet Deposition," Accepted for publication by the *J. Am. Stat. Assoc.*.
16. Watson, C. R. and A. R. Olsen. "Acid Deposition System (ADS) for Statistical Reporting. System Design and User's Code Manual," Environmental Protection Agency, U.S. EPA Report-600/8-84-023 (1984).
17. Sisterson, D. L., B. C. Bowersox, and A. R. Olsen. "Deposition Monitoring: Methods and Results, *National Acid Precipitation Assessment Program, State of Science & Technology,* No. 6, Washington D.C. (1990).
18. Integrated Assessment: Questions 4 & 5, Results and Comparisons of Illustrative Future Scenarios," National Acid Precipitation Assessment Program, Washington D.C. (1990).
19. An Assessment of Atmospheric Exposure and Deposition to High Elevation Forests in the eastern United States," Environmental Protection Agency, U.S. EPA Report-600/3-90/059 (1990).

CHAPTER 8

National Air Quality and Emissions Trends Report

Barbara A. Beard and Warren P. Freas

INTRODUCTION

This chapter contains graphs from the 18th annual U.S. Environmental Protection Agency (EPA) Trends Report[1] documenting air pollution trends in the United States. The air pollutants included in the report are those which have National Ambient Air Quality Standards (NAAQS) established by the EPA. There are two types of NAAQS, primary and secondary. Primary standards are designed to protect public health while secondary standards protect public welfare such as effects of air pollution on vegetation, materials, and visibility.

There are six pollutants that have NAAQS: particulate matter (formerly as total suspended particulate [TSP] and now as PM-10 which emphasizes the smaller particles), sulfur dioxide (SO_2), carbon monoxide (CO), nitrogen dioxide (NO_2), ozone (O_3), and lead (Pb). It is important to note that the discussions of ozone in this report refer to ground level, or tropospheric, ozone and not to stratospheric ozone.

Two kinds of trends are tracked: *air concentrations*, which are based on actual direct measurements of pollutant concentrations at selected sites throughout the country; and *emissions*, which are based on the best available engineering calculations. Both trend assessments contain possible uncertainties. Although the ambient air pollution measurements are actual measurements, uncertainties can be introduced by changes in monitoring methodology and monitor siting. These uncertainties can be reduced by restricting the trend assessment to periods with consistent ambient monitoring methodologies and monitoring locations with historic records. The emission estimates have the advantage that these engineering calculations can be computed back in time using consistent methodologies; however, it should be recognized that these estimates of total emissions released in the air are not actual measurements. Because of the possible uncertainties in both assessments, the Trends Report uses both approaches so that one approach serves as an independent consistency check on the other.

The Trends Report also provides (1) a detailed listing of selected 1990 air quality summary statistics for every metropolitan statistical area (MSA) in the nation and (2) maps highlighting the largest MSAs.

The analysis presented here illustrates some of the complications that are often encountered in the analysis of environmental data. The characteristics of the monitoring network changed in 1980 making comparisons before and after that date difficult. Recently the definition of total suspended particulates changed to reflect respirable particles. This different definition makes temporal intercomparisons difficult if not impossible. There is considerable variation in the reported values because of spatial differences in monitor location, meteorologic variance, and monitoring imprecision. Thus, it is somewhat misleading to quote only the changes in the mean values without also including some measure of the overall variation. This problem is also reflected in the relationship (or lack thereof) between summary peak values and the means. To begin to address these problems, the air quality trends are presented using boxplots developed by Tukey.[2] The boxplots have the advantage of displaying simultaneously the 5th, 10th, 25th, 50th (median), 75th, 90th, and 95th percentiles of the data; and the composite mean. Thus, the use of boxplots facilitates the comparison of trends among the cleaner sites (lower percentiles), the higher sites (upper percentiles), and the composite mean. Although not shown in this chapter, the Trends Report also presents 95% confidence intervals about the composite means that can be used to make comparisons between years. The confidence intervals for the composite averages of the annual mean and second maxima air quality concentrations were computed from a two-way analysis of variance followed by an application of the Tukey studentized range.[3] Confidence intervals for the composite averages of the number of estimated exceedances were computed by fitting Poisson distributions to the exceedances each year and then applying the Bonferroni multiple comparisons procedure.[4-6] A more complete discussion of these procedures can be found in the report.

A landmark event for air pollution control in the United States occurred in November 1990, with the passage of the Clean Air Act Amendments. This has increased the interest in air pollution trends. While it is much too early for this act to have influenced these trends, some provisions are discussed briefly in the report because of the major role that the act will play in dictating future air quality and emission trends in the United States.

SOME PERSPECTIVE

The 10-year time period from 1981 to 1990 used to assess recent trends was selected to reduce the uncertainties because of the many changes in monitoring networks that occurred in the early 1980s due to more standardized air monitoring regulations. However, it is important not to overlook some of the earlier control efforts in the air pollution field. Emission estimates are useful in examining longer term trends. Between 1970 and 1990, lead clearly shows the most impressive decrease (–97%) but improvements are also seen for total particulate (–59%), sulfur oxides (–25%), carbon monoxide (–41%), and volatile organic compounds (–31%). (See Figure 1.) Only nitrogen oxides did not show improvement with emissions estimated to have increased 6%, due primarily to increased fuel combustion by stationary sources and motor vehicles. It is also important to realize that many of these reductions occurred even in the face of growth of emissions sources. More detailed information is contained in a companion report.[7]

While it is important to recognize that progress has been made, it is also important not to lose sight of the magnitude of the air pollution

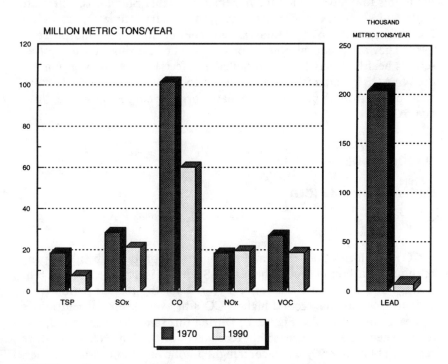

FIGURE 1. Comparison of 1970 and 1990 emissions.

problem that still remains. About 74 million people in the United States reside in counties which did not meet at least one air quality standard based on data for 1990 (Figure 2). The 63 million people living in counties that exceeded the ozone standard in 1990 is 4 million fewer than in 1989. The 1990 estimates for carbon monoxide and PM-10 are substantially lower than population totals for 1989. These population estimates are based only on a single year of data, 1990, and consider only counties with monitoring data for that pollutant. There are other approaches that would yield different numbers. For example, it is estimated that 140 million people live in ozone nonattainment areas based on the EPA October 1991 designations. This is because ozone nonattainment decisions are based on 3 years of data, rather than just 1 year, to reflect a broader range of meteorologic conditions. It was felt that 3-year averages would alleviate some of the problems encountered when 1 year was unusually high or low. Also, nonattainment boundaries may consider other air quality related information, such as emission inventories and modeling, and may extend beyond those counties with monitoring data to more fully characterize the ozone problem and to facilitate the development of an adequate control strategy.

Finally, it should be recognized that the report focuses on those six pollutants that have National Ambient Air Quality Standards. There are other pollutants of concern. According to industry estimates, more than 2.4 billion lb of toxic pollutants were emitted into the atmosphere in 1988. They are chemicals known or suspected of causing cancer or other serious health effects (e.g., reproductive effects). Control programs for the NAAQS pollutants can be expected to reduce these air toxic emissions by controlling particulates, volatile organic compounds, and nitrogen oxides.

MAJOR FINDINGS

Carbon Monoxide (CO)

Trends

CO is a colorless, odorless, and poisonous gas produced by incomplete burning of carbon in fuels. The NAAQS for CO is not to exceed an 8-hr average of 9 ppm more than once a year. Over the 10-year period from 1981 to 1990, annual second highest CO 8-hr average air concentrations decreased by 29%. (See Figure 3.) During this same 10-year period, CO experienced an 87% decrease in exceedances of the 8-hr NAAQS. More recently (1989 to 1990), the second highest nonoverlapping 8-hr CO average decreased 8%.

pollutant

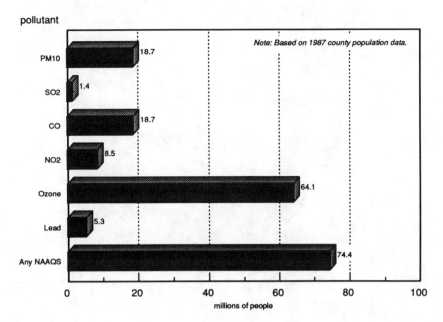

FIGURE 2. People in counties with 1990 air quality above primary National Ambient Air Quality Standards.

Carbon monoxide emissions have decreased 41% since 1970. Progress continued through the 1980s with a 22% reduction in total emissions. This progress occurred despite continued growth in miles of travel in the United States. Transportation sources account for approximately two thirds of the national CO emissions. Emissions from highway vehicles decreased 37% during the 1981–1990 period, despite a 37% increase in vehicle miles of travel. (See Figure 4.) Estimated nationwide CO emissions decreased less than 1% between 1989 and 1990, with forest fire activity in 1990 offsetting the 7% decrease in CO emissions from highway vehicles.

Status

In October 1991, EPA designated 42 areas as nonattainment for CO.

1990 Clean Air Act

The CO nonattainment areas have specific planning and implementation requirements specified in Title I of the act that vary depending on the magnitude of the CO problem. In addition, Title II of the act, which deals with mobile sources, includes a variety of provisions to help reduce

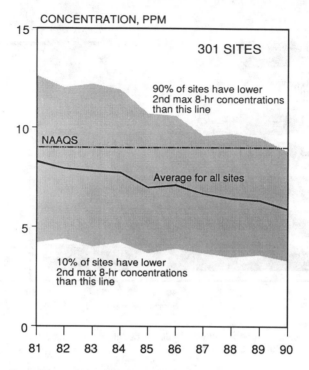

FIGURE 3. Carbon monoxide trend from 1981 to 1990 (annual second maximum 8-hr average).

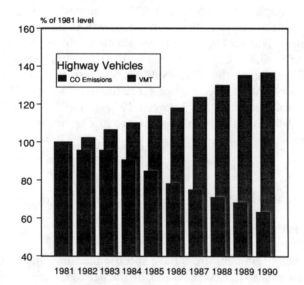

FIGURE 4. Comparison of trends in total national vehicle miles traveled and national highway vehicle emissions from 1981 to 1990.

CO levels including a winter time oxygenated fuels program for CO nonattainment areas, increased application of vehicle inspection and maintenance programs, and a tail pipe standard for CO emissions under cold temperature conditions.

Lead (Pb)

Trends

The NAAQS for lead is a maximum quarterly average of 1.5 $\mu g/m^3$ which is not to be exceeded. From 1981 to 1990, maximum quarterly average lead air concentrations decreased 85%. (See Figure 5.) In recent years (1989–1990), maximum quarterly average lead concentrations decreased 12%.

Lead gasoline additives, nonferrous smelters, and battery plants are the most significant contributors to atmospheric Pb emissions. Total lead emissions have dropped 97% since 1970 due principally to reductions in ambient lead levels from automotive sources. For the past 10 years (1981–1990), total lead emissions have decreased 87% and lead

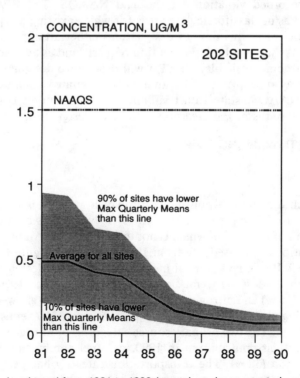

FIGURE 5. Lead trend from 1981 to 1990 (annual maximum quarterly average).

emissions from transportation sources have decreased 95%. From 1989 to 1990, total lead emissions decreased 1% and lead emissions from transportation sources experienced no change. The drop in Pb consumption and subsequent Pb emissions was brought about by the increased use of unleaded gasoline in catalyst-equipped cars (97% of the total gasoline market in 1990) and the reduced Pb content in leaded gasoline.

Status

In 1990, the reduction of exposure to lead became a top priority objective for the EPA. Among other things, the EPA identified 29 stationary sources with potential problems. An assessment of the compliance status, ambient monitoring availability, and state implementation plan (SIP) adequacy of these sources was completed.

1990 Clean Air Act

The amendments, for the first time, authorize the EPA to designate areas nonattainment, attainment, or unclassifiable for the lead NAAQS. As such, the EPA has designated as nonattainment 12 areas which have recently recorded violations of the lead NAAQS. The EPA has also designated as unclassifiable nine areas for which existing air quality data are insufficient at this time to designate as either attainment or nonattainment. As states submit designation requests and as ambient monitoring data become available, the EPA will proceed to designate additional lead areas as appropriate. Once an area is designated nonattainment for the lead NAAQS, states must submit revised pollution control plans within 18 months of the area nonattainment designation.

Nitrogen Dioxide (NO$_2$)

Trends

Nitrogen dioxide is a poisonous, brownish gas that is a precursor to ozone and to acidic precipitation. The NO$_2$ NAAQS is an annual arithmetic mean of 0.053 ppm which is not to be exceeded. Annual mean NO$_2$ concentrations decreased 8% from 1981–1990. (See Figure 6.) In recent years (1989–1990), NO$_2$ annual mean concentrations decreased 6%.

Nitrogen oxide (NO$_x$) emissions increased 6% since 1970 but both emissions (–6%) and nitrogen dioxide air quality (–8%) showed improvement during the 1980s. During the past 2 years, NO$_x$ emissions have decreased 1%. The two primary source categories of nitrogen oxide emissions, and their contribution in 1990, are fuel combustion (57%) and transportation (38%). The transportation category has decreased 24% while fuel combustion emissions are estimated to have increased by 12%.

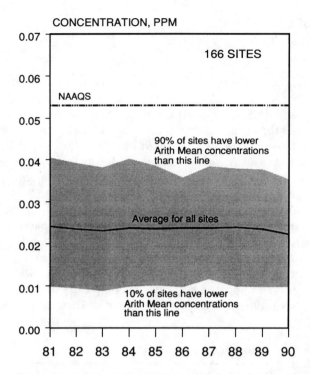

CONCENTRATION, PPM

FIGURE 6. Nitrogen dioxide trend from 1981 to 1990 (annual arithmetic mean).

Status

In October 1991, the EPA designated only one area as nonattainment for NO_2. Los Angeles, CA, which reported an annual mean of 0.056 ppm in 1990, is the only urban area that has recorded violations of the annual NO_2 NAAQS of 0.053 ppm during the past 10 years.

1990 Clean Air Act

Although Los Angeles is the only nonattainment area for nitrogen dioxide, the Clean Air Act Amendments of 1990 recognized the need for nitrogen oxide controls due to the contributing role of the gas in other problems including ozone (smog) and acid rain. The EPA has already issued final tighter tail pipe standards for NO_x as required under the new amendments. Future ozone (smog) control plans will address further NO_x controls, and the acid rain provisions of the act call for a 2 million ton NO_x reduction from affected utilities.

Ozone (O₃)

Trends

Ozone is a photochemical oxidant and the major component of smog. The NAAQS for O_3 is attained when the expected number of days per year with maximum hourly average concentrations above 0.12 ppm is equal to or less than one. Annual second highest daily maximum 1-hr O_3 concentrations decreased 10% from 1981 to 1990. (See Figure 7.) O_3 also experienced a 51% decrease in exceedances during this 10-year period. In the past 2 years, 1989–1990, second highest daily maximum 1-hr O_3 air concentrations decreased 1%.

Ground level ozone, the primary constituent of smog, has been a pervasive pollution problem for the United States. Ambient trends during the 1980s were influenced by varying meteorologic conditions. Relatively high 1983 and 1988 ozone levels are likely attributed in part to hot, dry, stagnant conditions in some areas of the country. Both 1989 and 1990 levels showed improvement, but the complexity of the ozone problem warrants caution in interpreting the data. There have been recent

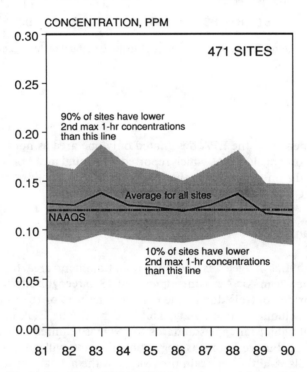

FIGURE 7.   Ozone trend from 1981 to 1990 (annual second daily maximum hour).

control measures, such as lower Reid vapor pressure (RVP) for gasoline resulting in lower fuel volatility and lower NO_x and VOC emissions from tail pipes. Emission estimates for volatile organic compounds (VOCs), which contribute to ozone formation, are estimated to have improved by 31% since 1970 and 12% since 1981. The past 2 years show a 1% increase in VOC emissions. However, these volatile organic compound (VOC) emission estimates represent annual totals. While these annual emission totals are the best national numbers now available, ozone is predominantly a warm weather problem and seasonal emission trends would be preferable. NO_x emissions, the other major precursor factor in ozone formation, decreased 6% between 1981 and 1990.

Status

In October 1991, the EPA designated 98 areas as nonattainment for O_3.

1990 Clean Air Act

This act expanded the framework for designating areas as attainment or nonattainment for ozone by further classifying areas based on the magnitude of their problem. Ozone nonattainment areas are now classified as marginal, moderate, serious, severe, or extreme. This allows more flexibility in the required control program. The act includes a variety of new requirements for cars and other sources of ozone precursors, including the introduction of cleaner (reformulated) gasoline beginning in 1995 into the nine U.S. cities with the worst ozone problems.

Particulate Matter

Trends

Air pollutants called particulate matter include dust, dirt, soot, smoke, and liquid droplets directly suspended in the air. The former TSP NAAQS was attained with an annual geometric mean of 75 $\mu g/m^3$ or less. In 1987, the EPA replaced the earlier TSP standard with a PM-10 standard. The current PM-10 NAAQS is attained with an annual arithmetic mean of 50 $\mu g/m^3$ or less. (PM-10 includes only those particles with aerodynamic diameters less than 10 μm. This focuses on the smaller particles likely to be responsible for adverse health effects because of their ability to reach the lower regions of the respiratory tract.) Ambient monitoring networks are being revised to measure PM-10 rather than TSP, and this is the reason for focusing on TSP for trends extending longer than 5 years. Over the 9-year period of 1982–1990, TSP experi-

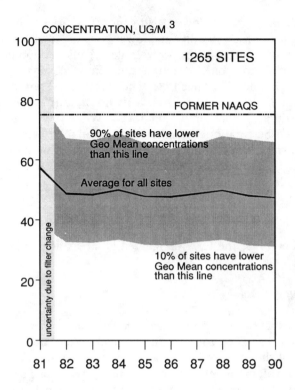

FIGURE 8.   Total suspended particulate from 1981 to 1990 (annual geometric mean).

enced a 3% decrease in air concentrations. (See Figure 8.) The 10-year change for TSP is not given because the 1981 data is affected by a change in filters. Ambient air concentrations decreased 3% for TSP and 8% for PM-10 during the last 2 years, 1989–1990.

Total particulate (TP) emissions from historically inventoried sources have been reduced 59% since 1970. During the 1980s, TP emissions decreased 6% while TP emissions increased 6% from 1982 to 1990. In the past 2 years, TP emissions increased 4%. Although PM-10 trends data are limited, emissions estimates increased 5% between 1989 and 1990. The PM-10 portion of TP emissions is estimated to have increased 7% since 1985 due to increases from transportation sources and forest fires. Nationally, fugitive sources provide six to eight times more tonnage of PM-10 emissions than historically inventoried sources.

Status

In October 1991, the EPA designated 70 areas as nonattainment for PM10. National average TSP levels in 1990 were the lowest of the past decade. Comparing 1989 and 1990, most of the country experienced an increase in precipitation and a decrease in TSP and PM-10.

1990 Clean Air Act

This act focuses attention on nonattainment of PM-10 health based standards. The acid rain provisions of the act address visibility impairment caused by fine (<2.5 μm) particles.

Sulfur Dioxide (SO$_2$)

Trends

Ambient sulfur dioxide results largely from stationary source coal and oil combustion, refineries, pulp and paper mills and nonferrous smelters. The two SO$_2$ NAAQS focused on are the annual arithmetic mean of 0.03 ppm and the 24-hr level of 0.14 ppm. SO$_2$ annual mean concentration decreased 24% from 1981 to 1990. (See Figure 9.) For this same 10-year period, annual second highest 24-hr SO$_2$ concentrations decreased 30% and the 24-hr exceedances decreased 87%. Recent changes show a 7% decrease in the SO$_2$ arithmetic mean concentration from 1989 to 1990.

SO$_x$ emissions decreased 25% since 1970. During the 1980s, emissions improved 6% while average air quality improved by 24%. This difference occurs because the historic ambient monitoring networks were population oriented while the major emission sources now tend to be in less populated areas. The exceedance trend is dominated by source-oriented sites. The 1981–1990 decrease in emissions reflects reductions at coal-fired power plants. The 1989–1990 emissions increase of 2% is due to increases from fuel combustion.

Status

Almost all monitors in U.S. urban areas meet the EPA ambient SO$_2$ standards. Dispersion models are commonly used to assess ambient SO$_2$ problems around point sources because it is frequently impractical to operate enough monitors to provide a complete air quality assessment. Currently, there are 50 areas designated as nonattainment for SO$_2$. Current concerns focus on major emitters, total atmospheric loadings, and possible need for a shorter term (i.e., 1-hr) standard. Of all national SO$_x$ emissions 70% are generated by electric utilities (92% of which come from coal-fired power plants).

1990 Clean Air Act

The acid rain provisions include a goal of reducing SO$_x$ emissions by 10 million tons relative to 1980 levels. The focus in this control program is

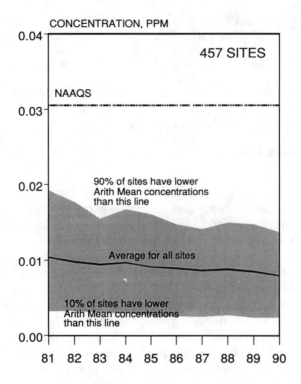

FIGURE 9. Sulfur dioxide trend from 1981 to 1990 (annual arithmetic mean).

innovative market-based emission allowances which will provide affected sources flexibility in meeting the mandated emission reductions. This is the first large-scale regulatory use of market-based incentives by the EPA. These reductions will improve visibility in the East by substantially reducing SO_x emissions. These emissions are transformed into fine acid sulfate aerosol, the main cause of regional visibility impairment.

SUMMARY

The air quality trends statistics in the annual Trends Report place emphasis on peak values and annual means related to the ambient air quality standards. These standards are designed to provide health protection against peak and long-term concentrations. While trend statistics based on peak values may be readily understood, there is the concern that they may be too variable to be used as trend indicators. The just released 1991 Trends Report[8] examines other possible indicators to check for consistency with these peak value trends. Some of the indicators considered include: maximum, second maximum, mean, 95th, 90th, 70th, and 50th percentiles. Although the indicators do not show complete agreement, there is general consistency in the direction of the trends. Examination of trends for these more robust statistics provides support for the peak value results and is a useful tool for these analyses. However, although statistical robustness is desirable, it is important to remember that the goal is to use trend indicators that are relevant. Trends in short term statistics should not be overlooked, but should be supplemented by analyzing a broader range of indicators.

REFERENCES

1. "National Air Quality and Emissions Trends Report, 1990," Office of Air Quality Planning and Standards, U.S. EPA Report-450/4–91–023 (November 1991).
2. J.W. Tukey, *Exploratory Data Analysis*, Addison-Wesley Publishing Company, Reading, MA, 1977.
3. B.J. Winer, *Statistical Principles in Experimental Design*, McGraw-Hill, New York, 1971.
4. N.L. Johnson and S. Kotz, *Discrete Distributions*, John Wiley and Sons, New York, 1969.
5. R.G. Miller, Jr., *Simultaneous Statistical Inference*, SpringerVerlag, New York, 1981.
6. A. Pollack, W.F. Hunt, Jr., and T.C. Curran, "Analysis of Variance Applied to National Ozone Air Quality Trends," presented at the 77th Annual Meeting of the Air Pollution Control Association, San Francisco, CA, June 1984.
7. "National Air Pollutant Emission Estimates, 1940–1990," Office of Air Quality Planning and Standards, U.S. EPA Report-450/4–91–026 (November 1991).
8. "National Air Quality and Emissions Trends Report, 1991," Office of Air Quality Planning and Standards, U.S. EPA Report 450-R-92–001 (October 1992).

SECTION III

Models and Data Interpretation

Daniel Krewski

INTRODUCTION

Data interpretation has not been widely done for environmental statistical data and information because generally the data are incomplete, have not been collected for the purpose of assembling the big picture, are spatially nonrepresentative, and as a rule are incompatible. However, in some cases such as for primary air pollutants, some interpretation is possible as shown in Chapter 11 by Schwartz. He asks and starts to answer such important questions as: are we concentrating our monitoring and regulatory efforts in the right areas? He suggests that it is cost effective to be concentrating resources on long-term effects, especially those due to particulates. He discusses some of the problems inherent in such analysis.

Some data interpretation is simple at first glance, but much more complex and difficult to interpret when looked at completely. An example is the description of soil quality as described in Chapter 12 by Cole. The pioneers could judge the soil by the vegetation. The composition of the currently available data involves such areas as: plant taxonomy, invertebrate biology and a wide range of disciplines and microscopic and macroscopic spatial variability, and sampling strategies and a wide range of analytical methodologies. Cole discussed the problems involved in integrating the diverse and complex data relating to soil quality.

In general, the data and information relating to the environment are too incomplete to be adequately analyzed and integrated. Methods for filling in the data and information gaps in any effort to describe the overall state of the environment include: professional judgment (see Chapter 1), guessing, interpreting, and modeling. The most familiar models are to be found in the areas of fate and transport in describing exposure and in models that fit dose-response data in describing and

179

estimating effects on human health and the environment. Two chapters in this section describe the different problems encountered in developmental endpoints (Chapter 9) and in broadly determining the human health potencies (Chapter 10). Some problem areas discussed include: thresholds, use of parametric and nonparametric statistical techniques, risk assessment, estimation of effects below the no observed adverse effect level (NOAEL), and time-to-tumor response and competing causes. Also discussed are more general questions such as: why are cancer rates increasing? With the plethora of chemicals produced there is a need to screen them to determine which ones are among the most important. Goddard and Krewski (Chapter 10) present a technique to develop such an approach in describing the potencies of chemicals as they can impact on human health.

The area of models and data interpretation involves a number of problems that need to be addressed in the overall determination of the state of the environment. It is, of course, most desirable to have actual data and information. However, in the meantime we will have to continue to rely on models. Perhaps in many areas there will be a long-term reliance on models. Thus it is most desirable to strengthen and improve all the models that describe environmental statistics and information.

CHAPTER 9

Statistical Issues for Development Toxicity Data

David W. Gaylor

INTRODUCTION

One of the concerns of modern society is the safe use of chemicals. Obviously, mankind has benefited from the development and use of new chemical products. Considerable attention has been given to potential cancer risk to humans due to exposure to chemicals used in food, drug, cosmetic, household, and industrial products and to contaminants in food, water, and air. Since the developing fetus may be particularly sensitive to chemical exposures experienced by the pregnant mother during gestation, attention has been given to limiting exposure to certain chemicals during pregnancy. Toxicity may be expressed as variations in development or behavior, morphological malformations, and fetal death. It is important to determine which chemicals are developmental toxicants and then to establish "safe" limits for their use. High exposure levels may produce profound health effects and should be avoided. The difficult issue is establishing exposure levels for developmental toxicants that result in negligible or no risk. Human data seldom are available to establish the relationship between the incidence of disease and chemical dose. Hence, animal bioassay data generally are employed to set safe exposure levels.

Animal bioassays are conducted to evaluate the toxicity of chemicals on the developing fetus. Typically, chemicals are administered to pregnant dams during a period of gestation. The pregnant dams generally are sacrificed near term, and the fetuses are examined for malformations and death. Chemicals are administered to the pregnant dam as the experimental unit. Since the individual fetal responses within a litter are likely to be correlated, failure to consider this correlation in statistical analyses may result in incorrect inferences. Hence, the sample size for a treatment group is the number of pregnant dams and not the number of fetuses.

181

The analyses discussed in this chapter focus on quantal data which are generally expressed as proportions of affected implants or fetuses per litter. For example, embryo lethality is usually measured by the proportion of implants in a litter that are dead and/or resorbed. Birth defects are generally measured by the proportion of live fetuses in a litter which possess a particular malformation among the live fetuses examined for that particular malformation. For malformations observable by external examination of a fetus (e.g., missing limb) all of the live fetuses in a litter usually are examinable. Detection of internal malformations require dissection of fetuses. Detection of certain skeletal malformations may require the use of special stains. Thus, care must be used to consider the number of fetuses in a litter examined for a particular type of malformation when computing the proportion of fetuses in a litter with each particular type of malformation.

For continuous data (e.g., fetal body weight) standard parametric and/or nonparametric statistical techniques can be employed. Such data are common for developmental measurements and behavioral tests conducted on offspring following birth to dams administered toxicants during gestation. The nesting of litters within treatment groups and fetuses within litters must be taken into account in statistical analyses.

Two major attributes of risk assessment will be addressed in this chapter: hazard identification and dose-response modeling. Various statistical techniques will be presented for testing for the existence of toxic effects during development. Dose-response modeling techniques will be discussed for estimating the incidence of developmental effects as a function of exposure to toxicants.

HAZARD IDENTIFICATION

The discussion in this section focuses on binomial responses where the fetus either possesses an attribute or it does not, e.g., malformed or normal. The measure of interest for the experiment unit (litter) is the proportion of abnormal fetuses. In general, we will be considering bioassay data where pregnant animals are administered different dosages of a potentially toxic substance. Suppose there are t treatments (dose) groups administered dosages of $d_1 < d_2 < \ldots < d_t$ plus control animals with $d_0 = 0$. Let $p_{ij} = x_{ij}/n_{ij}$ represent the proportion of affected implants or fetuses in the jth litter of the ith treatment group, where x_{ij} is the number of affected implants or fetuses out of n_{ij} examined for a particular defect.

Frequently, a large portion of litters will exhibit no abnormal fetuses for a particular anomaly, i.e., $p_{ij} = 0$, with a decreasing number of litters exhibiting increasing values of p_{ij}. Classical statistical procedures based on the normal (Gaussian) distribution are inappropriate for such skewed data. In such cases, it is recommended that nonparametric (distribution-free) statistical techniques are employed. The Mann-Whitney-Wilcoxin U test[1] may be employed to determine whether the proportion of abnormal fetuses in a treated group is statistically significantly greater than for the control group. If it is desired to maintain an experimentwise probability of a false positive for a particular type of anomaly at α (e.g., $\alpha = 0.05$), then the level of significance required for each comparison is $0.05/t$ (Miller[2]). For example, if five treatment groups are compared with the controls, then a P value of $0.05/5 = 0.01$ is required for each comparison in order to maintain an overall false positive rate of 0.05. An alternative nonparametric analysis is provided by Shirley[3] which compares the proportions of abnormal fetuses of increasing dose levels with the controls.

The Jonckheere[4] nonparametric test may be employed to determine whether there is a statistically significant dose-response trend for increasing proportions p_i of abnormal fetuses with increasing doses d_i.

Williams[5] proposed a parametric method for the statistical analysis of bioassay data. Kodell et al.[6] employs a beta-binomial distribution which assumes fetuses within the same litter behave independently according to a binomial distribution, and the probability of an abnormal fetus is assumed to vary among litters according to a beta distribution. The intralitter correlation coefficient is assumed to be constant within a dose group, but is allowed to vary across dose groups. The expected probability of an abnormal fetus for a litter of size n_{ij} at dose d_i is:

$$P(d_i,n_{ij}) = 1 - \exp\{-[\alpha + \theta_1(n_{ij} - \bar{n})]\}$$

for d_i less than or equal to a threshold dose of τ and $\bar{n}$ is the average litter size over all dose groups. For doses above the threshold dose, i.e., $d > \tau$, a Weibull dose-response is assumed:

$$P(d_i,n_{ij}) = 1 - \exp\{-[\alpha + \theta_1(n_{ij} - \bar{n}) + (\beta + \theta_2(n_{ij} - \bar{n}))(d_i - \tau)^w]\}$$

where $\alpha \geq 0, \beta \geq 0, \tau \geq 0, w \geq 1, [\alpha + \theta_1(n_{ij} - \bar{n})] \geq 0$, and $[\beta + \theta_2(n_{ij} - \bar{n})] \geq 0$ for all n_{ij}. The parameters can be estimated by the methods of maximum likelihood, and a test for a statistically significant dose-response trend can be performed. It is generally assumed that a threshold dose exists for most biological processes, other than cancer, below which the body can cope and no deleterious effects occur. However, estimation of a threshold dose from quantal bioassay data generally will have wide confidence limits.

Zeger and Liang[7] proposed methods for the statistical analysis of correlated binary data based on generalized estimating equations. Let the probability of abnormal fetuses in all m_i litters in the ith dose group be:

$$p_i = F(a + bd_i)$$

where F is a twice differentiable cumulative distribution function. The expected value and variance of x_{ij} are:

$$E(x_{ij}) = n_{ij}\, p_i$$
$$V(x_{ij}) = n_{ij}\, p_i\, (1 - p_i)\phi$$

where $\phi > 1$ accounts for extrabinomial variation. Estimates of the parameters a and b are obtained by solving the generalized estimating equations:

$$\Sigma\ \Sigma\ n_{ij}(\hat{p}_{ij} - p_i(a,\ b)) = 0$$

$$\Sigma\ \Sigma\ n_{ij}d_i(\hat{p}_{ij} - p_i(a,\ b)) = 0$$

which do not depend on ϕ:

where $\hat{p}_{ij} = x_{ij}/n_{ij}$.

Lefkopoulou et al.[8] show this leads to a generalized score statistic to test the null hypothesis of no dose-response trend b = 0:

$$T_{LMR} = \Sigma\ x_i(d_i - \bar{d})/[\Sigma v_i(d_i - \bar{d})^2]^{1/2}$$

where $\bar{d} = \Sigma n_i d_i/\Sigma n_i$, $V_i = \Sigma(x_{ij} - n_{ij}\,\hat{p})^2$, and $\hat{p} = \Sigma\Sigma x_{ij}/\Sigma\Sigma n_{ij}$. Asymptotically T_{LMR} is distributed as a standard normal deviate with mean 0 and variance 1.0.

Rao and Scott[9] suggested a method for the standard analysis of clustered binary data based on a concept of the "design effect" which leads to "effective numbers of abnormal fetuses" and "effective sample sizes." This method makes no specific assumption about the form of the intracluster correlation. The design effect is defined as the ratio of a consistent estimate of the variance $\hat{v}_i$ of $\hat{p}_i$ to the binomial variance:

$$\hat{D}_i = \hat{v}_i/[\hat{p}_i(1 - \hat{p}_i)/n_i]$$

where

$$\hat{v}_i = m_i\ \Sigma(x_{ij} - n_{ij}\,\hat{p}_i)^2/(m_i - 1)(n_i)^2$$

For small sample sizes, Scott and Smith[10] suggest using the modified estimator:

$$\tilde{v}_i = (m_i - 1)\ \hat{v}_i/(m_i - 3)$$

The effective number of abnormal fetuses is:

$$\tilde{x}_i = \Sigma x_{ij}/\hat{D}_i$$

and the effective sample size is:

$$\tilde{n}_i = \Sigma n_{ij}/\hat{D}_i$$

Since the intralitter correlation coefficient is not expected to be negative, if $\hat{D}_i$ is less than one, it is set equal to one. Fung et al.[11] propose using the Rao-Scott effective numbers $\tilde{x}_i$ and $\tilde{n}_i$ in the Cochran-Armitage test for trend:

$$T_{RS} = \Sigma \tilde{x}_i (d_i - \bar{d})/[\bar{p}(1 - \bar{p}) \Sigma \tilde{n}_i (d_i - \bar{d})^2]^{1/2}$$

where

$$\bar{p} = \Sigma \tilde{x}_i/\Sigma \tilde{n}_i$$

$$\bar{d} = \Sigma \tilde{n}_i d_i/\Sigma \tilde{n}_i$$

This adjusted Cochran-Armitage is asymptotically normally distributed with mean 0 and variance 1.0.

Chen et al.[12] discuss a Dirichlet-trinomial distribution for modeling data from developmental studies. They consider the three endpoints for each litter: the number of dead/resorbed fetuses, the number of malformed fetuses (generally for a specific anomaly), and the number of normal fetuses. In the previous discussion deaths and malformations are analyzed separately. The Dirichlet-trinomial distribution provides a procedure for the analysis of multiple endpoints simultaneously. The Dirichlet-trinomial model is a generalization of the beta-binomial model Kodell et al.[6] used for handling litter effects. Chen et al.[12] provide likelihood ratio tests for differences in death rates, malformation rates, and normal rates among dosed and control animals.

DOSE-RESPONSE MODELING

Typically, allowable daily intakes for potential developmental toxicants are determined by establishing a no observed adverse effect level (NOAEL) and then dividing by safety factors to account for uncertainty in extrapolating from animals to humans, variability among human sensitivities to toxic agents, and inadequate data. Hopefully, this process results in no or negligible risks to humans at these doses. Various authors, e.g., Crump[13] and Dourson et al.[14], have discussed the short-comings of using the NOAEL. Namely, the NOAEL is ill-defined, some-what subjective, and highly dependent on the dose spacing, and does not encourage or reward better experiments. That is, experiments with lower ability to detect toxic effects result in high NOAELs.

Crump,[13] Dourson et al.,[14] and Kimmel and Gaylor[15] proposed replacing the NOAEL by a dose corresponding to a low estimable incidence of an adverse effect in the range of 1–10% above the background incidence. To account for uncertainty in the experimental estimate of a dose corresponding to a specified incidence in the range of 1–10%, the lower confidence limit is suggested as a "benchmark" to replace the NOAEL. Better experiments generally result in tighter confidence limits which appropriately result in higher benchmark doses. Also, Gaylor[16] shows from a survey of 120 experiments on developmental toxicants in animals that the risk of either dead/resorbed or abnormal fetuses at the NOAEL exceeds 1% in about one fourth of the cases. This suggests that a benchmark dose with a risk of 1% would reduce higher risks frequently associated with NOAELs. Most plausible dose-response models which fit the experimental data adequately will provide similar estimates of doses for risks above 1%. Thus, the choice of a dose-response model for establishing a benchmark dose is generally not critical. However, estimation of risks below the benchmark dose generally requires judicious choice of a dose-response model.

As discussed previously, Kodell et al.[6] provide a threshold, Weibull dose-response model which includes an adjustment for incidence due to litter size. They utilize the asymptotic distribution of the likelihood ratio to estimate confidence limits for low-dose risk extrapolation.

Ryan[17] provides a multivariate model for simultaneously considering the proportion of malformed fetuses among the live fetuses in a litter and the proportion of dead/resorbed fetuses among the implants for the same litter. At this point, a slight change in notation is required to accommodate multiple endpoints.

Let Θ_x denote the probability that an implanted embryo is resorbed or dies during the gestation period observed. Let Θ_y denote the conditional probability of observing a particular type of malformation, given that the fetus lives to be examined. Suppose there are t groups with doses $0 = d_0 < d_1 < d_2 < \ldots < d_t$ where the jth dam in the ith dose group has n_{ij} implantation sites. Let x_{ij}, y_{ij}, and z_{ij} denote the numbers of dead, malformed, and normal fetuses, respectively, in a litter. Note that $(x_{ij} + y_{ij} + z_{ij}) = n_{ij}$. If outcomes occur independently at each implantation site, the random variables can be modeled as a trinomial with probabilities of dead, malformed, and normal fetuses in the ith dose group given by $P_x(d_i)$:

$$P_x(d_i) = \theta_x(d_i)$$

$$P_y(d_i) = [1 - \theta_x(d_i)]\,\theta_y(d_i)$$

$$P_z(d_i) = [1 - \theta_x(d_i)][1 - \theta_y(d_i)]$$

Due to different sensitivities of dams to a dose d_i within a group, more variability in the data is expected than can be expected due to the trinomial distribution. The method of generalized estimating equations suggested by Ryan[17] for multinomial data can be used to account for the overdispersion induced by litter effects by inflating the standard multinomial variance by a scale factor for each dose group. Ryan[17] discusses techniques for estimating the probabilities and the overdispersion parameter.

To describe the effect of a dose on θ_1 and θ_2, Ryan[17] considers the class of dose response models:

$$\theta_k(d) = F_k[\alpha_k + \beta_k \log(d)]$$

where F is a cumulative distribution function. For example a log probit model corresponds to the cumulative normal distribution, $F(\log d) = \Phi$ (log d) and the logit model corresponds to $F(\log d) = [1 + \exp(-\log d)]^{-1}$. Ryan[17] shows that the estimate of a virtually safe dose (VSD), corresponding to a low level of risk, based on the trinomial model leads to a lower VSD than the minimum of the VSDs based on separate models for death and malformation. When one of the individual outcomes has a much higher probability than the other, the estimate of the VSD based on the trinomial model is close to the VSD corresponding to the most probable endpoint.

An alternative to the trinomial model is to collapse the categories of dead or malformed into a single category of abnormal and apply methods for correlated binary data. For certain classes of models these two approaches are equivalent. In general, the trinomial model allows for a broader class of dose-response models and leads to more precise estimates of the VSD.

Chen and Li[18] consider the simultaneous modeling of death and malformation in the presence of litter effects. They consider a double beta-binomial model and the trinomial model. Both models assume that the proportion dead and the proportion malformed are independent. The double beta-binomial uses one intralitter correlation to describe the litter effect for death and another correlation for the malformation. The trinomial uses a single parameter to describe the litter effects for both endpoints. The double beta-binomial can be fitted separately by two beta-binomial models, although parameter estimates may not converge when the number of deaths or malformations is zero. The quasi-likelihood method of generalized estimating equations does not make an assumption on the distribution, and estimates of the dose-response coefficients are generally consistent and asymptotically normal. Quasi-likelihood estimates are generally more stable than maximum likelihood estimates. The generalized estimating equations approach allows for modeling a dependency between the two endpoints. The linear-logistic dose-response function is used by Chen and Li[18] to model the probability P_i of an adverse effect at dose d_i

$$\log [P_i/(1 - P_i)] = \alpha + \beta d_i$$

where α and β are parameters. One advantage of using the double beta-binomial model is that the parameters for dead or malformation can be estimated by fitting two separate beta-binomial models for each. The maximum likelihood estimates can be calculated under the assumption that the intralitter correlation is the same for all dose groups or that the intralitter correlations may be different across dose groups. To avoid bias in the estimation of the dose-response parameters, Kupper et al.[19] suggest that the unequal correlation structure should be used in dose-response modeling. However, Williams[5] notes that allowing unequal intralitter correlation across dose groups may result in unreliable estimates of risk.

Hopefully, an understanding of the mechanism of action of a developmental toxicant can lead to a more valid dose-response model. If so, estimation of the incidence of adverse effects may be more accurate in the low-dose range where precise experimental observation is not achievable with a moderate number of animals.

For example, suppose a malformation is due to inadequate cell numbers as measured by fetal weight. Many mathematical growth curves are dominated by an exponential term for early growth, as would be expected during gestation. The exponential growth model is:

$$W_t = W_0 e^{bt}$$

where W_t = fetal weight at time t
W_0 = weight at some point in time during gestation
t = time elapsed since some starting time (not necessarily conception)
b = exponential growth rate constant

Suppose the exponential growth rate constant can be related to dose (d) by $b = b_c (1 - ad^g)$

where $\quad b_c$ = growth rate constant for control animals ($d = 0$)
$\qquad$ a > 0 and g > 0 = parameters

If the probability of a normal fetus is proportional to fetal weight to a power, then the probability of an abnormal fetus for dose d is:

$$P(d) = 1 - e^{-(u_0 + u_1 d^g)}$$

which is a Weibull model and $u_0 \geq 0$, $u_1 > 0$, and $g > 0$ are parameters.

A different relationship can be used to describe the effects of dose on the growth rate constant. A polynomial function often provides a reasonable approximation to a large variety of dose-response functions:

$$b = b_c - (a_1 D + a_2 D^2 + \ldots)$$

Substituting this expression for b in the exponential growth curve gives:

$$W_t = W_0 e^{[b_c - (a_1 D + a_2 D^2 + \ldots)]t}$$

As before, if it is assumed that the probability of normal fetal development of a structure is proportional to fetal weight raised to a power, a polynomial-exponential model is obtained for the probability of an abnormal fetus:

$$P(d) = 1 - e^{-(q_0 + q_1 d + q_2 d^2 + \ldots)}$$

where the q's are parameters.

CONCLUSIONS

Establishing "safe" doses of chemicals to a developing fetus is complicated because the exposure to a chemical occurs via the pregnant mother. The current procedure of setting an allowable daily intake by dividing a NOAEL by safety factors is based on the premise that the resulting dose will generally be below a threshold dose for a toxic response. This assumes that the safety factors employed are adequate to protect most of the population. Further, this process assumes the existence of a threshold dose for each individual below which no deleterious effects occur. Most developmental effects occur at some low level of incidence in individuals that are not exposed to the chemical in question. That is, endogenous and/or other exogenous factors are adequate to produce a spontaneous background incidence of embryolethality or malformations in unexposed control individuals. That is, threshold doses in these individuals have already been surpassed. Thus, if the chemical under investigation augments an existing process that is already producing deleterious effects, the population threshold has already been surpassed and no threshold dose exists for the added chemical. In this case, it is important to establish a dose-response curve to estimate risks at low doses below the NOAEL or benchmark dose in order to establish relatively safe exposures to developmental toxicants. In a few cases it may be possible to develop biologically based dose-response models which are valid at low doses. In general, it will only be possible to adapt default procedures that are thought to provide adequate protection. The validity of any process can only be determined if more human dose-response data are collected for comparison with laboratory animal results.

REFERENCES

1. Siegel, S. *Nonparametric Statistics for the Behavioral Sciences* (New York: McGraw-Hill Book Co., 1956).
2. Miller, R. G. *Simultaneous Statistical Inference*, 2nd ed. (New York: Springer-Verlag, 1981).
3. Shirley, E. "A Non-Parametric Equivalent of William's Test for Contrasting Increasing Dose Levels of a Treatment," *Biometrics* 33:386–389 (1977).
4. Jonckheere, A. R. "A Distribution-Free k-Sample Test Against Ordered Alternatives," *Biometrika* 41:133–145 (1954).
5. Williams, D. A. "Estimation Bias using the Beta-Binomial Distribution in Teratology," *Biometrics* 44:305–308 (1988).
6. Kodell, R. L., R. B. Howe, J. J. Chen, and D. W. Gaylor. "Mathematical Modelling of Reproductive and Developmental Toxic Effects for Quantitative Risk Assessment," *Risk Anal.* 11:583–590 (1991).
7. Zeger, S. L. and K. Y. Liang. "Longitudinal Data Analysis for Discrete and Continuous Outcomes," *Biometrics* 42:121–130 (1986).
8. Lefkopoulou, M., D. Moore, and L. Ryan. "The Analysis of Multiple Correlated Binary Outcomes: Application of Rodent Teratology Experiments," *J. Am. Stat. Assoc.* 84:810–815 (1989).
9. Rao, J. N. K. and A. J. Scott. "A Simple Method for the Analysis of Clustered Binary Data," *Biometrics* 48:577–585 (1992).
10. Scott, A. J. and T. M. F. Smith. "Interval Estimates for Linear Combinations of Means," *Appl. Stat. (Ser. C)* 20:276–285 (1971).
11. Fung, K. Y., D. Krewski, J. N. K. Rao, and A. J. Scott. "Tests for Trend in Developmental Toxicological Experiments with Correlated Binary Outcomes," *Risk Anal.* 13: in press (1993).
12. Chen, J. J., R. L. Kodell, R. B. Howe, and D. W. Gaylor. "Analysis of Trinomial Responses from Reproductive and Developmental Toxicity Experiments," *Biometrics* 47:1049–1058 (1991).
13. Crump, K. S. "A New Method for Determining Allowable Daily Intakes," *Fundam. Appl. Toxicol.* 4:854–871 (1984).

14. Dourson, M. L., R. C. Hertzberg, R. Hartung, and K. Blackburn. "Novel Methods for the Estimation of Acceptable Daily Intake," *Toxicol. Ind. Health* 1:23–41 (1985).

15. Kimmel, C. A. and D. W. Gaylor. "Issues in Qualitative and Quantitative Risk Analysis for Developmental Toxicology," *Risk Anal.* 8:15–20 (1988).

16. Gaylor, D. W. "Incidence of Developmental Defects at the No Observed Adverse Effects Level (NOAEL)," *Reg. Toxicol. Pharmacol.* 15:151–160 (1992).

17. Ryan, L. "Quantitative Risk Assessment for Developmental Toxicology," *Biometrics* 48:163–174 (1992).

18. Chen, J. J. and L.-A. Li. "Dose-Response Modelling of Trinomial Responses from Developmental Experiments," *Acad. Sin.* (1994).

19. Kupper, L. L., C. Portier, M. D. Hogan, and E. Yamaoto. "The Impact of Litter Effects on Dose-Response Modeling in Teratology," *Biometrics* 42:85–89 (1986).

CHAPTER 10

Measuring Carcinogenic Potency

M. J. Goddard, Daniel Krewski, and Yiliang Zhu

ABSTRACT

Laboratory studies have identified over 1000 chemical substances capable of causing cancer in rodents. In this article, statistical methods for measuring carcinogenic potency based on the results of long-term animal studies are discussed, including a Weibull model that takes into account curvature in the dose-response curve, intercurrent mortality, and (when available) cause of death information. The distribution of the carcinogenic potency of known chemical carcinogens has been proposed as a means of establishing a threshold of regulation for substances not yet tested for carcinogenic potential. The use of empirical Bayes methods to adjust for overdispersion in the empirical distribution of potency values is explored.

INTRODUCTION

Modern medicines, efficient transportation, high quality food, and sturdy and comfortable housing are a few of the many advances in technology that contribute to the high standard of living we enjoy today. Along with this improved quality of life, however, we face a new set of potential health hazards. The benefits we derive from these technological achievements are often readily appreciated; the health risks are usually less obvious.

The risk associated with toxic substances present in the environment depends on both the potency of that hazard and the level of exposure. An extremely potent toxicant may be of little concern if people are never exposed to it. Herein, our focus is on the potency of chemical carcinogens present usually at low levels in the environment.

Although both epidemiological and toxicological data can be used to estimate carcinogenic potency, many more chemicals have been shown to cause cancer in animals than humans. To date, only 58 chemicals or radiological agents, chemical mixtures, or industrial processes have been shown to cause cancer in humans,[1] whereas the carcinogenic potency database (CPDB) established by Gold[2] currently includes estimates of the carcinogenic potency of over 1000 chemicals.

Our goal in this chapter is to describe the distribution of carcinogenic potencies using empirical Bayes methods. We begin with a historic review of measures of potency (next section), culminating with the TD_{50}, loosely defined as the dose inducing an excess cancer risk of 50% (the "Recent Potency Measures: the TD_{50}" section). In the "Thesholds of Regulation" section, we examine the variation in the TD_{50} values for some chemical carcinogens selected from the CPDB, and indicate potential applications of this approach in establishing a threshold of variation for chemicals not yet tested for carcinogenic potential. Empirical Bayes methods designed to adjust for overdispersion in the distribution of TD_{50} values are introduced in the section entitled "Empirical Bayes Shrinkage Estimators." (Details of the technique appear in the appendix.) Our conclusions are presented in the "Summary and Conclusions" section.

HISTORIC POTENCY MEASURES

Quantitative measures of potency appear in the scientific literature dating back to the 1930s. Twort and Twort[3,4] summarized their cancer bioassay results by tabulating the percentages of tumor-bearing animals surviving each week. Later, this approach was modified to present only new tumors appearing weekly. Potency was also expressed in terms of the estimated time at which 25% of the animals developed tumors. These measures were evaluated using data from simple experiments involving a single group of animals subjected to exposure to some level of test agent. Twort and Twort[3,4] made use of time-to-tumor information, attempted to distinguish between benign and malignant tumors, and anticipated the need to allow for such modern concepts as intercurrent mortality.

Iball[5] noted the possible biasing effects of intercurrent mortality. He suggested using 100 times the percentage of live tumor-bearing animals divided by average latent period (the average time to the first appearance of a tumor) as an indicator of carcinogenic potency. This index also took into account both the time-to-tumor and intercurrent mortality.

Irwin and Goodman[6] suggested additional potency indices for the database assembled by Twort and Twort. Their measures were also based on time (and not dose). Specifically, Irwin and Goodman[6] estimated the times at which 25, 50, 75, and 100% of the surviving animals had developed tumors. They also determined the expected tumor-free period and noted that different potency measures tended to lead to similar potency rankings.

In the 1940s and 1950s, Finney[7] developed statistical methods for probit analysis of bioassay data and proposed the ratio of probit slopes as a measure of relative potency. Bryan and Shimkin[8] used these methods to estimate the relative potency of chemicals shown to be carcinogenic in skin painting experiments with mice.

Meselson and Russel[9] utilized an index of carcinogenic potency defined by:

$$C_1 = (\log_e 2)/D_{1/2}$$

where $D_{1/2}$ is the dose of a compound in milligrams per kilograms per day that produces a 50% tumor response rate after a 2-year exposure. Values of $D_{1/2}$ were determined from the relationship:

$$I(t, d) = 1 - \exp\left\{ - \frac{\log_e 2}{D_{1/2}} dt^4 \right\}$$

where $I(t,d)$ = tumor incidence at time t (in years)

d = daily dose rate (in milligrams per kilograms per day)

Jones et al.[10] defined carcinogenic potency as:

$$C_2 = \log_e \left(1 + \frac{100}{dt^3} \right)$$

where t is the time (in weeks) required to yield a 50% tumor response rate and d is the dose (in mols) at which time 50% of the animals had died.

Crouch and Wilson[11] used the slope (β) of the simple dose-response model $P(d) = \beta d$ as a measure of carcinogenic potency. This linear equation approximates the "one hit" model in which the probability of a tumor occuring at dose d is given by:

$$P(d) = 1 - \exp(-\beta d)$$

$$\approx \beta d$$

for small d.

A model based on cytokinetic effects was developed by Jones et al.[12] The details of this approach, which models biological processes, are much more involved than others considered here. The method is unique in that it permits comparisons between the potencies of chemical and radiological agents capable of causing tumors.

Clayson[13] suggested the simple index:

$$C_3 = 7 - \log_{10} D_{1/2}$$

with dose measured in micromoles per kilograms per week. The index was used primarily for summary data on the number of animals developing tumors during the course of the study and requires the adjustment of results to a uniform experimental period, usually 2 years. The constant "7" is an empirical adjustment designed to scale the index C_3 to lie in the range of 1–10 with higher values indicating more potent carcinogens.

For further material in potency measures, the reader is referred to the paper by Barr[14] in which epidemiology, bioassay, skin painting, *in vitro* studies, acute toxicity, and structure-activity relationships are discussed.

RECENT POTENCY MEASURES: THE TD$_{50}$

Most recent developments in measuring carcinogenic potency are related to the TD_{50} index described by Peto et al.[15] and Sawyer et al.[16] Formally, the TD_{50} is defined as the dose that will halve the proportion of tumor-free animals at a specified point in time. The latter investigators assumed a hazard function of the form

$$\lambda(t;\ d) = (1 + \beta d)\lambda_0(t) \tag{1}$$

where $\lambda_0(t)$ denotes the baseline hazard in the absence of exposure. Note that the hazard $\lambda(t;d)$ is assumed to be a linear function of dose d. The hazard function $\lambda (t;d)$ is related to the probability, $Q_0(t)$, that an unexposed animal has not been diagnosed with a tumor by age t in the absence of all other causes of death by:

$$Q_0(t) = \exp \left\{ -\int_0^t \lambda_0(x)dx \right\} \tag{2}$$

At a fixed time, T, the TD_{50} is derived from the combination of Equations 1 and 2:

$$TD_{50} = \frac{\log {}^1/_2}{\beta \log\{Q_0(T)\}}$$

The probability, $Q_d(t)$, that an animal at dose d has not been diagnosed with a tumor by age t in the absence of all other causes of death is related to $Q_0(t)$ by the proportional hazards model:

$$Q_d(t) = \{Q_0(t)\}^{1+\beta d}$$

Sawyer et al.[16] use time to event information (specifically, the survival times of tumor-bearing animals) and not just the counts of the number of animals with tumors at the end of the experiment to estimate the TD_{50}.

Ideally, the time of tumor occurrence would be used as the basis for statistical inference about the hazard function for tumor induction. Unfortunately, most tumors are usually unobservable in live animals so that the onset time for the tumor is generally unknown. Although tumors differ with respect to their lethality, one approach to this problem is to assume that the tumors are "rapidly lethal", *i.e.*, that the time of death corresponds closely to the tumor onset time. This is implicit in the work of Sawyer et al.[16] Portier and Hoel[17] point out that the TD_{50} is quite sensitive to this assumption.

Several approaches have been developed to calculate a potency measure that applies to unseen or "occult" tumors and also corrects for intercurrent mortality. Underlying these proposed models is the compartmental model shown in Figure 1. Animals that are initially tumor free eventually either develop a tumor or die tumorless. Those that do develop tumors either die as a result of the tumor or die from other causes.

Finkelstein and Ryan[18] presented an approach based on the procedures developed by Peto et al.[19] for combining tumor mortality and prevalence data. Their potency measure is a weighted combination of the slope of a proportional hazards model for tumor mortality and the slope of a logistic model for tumor prevalence. The approach enjoys the simplicity of calculation similar to that of the statistic proposed by Peto et al.[19] to test for increased tumor risk in animal carcinogenicity studies taking into account the time at which tumors are observed in experimental animals.

Bailer and Portier[20] explored the effects of assumptions about tumor lethality on statistical tests for carcinogenicity by simulation. They concluded that a simple survival-adjusted quantal response test appeared to be robust against misspecification of tumor lethality. Bailer and Portier[21] took their 1988 results on statistical tests and developed an index of

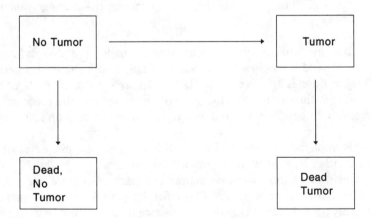

FIGURE 1. General compartmental model underlying recent potency measures.

tumorigenic potency. They used the TD_{50} adjusted for intercurrent mortality using a Weibull hazard model. This index is readily estimated using statistical software for quantal response data such as GLOBAL82 applied to the survival adjusted data.

Dewanji et al.[22] used three Weibull distributions to characterize the compartmental model in Figure 1. Specifically, the survivor functions for the time to tumor onset (X), the time to death from tumor (Y), and the time to death from competing risks (Z) are given by:

$$S_X(t,d) = \exp\{-(\alpha + \beta d^\delta)t^\gamma\}$$

$$S_Y(t,d) = \exp\{-(\alpha + \beta d^\delta)\rho t^\gamma\}$$

and

$$S_Z(t,d) = \exp\{-(\mu + \nu d^\eta)t^\kappa\}$$

respectively. The parameter ρ represents tumor lethality with $\rho = 0$ corresponding to nonlethal or incidental tumors and $\rho = 1$ reflecting rapidly fatal tumors. Values of ρ between zero and one represent tumors of intermediate lethality.

Various likelihoods can be constructed from this general model, depending on which data are available and which assumptions one makes about tumor lethality. These include the following special cases

    I. cause of death and tumor lethality unknown
   II. rapidly fatal tumors (*i.e.*, $S_Y \equiv S_X$)
  III. incidental tumors (*i.e.*, $S_Y \equiv 1$)
  IV. all causes of death (death from tumor, death from competing
       risk with tumor, and death from competing risk without tumor)
       known

The behavior of the model was studied using the 2-acetylaminofluorene (2-AAF) data described by Littlefield et al.[23] These data permitted TD_{50} estimates for liver and bladder tumors induced by 2-AAF in cases I–IV. Estimates of the TD_{50} were obtained both in the presence and absence of competing risks, and as a function of the time on test (Figure 2).

In this application, assumptions I–IV lead to similar estimates of the TD_{50} for either liver or bladder tumors. Note that although the TD_{50}s for both liver and bladder tumors change comparatively little between 20 and 36 months of testing the TD_{50}s for liver tumors drop markedly during this period. The TD_{50}s in the presence of competing risks are only slightly lower than those in the absence of competing risks.

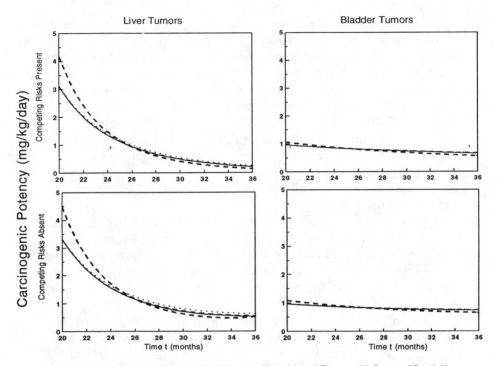

I. Cause of Death Unknown II. Rapidly Fatal Tumors III. Incidental Tumors IV. Cause of Death Known
Carcinogenic Potency of 2-Acetylaminofluorene as a Function of Time t

FIGURE 2. Potency measures as a function of time for liver and bladder tumors. (Based on various models described in Dewanji et al.[22])

THRESHOLDS OF REGULATION

Gold et al.[2] observed considerable variation in the potencies of chemical carcinogens, with TD_{50} values spanning more than a 10 million-fold range. To illustrate the variation in carcinogenic potencies, consider the empirical distribution of TD_{50} values for a sample of 191 studies in the CPDB. This distribution spans a range of about seven orders of magnitude and is approximately lognormal. This subset of the CPDB was selected by Krewski et al.[24] Data for rats, mice, and hamsters were used provided there were at least two dose groups in addition to the unexposed control group. Studies showing clear evidence of carcinogenicity were selected; where more than one study for a compound warranted inclusion, that study yielding the lowest potency was used. The TD_{50} was calculated based on a simple Weibull model fit to the proportion of animals developing tumors during the course of the study.

The use of the distribution of carcinogenic potencies as a means of establishing a threshold of regulation for chemicals not yet tested for carcinogenic potential has been discussed by Frawley,[25] Ramsey,[26] the National Research Council,[27] Rulis,[28] and Flamm et al.[29] An excellent discussion of the concept of a threshold of regulation based on a workshop on safety assessment procedures for indirect food additives is presented by Munro.[30]

The basis for this concept is the idea that for chemicals not yet tested for carcinogenicity a threshold of regulation might be determined. If the exposure of a new chemical is below this threshold, it is assumed not to pose an appreciable risk; and hence it does not require extensive regulatory attention.

Regulation of new chemicals could benefit in two ways by using regulatory thresholds. First, the expense and delay of extensive toxicological testing are avoided. Second, all chemicals presented for regulation receive identical treatment. The principal disadvantage of such an approach is the risk that the new chemical may be more potent than the threshold compound.

Rulis[28] and Flamm et al.[29] explored the application of the concept of a threshold of regulation for indirect food additives, such as compounds migrating from packaging material into food. Specifically, a set of 343 chemicals was selected to represent the variation in the potency of known chemical carcinogens.

In general, an index of potency is selected and the distribution of potencies for the set of chemicals is determined. This distribution of potencies is then transformed to a risk-equivalent exposure distribution. A threshold of regulation based on the risk-equivalent exposure distribution could then be determined by choosing a lower quantile of the distribution. Rulis[28] used linear extrapolation from the TD_{50} values reported in the CPDB to estimate the risk-specific dose corresponding to a lifetime risk of 10^{-6}. This simple approach to lifetime risk-specific dose provides results similar to those based on the linearized multistage model used by the U.S. Environmental Protection Agency[31] (*cf.*, Krewski[32]). Then, if the anticipated level of exposure did not exceed the threshold of variation, the need for further testing would be obviated.

Rulis[28] used the empirical distribution of TD_{50}s to determine the risk-equivalent exposure distribution. Since these TD_{50}s are estimated on the basis of bioassay data, they are subject to estimation error. By adopting Bayesian shrinkage estimators, it is possible to adjust for variability of the individual potency estimates and reduce the overall variation of the distribution of estimated potencies to correspond to that of the actual potencies.

EMPIRICAL BAYES SHRINKAGE ESTIMATORS

Ideally, a threshold of regulation would be based on the distribution of *actual* TD_{50}s rather than *estimates* of carcinogenic potency. As noted in the previous section, the distribution of estimated TD_{50}s will exhibit greater variation than the distribution of actual TD_{50}s because of the extra variation associated with estimation error. This overdispersion can be reduced using Bayesian methods as described in the appendix (*cf.*, Louis[33]).

These methods can be illustrated using the 191 chemicals discussed in the previous section. It should be noted that any threshold of regulation will depend very strongly on the database from which it was determined.[28] We make no representations that this particular data set is fully suitable for the the determination of a regulatory threshold. In selecting of a data set for use in practice it is important that the sample of chemicals should represent the class of compounds to which the threshold of regulation would apply. Limitations on the possible use of the CPDB in determining regulatory thresholds have been discussed by Munro.[30]

The distribution of logarithms of the actual TD_{50} values is characterized by its mean μ and variance τ^2. Estimates of these parameters obtained by fitting a normal distribution to the logarithm of the estimated TD_{50} values are $\hat{\mu} = 20.5$ (mg/kg/day) and $\hat{\tau}^2 = 168$ (mg/kg/day)^2. These estimates of μ and τ correspond to the distribution shown by the solid line in Figure 3. This distribution is subject to overdispersion due to the estimation error associated with the individual TD_{50} values.

The empirical Bayes shrinkage estimate of μ and τ^2 are $\hat{\mu}_B = 23.6$ (mg/kg/day) and $\hat{\mu}_B^2 = 41.3$ (mg/kg/day)^2. Louis[33] notes that the Bayes estimators tend to overadjust for overdispersion in the original distribution. The modified Bayes estimators based on Equations 9 and 10 in the annex yield $\hat{\mu}_{MB} = 23.6$ (mg/kg/day) and $\hat{\tau}_{MB}^2 = 59.3$ (mg/kg/day)^2. The effect of this modification is to reduce the degree of shrinkage somewhat, with $\hat{\tau}^2 < \hat{\tau}_{MB}^2 < \hat{\tau}_B^2$. Note that the Bayes estimates alter the mean of the original distribution with $\hat{\mu}_B = \hat{\mu}_{MB} > \hat{\mu}$. The three distributions are shown in Figure 3.

The effect of standard and adjusted Bayesian shrinkage estimates on selected lower percentiles of the distribution of carcinogenic potencies is illustrated in Table 1. In all cases, the adjusted Bayes value lies midway between the empirical value and the standard Bayes value.

If the centers of all three distributions coincided and one only considered thresholds corresponding to very low percentiles, then one would expect similar ranks of predictions to those in Table 1. This ranking is not universal: for percentiles close to the center of the distribution and for situations where the empirical and the Bayes approaches yield distributions with very different centers, the ranking may change.

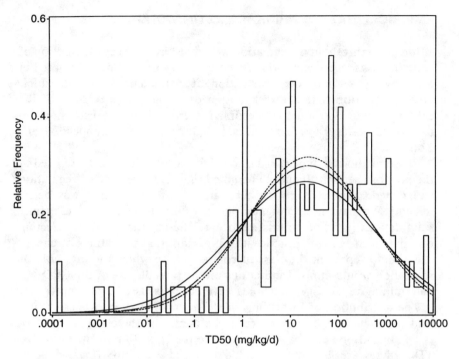

FIGURE 3.   Empirical and fitted curves for sample of 191 TD_{50}s. Solid curve: standard fit; dotted curve: standard Bayes fit; dash-dot curve: adjusted Bayes curve.

SUMMARY AND CONCLUSIONS

In this chapter we provide an overview of the many attempts to quantify carcinogenic potency. Even researchers attempting to measure potency earlier this century attempted to use time-to-event data and to allow for possible biasing effects of competing risks. The TD_{50} proposal in 1984 was based on statistical methods to deal with these facets of the data, and this measure has become a benchmark measure of carcinogenic potency.

Table 1.   Lower Percentiles of the Distribution of the Logarithms of the TD_{50} Based on Empirical, Standard Bayes, and Adjusted Bayes Methods

Percentile	1%	5%	10%	15%	20%	25%	30%	40%	50%
Empirical	0.01	0.07	0.25	0.58	1.14	2.02	3.39	8.60	20.5
Adjusted Bayes	0.02	0.15	0.46	0.98	1.78	2.98	4.72	10.83	23.6
Standard Bayes	0.03	0.19	0.55	1.13	2.06	3.34	5.16	11.31	23.6

Note: Entries expressed in milligrams per kilograms per day.

Regulatory agencies can use measures of carcinogenic potency to guide their decisions. A specific example is the concept of threshold of regulation. With this approach, carcinogencitiy testing would not be done if anticipated human exposure was less than the threshold of regulation. The distribution of carcinogenic potencies would be the basis for the determination of such a threshold value.

Each value in a distribution of potencies is a statistical estimate, usually derived from a single toxicological study. Thus, each value has an associated uncertainty. It is feasible to apply well-known Bayesian methods when combining the individual potencies and their uncertainties to estimate a distribution with less variability than the raw empirical distribution of potency estimates. The standard Bayesian approach, however, leads to an estimated distribution that has too little variability. To avoid this, we used an adjusted Bayesian method suggested by Louis[33] and applied the technique to a collection of potency estimates derived from the large database of carcinogenic studies maintained by Gold et al.[2] We observe that for percentiles between 1 and 40% the estimated threshold based on an empirical distribution is uniformly lower than that predicted using the adjusted Bayes estimator, which itself was less than the standard Bayes estimator.

There are many aspects of this approach which warrant further research. We have adopted a model assuming the prior distribution is normal, and we determined estimates of parameters for this distribution in the first stage. Alternative approaches include the possibility of using hyperparameter methods, for example, using "uniform ignorance" as prior distributions for μ and τ^2. From the data analytic point of view, there is also a need to explore the various values of predicted thresholds of regulation for different collections of chemicals using the standard and adjusted Bayes approaches described here.

REFERENCES

1. Tomatis, L., Aitio, A., Wilbourn, J. and Shuker, L. (1989) Human carcinogens so far identified, *Jpn. J. Cancer Res.*, 80, 795–807.
2. Gold, L., Sawyer, C., Magaw, R., Backman, G., de Veciana, M., Levinson, R.L., Hooper, N., Havender, W., Bernstein, L., Peto, R., Pike, M., and Ames, B. (1984) A carcinogenic potency database of standardized results of animal bioassays, *Environ. Health Perspect.*, 58, 9–322.
3. Twort, C. C. and Twort, J. M. (1930). The relative potency of carcinogenic tars and oils. *J. Hygiene*, 29, 373–379.
4. Twort, C. C. and Twort, J. M. (1933). Suggested methods for the standardization of the carcinogenic activity of different agents for the skin of mice. *Am. J. Cancer*, 17, 293–320.

5. Iball, J. (1939). The relative potency of carcinogenic compounds. *Am. J. Cancer*, 35, 188–190.

6. Irwin, J. O. and Goodman, N. (1946). The statistical treatment of the carcinogenic properties of tars (Part I) and mineral oils (Part II). *J. Hygiene*, 44, 362–420.

7. Finney, D. J. (1978). Statistical method in biological assay. 3rd ed. Charles Griffin & Co. Ltd., London.

8. Bryan, W. R. and Shimkin, M. B. (1943). Quantitative analysis of dose response data obtained with three carcinogenic hydrocarbons in strain C3H male mice. *J. Natl. Cancer Inst.*, 3, 503–531.

9. Meselson, M. and Russel, K. (1977). Comparisons of carcinogenic and mutagenic potency. In: *Origins of Human Cancer, Book C. Cold Spring Harbour Conference on Cell Proliferation*, Vol. 4. (H. Hiatt, J. R. Watson and J. A. Winston, Eds.). Cold Spring Laboratory, Cold Spring Harbour, pp. 1473–1481.

10. Jones, C. A., Marlins, P. J., Lijinksy, W., and Huberman, E. (1981). The relationship between the carcinogenicity and mutagenicity of nitrosamines in a hepatocyte–mediated mutagenicity assay. *Carcinogenesis*, 2, 1075–1077.

11. Crouch, E. and Wilson, R. (1981). Regulation of carcinogens. *Risk Anal.*, 1, 47–66.

12. Jones, T. D., Griffin, G. D., and Walsh, P. J. (1983). A unifying concept for carcinogenic risk assessments. *J. Theor. Biol.*, 105, 35–61.

13. Clayson, D. B. (1983). Trans–species and trans–tissue extrapolation of carcinogenicity assays. In: *Organ and Species Specificity in Chemical Carcinogenesis* (R. Langenbach, S. Newnow, and J. M. Rice, Eds.). Plenum, New York, pp 637–651.

14. Barr, J. T. (1985). The calculation and use of carcinogenic potency: A review, *Regul. Toxicol. Pharmacol.*, 5, 432–459.

15. Peto, R., Pike, M., Bernstein, L., Gold, L. S., and Ames, B. (1984). The TD_{50}: A proposed general convention for the numerical description of the carcinogenic potency of chemicals in chronic exposure animal experiments, *Environ. Health Perspect.*, 58, 1–9.

16. Sawyer, C., Peto, R., Bernstein, L., and Pike, M. (1984). Calculation of carcinogenic potency from long–term animal carcinogenesis experiments, *Biometrics*, 40, 27–40.

17. Portier, C. and Hoel, D. (1987). Issues concerning the estimation of the TD_{50}, *Risk Anal.*, 437–447.

18. Finkelstein, D. M. and Ryan, L. M. (1987). Estimating carcinogenic potency from a rodent tumorigenicity experiment, *Appl. Stat.*, 36, 121–133.

19. Peto, R., Pike, M., Day, N., Gray, R., Lee, P., Parish, S., Peto, J., Richards, S., and Wahrendorf, J. (1980). Guidelines for simple, sensitive significance tests for carcinogenic effects in long-term animal experiments, in *IARC Monographs on the Evaluation of the Carcinogenic Risk of Chemicals to Humans: Supplement of Long-term and Short-term Screening Assays for Carcinogens: A Critical Appraisal*, Lyon: IARC, pp. 331–425.

20. Bailer, A. J. and Portier, C. J. (1988). Effects of treatment-induced mortality and tumor induced mortality on tests for carcinogenicity, *Biometrics*, 44, 417–431.

21. Bailer, A. J. and Portier, C. J. (1993). An index of tumorigenic potency. *Biometrics*, 49, 357–365.
22. Dewanji, A., Krewski, D., and Goddard, M. J. (1993). A Weibull model for the estimation of tumorigenic potency, *Biometrics*, 49, 367–377.
23. Littlefield, N. A., Farmer, J. H., Gaylor, D. W., and Sheldon, W. G. (1980). Effects of dose and time in a long-term, low-dose carcinogenic study. *J. Environ. Pathol. Toxicol.*, 3, 17–35.
24. Krewski, D., Gaylor, D. W., Soms, A. P., and Szyszkowicz, M. (1993). Correlation between carcinogenic potency and the maximum tolerated dose: implications for risk assessment, in: *Issues in Risk Assessment*, National Academy Press, pp. 111–171.
25. Frawley, J. P. (1967). Scientific evidence and common sense as a basis for food-packaging regulations, *Food Comest. Toxicol.*, 5, 293–308.
26. Ramsey, L. L. (1966). The food additive problem of plastics used in food packaging. Presented at the National Technical Conference of the Society of Plastics Engineers, November 4–6.
27. National Academy of Sciences (NRC) (1969). *Guideline for estimating toxicologically insignificant levels of chemicals in food*, Food Protection Committee, Food and Nutrition Board, National Academy of Sciences, National Research Council, Washington, DC.
28. Rulis, A. M. (1986). *De minimis* and the threshold of regulation. In *Food Protection Technology* (C. W. Felix, Ed.,) Lewis Publishers, Chelsea, MI. pp 29–37.
29. Flamm, W. G., Lake, L. R., Lorentzen, R. J., Rulis, A. M., Schwartz, P. S., and Troxell, T. C. (1987). Carcinogenic Potencies and Establishment of a Threshold of Regulation for Food Contact Substances. In *De Minimis Risk* (C. Whipple, Ed.) Plenum, New York. pp. 87–92.
30. Munro, I. (1990). Safety assessment procedures for indirect food additives: an Overview. Report of a workshop, *Regul. Toxicol. Pharmacol.*, 12, 2–12.
31. U. S. Environmental Protection Agency (1986). Guidelines for carcinogen risk assessment, *Fed. Regist.*, 51, 33992–34003.
32. Krewski, D. (1990). Measuring carcinogenic potency *Risk Anal.*, 10, 615–617.
33. Louis, T. (1984). Estimating a population of parameter values using Bayes and empirical Bayes methods, *J. Am. Stat. Assoc.*, 79, 393–398.

APPENDIX

Let θ represent the logarithm of the potency of a chemical carcinogen, i.e.,

$$\theta = \log_{10} TD_{50}$$

and let y be an estimator of θ. These estimates are assumed independent, with $y_i \sim N(\theta_i, \sigma_i^2)$. Asymptotic normality of each y_i will hold under quite general conditions using classical methods of estimating potency such as maximum likelihood (cf., Dewanji et al.[22]). Here, θ_i denotes the actual potency of the i th carcinogen and σ_i^2 denotes the variance of the estimate y_i of θ_i.

Since the distribution of logarithms of the TD_{50} is approximately lognormal (Krewski et al.[32]) we assume that $\theta_1, \ldots, \theta_n$ are independent identically distributed random variables with $\theta_i \sim N(\mu, \tau^2)$. Our objective is to make inferences about this distribution of potencies based on the estimated values $y_1, \ldots, y_n$ in our sample.

With this framework, the posterior distributions of the θ_i given the estimated potency y_i are normal distributions with mean and variance given by:

$$E[\theta_i | y_i] = D_i y_i + (1 - D_i)\mu$$

and

$$Var[\theta_i | y_i] = D_i \sigma_i^2$$

where $D_i = \tau^2 (\tau^2 + \sigma_i^2)^{-1}$ $(i = 1, \ldots, n)$.

The posterior expectation of the sample mean of the θ_is $(\bar{\theta}. = n^{-1} \Sigma \theta_i)$ given the observation $y = (y_1, \ldots, y_n)$ is:

$$E[\bar{\theta}. | y] = \mu + n^{-1} \sum_{i=1}^{n} D_i (y_i - \mu) = \bar{y}_D + (1 - \bar{D})\mu \qquad (3)$$

where $\bar{D} = n^{-1} \Sigma D_i$ and $\bar{y}_D = n^{-1} \Sigma D_i y_i$. The posterior expectation of the sample variance of the θ_is is:

$$\frac{E[\Sigma_{i=1}^{n}(\theta_i - \bar{\theta}.)^2 | y]}{n - 1} = \frac{\Sigma_{i=1}^{n} D_i \sigma_i^2}{n} +$$

$$\frac{\Sigma_{i=1}^{n} \{D_i y_i - \bar{y}_D - \mu(D_i - \bar{D})\}^2}{n - 1}. \qquad (4)$$

Given μ and τ^2, Bayes estimates of the θ_i are given by:

$$\hat{\theta}_i = (1 - D_i)\mu + D_i y_i. \tag{5}$$

An estimate of the mean μ of the potency distribution based on these estimates is:

$$\hat{\theta}. = n^{-1} \sum_{i=1}^{n} \hat{\theta}_i = (1 - \bar{D})\mu + \bar{y}_D. \tag{6}$$

While the Bayes estimator $\hat{\theta}.$ in Equation 6 is equal to the expectation $E[\bar{\theta}.|y]$ in Equation 3, Louis[33] notes that the sample variance of the estimates in Equation 5:

$$\frac{\sum_{i=1}^{n} (\hat{\theta}_i - \hat{\theta}.)^2}{n - 1} + \frac{\sum_{i=1}^{n} \{D_i y_i - \bar{y}_D - \mu(D_i - \bar{D})\}^2}{n - 1} \tag{7}$$

is less than the expectation of the posterior sample variance from Equation 4. Since these Bayes shrinkage estimators (Equation 5) tend to offer too much shrinkage, a modified Bayes approach was proposed by Louis.[33]

Under weighted squared error loss $\Sigma w_i(\hat{\theta}_i - \theta_i)^2$ with weights $w_i \propto (D_i\sigma_i^2)^{-1}$, optimally adjusted Bayes estimators are given by:

$$\hat{\theta}_i^q = A_i y_i + (1 - A_i)\zeta \tag{8}$$

which when averaged yield an adjusted estimator for the population mean:

$$\hat{\theta}_{\cdot}^q = \zeta + n^{-1} \sum_{i=1}^{n} A_i(y_i - \zeta) = (1 - \bar{A})\zeta + \bar{y}_A. \tag{9}$$

Here $A_i = D_i/(1 + \lambda D_i\sigma_i^2)$, $\bar{y}_A = n^{-1} \Sigma A_i y_i$, and $\bar{A} = n^{-1} \Sigma A_i$.

We determine λ and ζ so that $\hat{\theta}_{\cdot}^q$ and $(n - 1)^{-1} \Sigma(\hat{\theta}_i^q - \hat{\theta}_{\cdot}^q)^2$ are equal to the posterior expectations of the sample mean (Equation 3) and variance (Equation 4), respectively. That is, set:

$$(1 - \bar{A})\zeta + \bar{y}_A = (1 - \bar{D})\mu + \bar{y}_D \tag{10}$$

and

$$\frac{\sum_{i=1}^{n} \{A_i y_i - \bar{y}_A - \zeta(A_i - \bar{A})\}^2}{n-1} =$$

$$\frac{\sum D_i \sigma_1^2}{n} + \frac{\sum_{i=1}^{n} \{D_i y_i - \bar{y}_D - \mu(D_i - \bar{D})\}^2}{n-1}. \tag{11}$$

Since the population mean μ and variance τ^2 involved in the prior distributions are usually unobservable, the general Bayes approach requires the specification of a prior distribution for μ and τ^2. In particular, the empirical Bayes method assumes the prior is a probability mass concentrating on some estimated value of μ and τ^2. For instance, μ and τ^2 can be estimated by maximum likelihood based on the likelihood $P(y \mid \mu, \tau^2)$ obtained by integrating out $\theta_1, \ldots, \theta_n$ from $P(y_i \mid \theta_i)P(\theta_i \mid \mu, \tau^2)$. This approach results in the estimator:

$$\hat{\mu} = \frac{\bar{y}D}{\bar{D}} \tag{12}$$

where τ^2 is replaced by the estimator $\hat{\tau}^2$ obtained by solving the equation:

$$\sum_{i=1}^{n} \frac{1}{\sigma_i^2 + \tau^2} = \sum_{i=1}^{n} \left\{ \frac{y_i - \hat{\mu}}{\sigma_i^2 + \tau^2} \right\}^2. \tag{13}$$

With μ and τ estimated and the σ_i^2 assumed known and fixed, the right-hand sides of Equations 10 and 11 can be calculated and only the left-hand sides depend on λ and ζ. Equation 10 can be simplified by substituting $\hat{\mu}$ from Equation 12:

$$\hat{\zeta} = \frac{\hat{\mu} - \bar{y}_A}{1 - \bar{A}} \tag{14}$$

Using the data described in the main body of the chapter we have obtained the following results. In the first stage, values of $\hat{\mu}$ and $\hat{\tau}^2$ were calculated by iteratively solving equations 12 and 13 yielding $\hat{\mu} = 1.37$ (mg/kg/day) and $\hat{\tau}^2 = 1.76$ (mg/kg/day)^2. The standard unadjusted posterior estimates (Equation 5) were determined; and values of the sample mean and variance of these were 23.6 (mg/kg/day) and 41.3 (mg/kg/day)^2, respectively. In the second stage, Equations 10 and 11 were iteratively solved to yield $\hat{\zeta} = 1.39$ (mg/kg/day) and $\hat{\lambda} = -0.41$. The sample mean and sample variance of the adjusted values given by Equation 8 are 23.6 (mg/kg/day) and 59.3 (mg/kg/day)^2, respectively. Figure 3 displays the empirical density of the raw potency values along with three fitted curves, corresponding to the empirical (solid curve), standard Bayes (dashed curve), and the adjusted Bayes (dash-dot curve) models.

CHAPTER 11

Environmental Pollution and Human Health: An Epidemiology Perspective

Joel Schwartz

INTRODUCTION

A Martian, newly arrived in the United States, would have little trouble identifying what the inhabitants regard as their most serious air pollution problems. Short-term (1-hr) excursions of ozone and carbon monoxide are clearly the largest threats to public health. Exposure to airborne particles and sulfur oxides are a minor problem, as is long-term average exposure to all air pollutants, with the possible exceptions of ozone and carcinogenic organic compounds. These messages are clear in the dollar amounts spent on controlling short-term peaks in ozone and carbon monoxide; in the effort and attention given to designing state plans to reduce maximum hourly exposure to those pollutants; in the research agenda adopted by the U.S. Environmental Protection Agency (EPA); and in its trends reports, which track not average ozone levels but days above 120 ppb.

Should the Martian look further and examine the basis for these concerns, he would be forced to two conclusions: (1) the behavior of Americans is completely inscrutable and (2) the health effects of air pollution are entirely inconsequential. This is because these standards are based on studies in exposure chambers, under unrealistic scenarios, producing effects of little consequence. The EPA justifies a vast array of pollution control strategies for ozone with chamber studies showing completely reversible decreases in lung function during prolonged exercise. The decreases are generally not detectable to the subjects, and the number of people engaged in vigorous outdoor exercise at 3 p.m. on a day with high ozone levels is not large. Is this really the major public health threat from air pollution?

Further research into studies the EPA *ignores* in setting its standards would leave the Martian even more puzzled. For the epidemiology

209

literature is replete with dozens of studies associating air pollution with increased respiratory symptoms, need for asthma medication, asthma attacks, emergency room visits, hospitalization for respiratory illness, and early mortality. Long-term exposure to air pollution has been linked to the development of chronic lung disease and long-term reductions in lung function. The air pollutant most commonly associated with all of these outcomes is airborne particles, not ozone. However, ozone is sometimes also associated with these outcomes, which are far more serious than the effects measured in the chamber studies. These outcomes appear to be occurring in persons who are already acutely or chronically ill because of other factors, and whose illness is exacerbated by air pollution. These findings would not be surprising to the Martian, who majored in earth history. They repeat results that have been investigated for centuries.

Investigations of the potential health effects of air pollution date to the *Fumifugium* of Evelyn[1] (in 1661) if not earlier. Evelyn argued that the increased rates of respiratory disease in London were associated with increased coal combustion. Graunt[2] (in 1662) linked variations in weekly bills of mortality in London to variations in the "airs." The rapid industrialization of the Western world worsened air quality in major urban areas, and furthered early investigations of these issues. Studies grew explosively after World War II and were critical in the establishment of the clean air laws in the 1950s and 1960s in the United States and Great Britain. These and subsequent studies have established air pollution as an important potential modulator of health. The major endpoints examined and some important findings are discussed briefly below.

In contrast to many studies of cancer, the outcomes associated with air pollutants such as ozone or airborne particles are prevalent and highly nonspecific. Not surprisingly, the relative risks are low. However, in contrast to airborne carcinogens from chemical plants or exposures from hazardous waste sites, the exposed population is very large. Hence the attributable risks can be quite substantial. Indeed the magnitude of these risks dwarfs those associated with many other pollution exposures.

For example, recent animal studies of the effects of long-term exposure to ozone at concentrations similar to those seen in Los Angeles, showed chronic lung damage which increased monotonically with cumulative dose and showed no evidence of a threshold. If these effects are also occurring in humans, as seems likely, then approximately 200 million people are exposed to ozone concentrations that may produce chronic lung damage every summer in the United States. Moreover, pretreatment with ozone for a short period at commonly occurring concentrations halved the dose of allergens required to trigger an asthmatic response in allergic asthmatics.[3]

Recent studies in 12 cities have reported associations of airborne particles with small increases in the risk of mortality. Again, no evidence of a threshold was found, and the relationships reported suggest that airborne particles are responsible for about 70,000 early deaths in the United States each year. In a recent community study, the closing of a steel mill for a year was associated with over a 40% decrease in hospital admissions for asthma for children.[4] The next year the mill reopened, and hospitalization rates climbed back to their previous level. Another recent study found acute bronchitis rates in children were about twice as high in a town with particle concentrations at the U.S. air quality standard as they were in a community with near background concentrations, with intermediate communities falling in the middle.[5] Acute respiratory symptoms in school children[6] and preschool children[7] similarly show major changes across concentrations normally seen in the U.S. urban population. Given the large number of people exposed, the estimated number of attributable cases is large.

These studies illustrate that while progress has been made in controlling air pollution, the pollution remaining may have important public health consequences. The recent Clean Air Act Amendments (CAAA) focused attention on this issue. Unfortunately, the current law and regulatory structure may be misdirecting society's resources in this regard.

For example, attention is focused on airborne carcinogens, short-term effects of ozone, and short-term effects of carbon monoxide in the current regulatory regime. However, there is little evidence that airborne carcinogens are of much concern; and most health scientists are more concerned about the potential effects of chronic ozone exposure, than of the short-term effects of exceeding the 1-hr standard. Regulatory attention is on meeting that standard, even when less expensive measures might reduce long-term average exposure by a greater amount. Moreover, there is little attention paid to lowering chronic ozone exposure in locations that are meeting the short-term ozone standard. Few researchers consider carbon monoxide a major problem, whereas there is growing evidence that airborne particles have potentially major health impacts. However, since few locations currently exceed our particle standards and many more exceed our CO standard, attention is focused on the less important risk. A better understanding of the health risks of air pollution is clearly needed.

ACUTE HEALTH EFFECTS

Recent exposure to air pollution has been associated with a number of acute outcomes from reversible reductions in lung function, to increased respiratory symptoms and illness, to increased hospitalization and emergency visits, to increased death rates from respiratory and cardiovascular disease. They are discussed briefly below in increasing order of severity.

Short-Term Changes in Lung Function

Exposure of exercising volunteers to ozone in chambers at concentrations below current air quality standards has been associated with decrements in lung function.[8] While these studies involve durations and exercise levels that may not be commonly experienced by the general population, studies of children in summer camps[9] have confirmed the effect in children — at least in the summer months — when they are mostly outdoors during the day. Smaller, but still significant effects have been reported in school children during the school year, when outdoor activity is less.[10] Studies of lavage fluid from the lungs of volunteers have also provided evidence of inflammatory processes in the lung following such ozone exposure.

Acute episodes of moderately elevated particle concentrations have also been associated with pulmonary function deficits.[11,12] Subsequent studies looking at daily time series rather than episodes have also associated PM-10 with short-term changes in peak flow at concentrations below current air quality standards.[13,14]

Acute exposure to short duration peaks of SO_2 have also been associated with asthmatic responses in exercising asthmatics, although the combination of exercise and concentration required suggest that this is a less frequent occurrence than those cited above.

Increases in Respiratory Symptoms

The chamber studies of ozone exposure cited above also provided evidence of increases in respiratory symptoms, although the symptoms seem short lived and qualitatively different from those reported in diary studies. Similar findings in diary studies of populations have not been common. One exception is Krupnick et al.[15] Ozone exposure was associated with increased duration of respiratory episodes in one diary.[16]

In contrast to the findings for ozone, a number of studies have associated daily symptom incidence or duration with exposure to airborne particles. These include results from the Six City study diary,[17] where particle concentrations never exceeded 75% of the current air quality standard. Other similar findings come from Switzerland[7] and Provo, Utah.[18,19]

Increases in Respiratory Illness

Exposure to airborne particles has been associated with increased rates of bronchitis in children[5] at concentrations below current standards, and also with increased rates of croup attacks in children.[20] Chronic respiratory symptoms were also associated with PM-10 exposures in a study by

Vedal and co-workers.[21] These effects do not appear to be limited to children. For example, Ostro and Rothschild[22] have reported an association between airborne particles and ozone and respiratory illness severe enough to restrict activity in adults. School absences have also been associated with PM-10 concentrations.[23] These associations show dose-dependent increases with exposure with no evidence of a threshold. This is illustrated in Figure 1, using data from Dockery et al.[5]

Increased Hospitalization for Respiratory Illness

Bates and Sizto[24] reported that both ozone and sulfate exposure were associated with increased hospitalization for respiratory illness in Ontario. Pope,[4] as mentioned above, found a sharp change in hospitalization rates for children when a steel mill closed and then reopened. The only major air pollutant in the valley was respirable particles. In a follow-up study in Provo and Salt Lake City looking at fluctuations in hospitalization continuously over time, a significant association was found with PM-10 in both cities. Hospital admissions were also increased in the German Smog episode of 1985[25] in response to sharp increases in both TSP and SO_2. Recently, Sunyer and co-workers[26] have reported that particles and SO_2 were associated with hospitalization for respiratory illness in Barcelona. Hospital emergency room visits were also associated

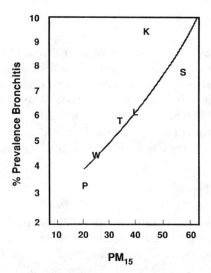

FIGURE 1. Shows the adjusted rate of acute bronchitis in children vs the PM-15 concentration ($\mu g/m^3$) in each of the cities in the Harvard Six City study. The rates in each city have been adjusted for individual risk factors. (P = Portage, T = Topeka, W = Watertown, K = Kingston, L = St. Louis, and S = Steubenville.)

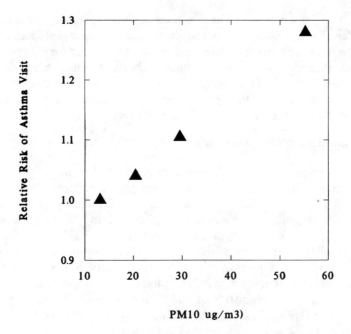

FIGURE 2.   Shows the relative risk of an asthma emergency visit in Seattle vs quintiles of daily PM-10 concentration. These risks are after controlling for age, hospital, season, and weather. The risk in the lowest quintile was defined as one.

with sulfates and SO_2 in a study in Vancouver,[27] and respirable particles were associated with hospital emergency room visits in Israel.[28] Cody and co-workers[29] reported an association between ozone levels and hospital emergency visits for asthma. Emergency room visits for asthma were also associated with PM-10 in a recent study in Seattle.[30] Again little evidence is seen for a threshold in these studies. This is illustrated in Figure 2 from Schwartz et al.[30] Note the daily ambient air quality standard for PM-10 is 150 $\mu g/m^3$.

Increased Mortality Rates

The dramatic events of London in 1952, where the death rate more than doubled during an air pollution episode, left little doubt that particulate-based smog at high concentrations could increase mortality. In the 4 days ending December 5, 1952, the mortality rate was 193 people per day in the London Administrative County. In the next 4 days, the daily death rate increased to 406 deaths per day.[31] In the 160 Great Towns of England excluding London, death rates changed little during this period (4585 to 4749). No evidence of an infectious epidemic was found, and the deaths began to rise with a lag of about a day following the dramatic increase in air pollution. The average increase in particle concentration associated with this episode was about 1200 μg/m^3 in the London Administrative County, however. Whether smaller changes in particle concentrations are associated with smaller increase in mortality remains a critical issue.

Schwartz and Dockery[32] examined the correlation between daily deaths in Philadelphia and daily concentrations of total suspended particles (TSP) during the years 1973–1980. After controlling for the year of study, a continuous time trend, the daily temperature, the daily humidity, an indicator for very hot days, an indicator for very humid days, and the indicator variables for season, TSP was a highly significant predictor of daily mortality. While SO$_2$ was also a significant predictor of daily mortality, when both pollutants were considered simultaneously, TSP remained significant with only a minor reduction in its estimated effect size; on the other hand, SO$_2$ became insignificant with a substantial reduction in its regression coefficient. TSP was highest during the warm months in this study, while mortality peaked in the winter; this suggests that inadequate control for seasonal factors would only bias downward the estimated pollution effect. When quintiles of TSP were used instead of the continuous pollution variable, a dose-dependent increase in mortality risk with increasing TSP was seen with no evidence of a threshold. This is illustrated in Figure 3. Schwartz and Dockery[32] also examined age and cause-specific risks of particulate exposure. The relative risk was highest in persons aged 65 and older and for deaths from chronic obstructive pulmonary disease (COPD), pneumonia, and cardiovascular disease. The pattern of much larger relative risks in the elderly and for COPD deaths, followed by pneumonia and cardiovascular disease, parallels what was seen in London in 1952. Schwartz,[33] in a more detailed analysis of cause and age at death, found the pattern in Philadelphia almost identical to that seen in the London 1952 episode. He also reported that most of the increase in cardiovascular mortality was associated with respiratory complications.

What makes this finding particularly interesting is its consistency with other recent findings. Figure 4 shows the estimated relative risk of exposure to 100 μg/m^3 TSP for all cause mortality in recent studies in Steubenville, OH;[34] Philadelphia, PA;[32] Detroit, MI;[35] Provo, UT;[19] St. Louis, MO, and Eastern Tennessee;[36] Minneapolis-St. Paul, MN;[33] and Birmingham, AL.[37] In addition, the relative risk was also estimated from the London episode of 1952.

These studies—showing similar slopes in areas with different mean temperatures and climatic conditions, and with both winter and summer peaking air pollution—make a strong case for the association. Other studies have used somewhat different methodologies; thus their effect

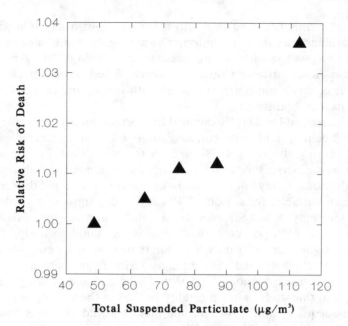

FIGURE 3. Shows the relative risk of death in Philadelphia by quintiles of daily TSP concentration, after controlling for weather and season. The risk in the lowest quintile was defined as one.

size estimates are not directly comparable. However, Fairley[38] and Schwartz and Marcus[39] both reported optical measures of airborne particles were associated with daily mortality. Qualitatively, the effect size estimates seemed similar. Hatzakis and co-workers[40] reported an association in Athens that was primarily with SO_2 and not with particulate matter. Reanalysis of that data using only the winter season[41] reported that the principal association was with particulate matter and not SO_2. Kinney and colleagues[42] have reported associations with both ozone and k_m, an optical measure of particles, with daily mortality in Los Angeles. The ozone association was stronger in that study. The poorer measure of particle exposure and high correlation between the two pollutants in Los Angeles make it difficult to assess the relative importance of the two pollutants in those data.

CHRONIC EFFECTS OF AIR POLLUTION

Decreases in Lung Function

Spektor and co-workers[43] have recently reported that long-term exposure of children to particulate air pollution was associated with lower

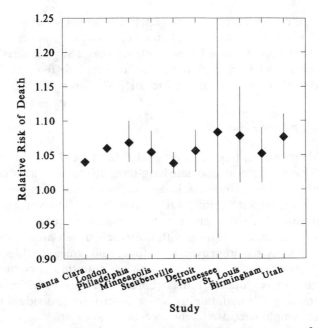

FIGURE 4. Shows the relative risk of death associated with a 100 μg/m³ increase in TSP in each of the cities where regression analyses were done. The vertical bars show the 95% confidence intervals.

lung function. Chronic exposure to both particles and ozone was associated with lower lung function in children in the NHANES II survey.[6] The ozone effect was stronger in that study. Chestnut and colleagues[44] have reported long-term TSP exposure was associated with lung function decrements in adults. The most striking effects have come from animal studies, however, where chronic exposure to ozone at Los Angeles levels has produced chronic damage to the small airways of the lung. This damage increases monotonically with cumulative dose, suggesting there is no threshold for the effect. While similar data are not available for humans, the findings in primates are particularly worrisome.

Increases in Chronic Respiratory Symptoms

A number of studies have associated differences in long-term exposure to air pollution with increased rates of chronic respiratory illness. For example, Euler et al.[45,46] have reported that cumulative exposure to TSP was associated with increased rates of chronic bronchitis in a prospectively followed cohort of Seventh Day Adventists in California. The association held when ozone exposure was also examined. A weaker association was found with ozone exposure, which was not significant after controlling for particles. Schwartz[47] confirmed an association between TSP exposure and chronic bronchitis in never smokers. Portney and Mullahy[48] have likewise found evidence that TSP was a risk factor for bronchitis. Detels et al.[49] have also reported differences between communities differentially exposed to air pollution in the Los Angeles area.

Chronic Mortality Studies

A number of studies (e.g., Lipfert et al,[50] Evans et al.,[51] Ozkaynak and Thurston[52] have sought to associate long-term differences in air pollution concentrations with differences in age adjusted mortality rates across major urban areas. Such studies suffer from the difficulty in adequately controlling for other risk factors such as smoking, problems which are avoided by time series analyses within a single urban area. In general, their methodology has improved over time, and while considerable uncertainty remains, some association with air pollution is certainly suggested. Prospective cohort studies represent a way to estimate these chronic effects, while maintaining better control for individual risk factors. Large sample sizes are likely necessary, however.

Opportunities and Needs

While the studies above and others make a convincing case that air pollution can be a major public health hazard, many unanswered questions remain. A critical issue is the effects of chronic exposure to ozone on humans. The combination of animal data on long-term impact with findings of inflammatory processes in short-term human exposure makes it likely that there is some cumulative damage. What remains unknown is whether that damage proceeds at such a slow pace as to result in little impact of chronic respiratory morbidity, or whether it is an important risk factor. It is interesting that the existing epidemiology indicates that particles instead of ozone comprise the main air pollution risk factor for this outcome. More studies of the impact of long-term exposure to both pollutants are clearly needed, as is a reversal of the ozone bias at the EPA. These studies will of necessity require large sample sizes. However, National Institutes of Health (NIH) studies on cancer and heart disease have recently involved sample sizes of 100,000 (the Nurses Health study) and 50,000 (the Physicians Health study) showing these sizes are possible if the sufficient priority is given to the study.

The use of better measures of population ozone exposure, such as the recently developed ozone badges,[53] will allow better identification of acute symptomatic responses to ozone. While the case for lung function changes is clear, the evidence that acute morbidity is invoked remains weak at present.

The association of airborne particle exposure with acute health endpoints from lung function changes to early mortality has been well documented by now. What is less clear is whether some inhalable particles are more potent and others less important in producing these effects. Some studies of acid aerosols have been funded recently, although most have not yet fully reported. More studies of speciation are needed, and the role of the organic particulate components has been little examined to date. More detailed studies to determine the characteristics of the subjects who are, for example, turning up in hospitals or dying when particle concentrations increase are needed. This would allow us to identify the potentially sensitive groups, and provide clues toward a better understanding of the mechanism. Likewise, heterogeneity in response to ozone exposure has been well documented for acute lung function changes, but efforts to classify the sensitive population have failed so far.

REFERENCES

1. Evelyn, J. Funifugium, or the inconvenience of the aer and smoke of London dissipated. In: Lodge, J. P., Jr. Ed. *The Smoke of London: Two Prophecies.* Elmstead NY: Maxwell Reprint Co. 1969.
2. Graunt, J. *Natural and Political Observations Made upon the Bills of Mortality, London 1662.* Baltimore: Johns Hopkins Press, 1939.
3. Molfino, N. A., Wright, S. C., Katz, T., Tarlos, S., Silverman, F., McClean, P. A., Szalai, J. P., Raizenne, M., Slutsky, A. S., and Zamel, N., Effect of low concentrations of ozone on inhaled allergen response in asthmatic subjects, *Lancet*, 1991, 338(8761):199-203.
4. Pope, C. A., III. Respiratory disease associated with community air pollution and a steel mill, Utah valley. *Am. J. Public Health* 1989, 79: 23-28.
5. Dockery, D. W., Speizer, F. E., Stram, D. O., Ware, J. H., Spengler, J. D., and Ferris, B. G., Jr. Effects of inhaled particles on respiratory health of children. *Am. Rev. Respir, Dis.* 1989, 39:587-594.
6. Schwartz, J. Lung function and chronic exposure to air pollution: a cross-sectional analysis of NHANES II. *Environ. Res.* 1989; 50:309-321.
7. Braun-Fahrlander, C., Ackermann-Liebrich, U., Schwartz, J., Gnehm, H. P., Rutishauser, M., and Wanner, H. U. Air pollution and respiratory symptoms in preschool children. *Am. Rev. Respir. Dis.*, 1992, 145:42-47.
8. Horstman, D. H., Folinsbee, L. J., Ives, P. J., Abdul-Salaam, S., and McDonnell, W. F. Ozone concentrations and pulmonary function response relationships for 6.6 hour exposures with five hours of moderate exercise to 0.08, 0.10, and 0.12 ppm. *Am. Rev. Respir. Dis.* 1989, 138:407-15.
9. Spektor, D. M., Lippmann, M., Lioy, P. J., et al. Effects of ambient ozone on respiratory function in active, normal children. *Am. Rev. Respir. Dis.* 1988, 137:313-20.

10. Brunekreef, B., Kinney, P. L., Ware, J. H., Dockery, D. W., Speizer, F. E., Spengler, J. D., and Ferris, B. G., Jr. Sensitive subgroups and normal variation in pulmonary function in response to air pollution episodes. *Environ. Health Perspect.* 1991, 90:189–194.

11. Dockery, D. W., Ware, J. H., Ferris, B. G., Jr., Speizer, F. E., Cook, N. R., and Hermann, S. M. Change in pulmonary function in children associated with air pollution episodes. *JAPCA* 1982, 32: 937–42.

12. Dasen, W., Brunekreef, B., Hoek, G., et al. Decline in children's pulmonary function during an air pollution episode. *JAPCA* 1986, 36:1233.

13. Pope, C. A., Dockery, D. W., Spengler, J. D., and Raizenne, M. E. Respiratory health and PM_{10} pollution: a daily time series analysis. *Am. Rev. Respir. Dis.* 1991, 144:668–674.

14. Pope, C. A. III and Dockery, D. W. Acute health effects of PM10 pollution on symptomatic and asymptomatic children. *Am. Rev. Respir. Dis.* 1992, 145: 1123–28.

15. Krupnick, A. J., Harrington, W., and Ostro, B. D., Ambient ozone and acute health effects: evidence from daily data, *J. Environ. Econ. Manage.*, 1990, 18:1–18.

16. Schwartz, J., Air pollution and the duration of acute respiratory symptoms. *Arch. Environ. Health.* 1992, 42(2):116–122.

17. Schwartz, J., Dockery, D. W., Ware, J. H., Spengler, J. D., Wypij, D., Koutrakis, P., Speizer, F. E., and Ferris, B. G., Jr. Acute effects of acid aerosols on respiratory symptom reporting in children. *Air Pollut. Control Assoc. Paper* 1989, 89:92.1.

18. Pope, C. A., III. Respiratory hospital admissions associated with PM_{10} pollution in Utah, Salt Lake, and Cache valleys. *Arch. Environ. Health* 1991, 46:9–97.

19. Pope, C. A., Schwartz, J., and Ransom, M. Daily mortality and PM_{10} pollution in Utah Valley. *Arch. Environ. Health* 1992, 42(3):211–217.

20. Schwartz, J., Spix, C., Wichmann, H. E., and Malin, E. Air pollution and acute respiratory illness in five German communities. *Environ. Res.* 1991, 56:1–14.

21. Vedal, S., Blair, J., and Manna, B. Adverse respiratory health effects of ambient inhalable particle exposure. *Air Waste Manage. Assoc.* 1991, 56:91–810.

22. Ostro, B. D. and Rothschild, S. Air pollution and acute respiratory morbidity: an observational study of multiple pollutants. *Environ. Res.* 1989, 50: 238–247.

23. Ransom, M. R. and Pope, C. A. III. Elementary school absences and PM10 pollution in Utah valley. *Environ. Res.* 1992, 58: 204–219.

24. Bates, D. V. and Sizto, R. Hospital admissions and air pollution in Southern Ontario: the acid summer haze effect. *Environ. Res.* 1987, 50:238–247.

25. Wichmann, H. F., Mueller, W., Allhoff, P., Beckmann, M., Bocker, N., Csicsaky, M. J., Jung, M., Molik, B., and Schoeneborg, G., Health effects during a smog episode in West Germany in 1985, *Environ. Health Perspect.*, 1989, 79:89–99.

26. Sunyer, J., Anto, J. M., Murillo, C., and Saez, M. Effects of urban air pollution on emergency room admissions for chronic obstructive pulmonary disease. *Am. J. Epidemiol.* 1991, 134:277–286.

27. Bates, D. V., Baker-Anderson, M., and Sizto, R. Asthma attack periodicity: a study of hospital emergency visits in Vancouver. *Environ. Res.* 1990, 51:51–70.

28. Gross, J., Goldsmith, J. R., Zangwill, L., and Lerman, S. Monitoring of hospital emergency room visits as a method for detecting health effects of environmental exposures. *Sci. Total Environ.*, 1984, 32:289–302.

29. Cody, R. P., Weisel, C. P., Birnbaum, G., and Lioy, P. J. The effect of ozone associated with summertime photochemical smog on the frequency of asthma visits to hospital emergency departments. *Environ. Res.*, 1992, 58, 184–194.

30. Schwartz, J., Koenig, J., Slater, D., and Larson, T. Particulate air pollution and hospital emergency visits for asthma in Seattle. *Am. Rev. Respir. Dis.* 1993, 147–326.

31. Mortality and Morbidity during the London fog of December 1952, Her Majesty's Public Health Service Report No. 95 on Public Health and Medical Subjects, Her Majesty's Stationery Office, London, 1954.

32. Schwartz, J. and Dockery, D. W. Increased mortality in Philadelphia associated with daily air pollution concentrations. *Am. Rev. Respir. Dis.* 1992, 145:600–604.

33. Schwartz, J. Particulate air pollution and daily mortality: a synthesis. *Public Health Rev.* 1992, 19:39–60.

34. Schwartz, J. and Dockery, D. W. Particulate air pollution and daily mortality in Steubenville, Ohio. *Am. J. Epidemiol.* 1992, 135:12–19.

35. Schwartz, J. Particulate air pollution and daily mortality in Detroit. *Environ. Res.* 1991, 56:204–213.

36. Dockery, D. W., Schwartz, J., and Spengler, J. D. Air pollution and daily mortality: associations with particulates and acid aerosols. *Environ. Res.* 1992, 59:362–373.

37. Schwartz, J. Air pollution and daily mortality in Birmingham, AL. *Am. J. Epidemiol.* 1993, 137:1136–1147.

38. Fairley, D. The relationship of daily mortality to suspended particulates in Santa Clara county, 1980–1986. *Environ. Health Perspect.* 1990, 89:159–168.

39. Schwartz, J. and Marcus, A. Mortality and air pollution in London: a time series analysis. *Am. J. Epidemiol.* 1990, 131:185–194.

40. Hatzakis, A., Katsouyanni, K., Kalandidi, A., Day, N., and Trichopoulos, D. Short term effects of air pollution on mortality in Athens. *Int. J. Epidemiol.* 1986, 15:73–81.

41. Katsouyanni, K., Hatzakis, A., Kalandidi, A., and Trichopoulos, D. Short term effects of atmospheric pollution on mortality in Athens. *Arch. Hellen. Med.* 1990, 7:126–132.

42. Kinney, P. L. and Ozkaynak, H. Associations of daily mortality and air pollution in Los Angeles County. *Environ Res.* 1991, 54:99–120; Krupnick, A. J., Harrington, W., and Ostro, B. D. Ambient ozone and acute health effects: evidence from daily data. *J. Environ. Econ. Manage.* 1990, 18:1–18.

43. Spektor, D. M., Hofmeister, V. A., Artaxo, P., et al. Effects of heavy industrial pollution on respiratory function in children of Cubatao, Brazil: a preliminary report. *Environ. Health Perspect.* 1991, 94:51–54.
44. Chestnut, L. G., Schwartz, J., Savitz, D. A., and Burchfiel, C. M. Pulmonary function and ambient particulate matter: epidemiologic evidence from NHANES I. *Arch. Environ. Health*, 1991, 46:135–144.
45. Euler, G. L., Abbey, D. E., Hodgkin, J. E., and Magie, A. R. Chronic obstructive pulmonary disease symptoms effects of long term cumulative exposure to ambient levels of total suspended particulates and sulfur dioxide in California Seventh-Day Adventist residents. *Arch. Environ. Health* 1987, 42:213–222.
46. Euler, G. L., Abbey, D. E., Hodgkin, J. E., and Magie, A. R. Chronic obstructive pulmonary disease symptom effects of log term cumulative exposure to ambient levels of total oxidants and nitrogen dioxide in California Seventh-Day Adventist residents. *Arch. Environ. Health* 1988, 43:279–285.
47. Schwartz, J. Particulate air pollution and chronic respiratory disease. *Environ. Res.*, 1993, 62:7–13.
48. Portney, P. and Mullahy, J. Urban air quality and respiratory disease. *Reg. Sci. Urban Econ.* 1990, 20:407–418.
49. Detels, R., Sayre, J. W., Coulson, A. H., et al. The UCLA population studies of chronic obstructive respiratory disease. Respiratory effect of log term exposure to photochemical oxidants, nitrogen dioxide, and sulfates on current and never smokers. *Am. Rev. Respir. Dis.* 1981, 124:673–680.
50. Lipfert, F. W., Malone, R. G., Daum, M. L., et al. A statistical study of the macroepidemiology of air pollution and total mortality. Upton, NY, Brookhaven National Laboratories, 1988.
51. Evans, J. S., Tosteson, T., and Kinney, P. L. Cross-sectional mortality studies and air pollution risk assessment. *Environ. Int.* 1984, 10:55–83.
52. Ozkaynak, H. and Thurston, G. D. Association between 1980 U.S. mortality rates and alternative measures of airborne particle concentration. *Risk Anal.* 1987, 7:449–461.
53. Koutrakis, P., Wolf, J. M., Bunyaviroch, A., et al. Measurement of ambient ozone using a nitrite-coated filter. *Anal. Chem.* 1993, 65:209–214.

CHAPTER 12

Soil Quality as a Component of Environmental Quality

Michael A. Cole

INTRODUCTION

Soil is a critical support medium for terrestrial life. Historically, the kind and quantity of vegetation was used by pioneers to distinguish between high- and poor-quality soil. Plant biomass production or crop yields are still used by ecologists, agronomists, and regulators as a primary criterion of quality. For example, restoration of soil to its "original productivity" and topography after surface mining is a primary requirement of current regulations. There has been substantial recent interest in broadening the base for the soil quality concept to include aspects other than plant production because high biomass production based on external nutrient supplies can overestimate the intrinsic ability of the soil to support plant growth and ignores the value of biotic components other than the principal crop. Incorporation of environmental criteria into traditional land-use planning was discussed by van Lier[1] on both philosophical and practical levels. Swaminathan[2] proposed that crop yield should be defined as:

$$\text{YIELD} = (\text{output value} - \text{input value}) - \text{impact on environmental capital stocks.}$$

The suggestions of van Lier and Swaminathan are similar to arguments about who should pay for cleanup of historic industrial pollution. Both historic and current industrial and agricultural pollution have substantial off-site impacts. However, the historic trend in industry and agriculture has been to ignore off-site impacts; and where such impacts occur (nitrate and pesticide contamination of drinking water and siltation of reservoirs due to soil erosion, for example), the affected party,

0–87371–936–0/94/$0.00 + $.50

not the responsible party, has typically had to pay the costs of remedial action.

A wide variety of definitions and components of soil quality have been suggested by various researchers.[3] For the purposes of this discussion, an ecological rather than agricultural perspective will be used. In the context of overall ecosystem function, a high-quality soil provides conditions which allow development of a highly diverse flora and fauna and high annual biomass production. The soil furnishes ample supplies of critical nutrients in a sustainable manner from large intrinsic reserves of potentially limiting elements such as nitrogen and phosphorus. From an environmental perspective, stable ecosystems conserve nutrients and soil organic reserves are high.[4,5] Since stable systems are conservative, there are few adverse off-site impacts such as siltation of surface waters, eutrophication of surface waters from transported terrestrial nutrients, or contamination of groundwater by organic or inorganic chemicals originating in the soil.

INTRINSIC AND EXTRINSIC COMPONENTS OF SOIL QUALITY

The components of quality are the intrinsic physical and chemical properties, topographic relationships, biota, and potential off-site impacts of a specific soil. Various researchers have proposed a shorter list of predominantly intrinsic chemical and physical properties while ignoring the substantial impact that climate and other externalities have on soil. In contrast, pedologists incorporate components of geologic history, climate, placement in the landscape, and vegetation into their soil descriptions, as well as analytically determined chemical and physical properties when characterizing a soil.[6] Extrinsic factors such as climate are included on the grounds that these factors have a large influence on soil development from parent materials and they also affect the potential rate of soil degradation and biomass production. Temperature and available moisture are primary regulators of microbial activity, and moisture is a major controlling factor for plant biomass production. The inclusion of external properties such as climate and topographic relationships is beneficial because inaccurate conclusions can be made if quality is gauged only by intrinsic properties. A soil *in situ* as a past, present, and future entity is affected so much by its relationships to nonsoil properties of the environment that to exclude these properties is to deny the evaluator access to critical information that can have great impact on a high- *vs* low-quality decision. For example, many soils in semiarid regions would rate highly based on analytical properties; but in actuality, they have a limited biota and productivity because climatic conditions (lack of rainfall) prevent expression of the soil potential. Similarly, a permanently flooded soil has a limited flora and fauna because of the prevailingly anaerobic conditions that are inimicable to many organisms. In these cases, the single limitation of insufficient water (a function of climate) or excessive water (a function of climate and substratum geology) override the apparently high quality based on determination of intrinsic chemical and physical properties of these soils.

UTILITY OF PRESENT DATA FOR ASSESSING SOIL QUALITY

Most available data published on physical and chemical properties and microbiota of agricultural soils are not very useful for a national assessment of soil quality because of differences among disciplines and investigators in the purposes of gathering data about soil, sampling strategies, sample processing, and analytical methods . Although thousands of research papers and books on various aspects of soil properties and biology exist, very little of the work was intended to be used for absolute comparisons across time or location. Because of the heterogeneous nature of small soil samples, researchers have tended to process samples to obliterate small-scale heterogeneity and to remove many *in situ* components prior to analysis. Statements such as "root fragments, invertebrates, and undegraded plant materials were removed by sieving prior to analysis" are very common. Such treatment can dramatically change the apparent attributes of the soil being examined. For example, microbial populations are often ten to a hundred times higher on root systems than in surrounding soil,[7] and the species distribution is quite different.[8] In addition, the predominant microbes which colonize litter materials are often of different species or phylogenetic type than the predominant soil organisms.[9] Thus, removing and discarding root fragments and plant litter result in substantial quantitative and qualitative changes in the biological attributes of the samples. The research conducted with these samples has yielded valuable fundamental insights into soil processes and has resulted in the development of methods which would be applicable to studies of intact soils. However, the data have only limited utility for documenting the properties of the originally collected material such as species abundance, availability of organic substrates as potential carbon and energy sources, and nutrient fluxes. Since adequate methods for holistic soil analysis either exist or could be developed by adaptation of currently available methods, a program in which key properties of managed and unmanaged soils were compared would be highly desirable.

SYSTEMIC LIMITATIONS TO ASSESSING SOIL QUALITY

Establishing a valid database for soil properties is complicated by microscopic and macroscopic spatial variability in the physical, biotic, and chemical properties of soils and by wide seasonal fluctutations in some important properties. Microbial populations and activity can fluctuate widely within a few days, depending on soil moisture content and temperature. The best database is obtained by collecting numerous samples at a single time and by collecting samples at several times over a 1- to 2-year period. This stipulation creates the practical problem of extensive analytical time (and cost) being required to obtain information on a relatively small sample base.

A thorough botanical or faunal survey of relatively small areas requires several months of work by experienced specialists in vascular plant taxonomy and several subdisciplines in invertebrate biology, while a thorough microbial survey would entail a year or more of work to isolate and characterize the predominant microbes. Because of the labor-intensive nature of biotic surveys, relatively detailed studies would be feasible only if the focus were the impact on terrestrial biota of common land-use practices in several regions of the United States. Since the major land uses are cultivation of annual crops, managed and unmanaged grasslands, and various managed and unmanaged forestry systems, the number of prospective survey sites would be manageable. Such studies would provide a nucleus of data which would permit an assessment of management practices on soil quality. The studies conducted under the International Biological Program (IBP) in the 1970s are a good precedent for the studies proposed here. A preliminary assessment of impact on some biota such as higher plants, mammals, birds, and some insects and on soil chemical and physical properties can be made using published literature and various environmental impact documents which are required for projects such as dams and forestry management plans.

Accurate evaluation of quality will require a judicious weighting system for various soil physical and chemical properties, some of which are interactive and complex (i.e., composed of several characters) or which singly can be of overwhelming importance. As an example of interactive and conflicting influences, soils with a high clay content tend to accumulate larger amounts of organic matter than sandy soils do (which furnishes a positive value during soil quality evaluation); however, soils of high clay content also tend to be poorly drained (which can limit biotic diversity and thereby lower soil quality). A single factor such as very low or very high pH will severely restrict plant biomass production and diversity with substantial impacts on all other biota. When a single factor limits soil quality, a thorough survey of all soil quality characters is not very productive because other properties may fall within the same range as found in high-quality soils; however, they do not have a large influence on the overall behavior of the system.

SPECIFIC CRITERIA FOR ASSESSING SOIL QUALITY AND JUSTIFICATION FOR THEIR INCLUSION

Numerous investigators have proposed either a limited or extensive list of soil properties which contribute to soil quality. Table 1 contains a list of properties which reflect the components of the definition of a high-quality soil given in the introduction of this chapter. With a few exceptions, the proposed information can be obtained from existing data or extrapolations from existing data or by performing relatively simple analytical procedures. The list of traits is intended to serve for screening purposes so that selected ecosystems can be targeted for more detailed investigation. It should be noted that these criteria are more comprehensive than those used for economic valuation of land based on the ability of the soil to produce marketable biomass such as grain, hay, livestock, or wood products. In such valuation systems, no distinction is made between biomass produced from soil-derived or external nutrient sources (fertilizers) and biodiversity is not considered. Hence, the criteria that I propose will result in substantially lower soil quality values with highly productive agricultural soils than an economically based evaluation would.

Biotic Diversity

High biotic diversity in unmanaged ecosystems provides resilience to perturbation. The survival of the native prairie in the midwestern United States in the 1930s[10] and the failure of cultivated crops during the same period provide good examples of the benefits of diversity. In unmanaged systems, a high degree of biodiversity may also be perceived as an abstract, but desirable attribute, whereas a high degree of biodiversity in managed systems can have the practical benefit of prolonging the productive life span of a soil. In historic tropical agricultural systems, polycultures have allowed continuous crop production for several hundred years at one location,[11] in contrast to the 2- to 5-year production span that is possible in less diversified slash-and-burn systems. Traditional polycultures exhibit higher permanence, greater system stability on a short-term basis, and higher genetic diversity; and have better developed natural pest control systems than modern monocultures.[12] Therefore, for both esoteric and practical reasons, biodiversity should be included as a component of soil quality.

Table 1. Specific Criteria for Assessment of Soil Quality

Biotic diversity
Number of species, biomass, and/or number of individuals of principal
animal phyla inhabiting soil
Number of species and biomass of plants; annual and perennial
species are scored separately
Estimated number of species and biomass of bacteria and fungi in soil
and in litter layer (if present)

Geologic and topographic factors
Depth of topsoil
Drainage (rapid or slow), depth to water table
Ability of soil minerals to supply major and minor nutrient elements

Chemical properties
pH
Soluble salt content
Concentrations of toxic organic and inorganic chemicals
Organic matter and organic-N content: present value, historic value,
and ratio present value/historic value

Physical properties
Aggregate structure and stability

Biological activities and nutrient cycles
Biomass production and/or accumulated biomass: historic, current
without external nutrient sources, current with external nutrient
sources
Vegetative cover as percentage of total surface area
Carbon and nutrient additions from external organic materials
(manures, composts, wastes)
Net annual carbon mineralization
Net annual change in organic matter content
Net nitrogen (N) status: contributions of biotic processes (primarily
nitrogen fixation and denitrification) and biomass removal
Nitrogen input from external inorganic sources
Net nitrogen status: contributions of abiotic processes (primarily N from
precipitation and leaching losses)
Net annual change in nitrogen content

Off-site impacts
Annual soil erosion rate
Annual losses of organic matter, nitrogen, and phosphorus to surface
water
Annual losses of organic compounds and nitrogen to groundwater

Use of land for agricultural production or monoculture forestry plantations results in dramatic decreases in plant biodiversity when plant species abundance is reduced from several hundred plant species per square kilometer in undisturbed systems to a few crop and weed species per square kilometer in intensively managed agricultural systems. Because of the high degree of host plant dependency in insects, invertebrate diversity will always be adversely affected by annual crop agriculture and forestry plantations when compared to less intensively managed areas. Since animals have a substantial impact on soil properties,[13,14] the reduction in animal populations and diversity as host plant diversity decreases should be accompanied by changes in soil properties.

Natural forest systems display vertical stratification, which increases the effective habitat area per land area and also provides a wide variety of terrestrial and aerial sites for breeding and shelter of birds and mammals. This vertical stratification is achieved by a mixture of fallen woody material, low forbs and grasses at ground level, shrubs and young trees at intermediate levels, and mature trees at high levels. Conversion of the forests of eastern North America and floodplain areas to agricultural production has resulted in destruction of this stratified system as well as a diminution in overall biotic and faunal diversity.

The degree and importance of biodiversity of microbial populations in terrestrial systems has not been well documented.[15] In those microbes for which the main ecological function is nutrient cycling activities such as decay of organic materials, high species or genus diversity may not be particularly important because high percentages of bacteria and fungi isolated from soil are producers of the major enzymes required for degradation of cellulose and other polysaccharides, proteins, nucleic acids, lipids, and monomers derived from the respective polymers.[16]

Because of the short generation time of microbes compared to other organisms, population recovery after adverse impacts or environmental change such as drying, flooding, and addition of toxic organic and inorganic chemicals is usually rapid and long-term consequences are minimal when compared to the impact on higher organisms. An exception to this statement is the dramatic reduction in microbial diversity after chemical sterilization of soil.[17]

There are several specialized microbial groups such as the mycorrhizal fungi, symbiotic and nonsymbiotic dinitrogen fixers, and nitrifiers whose activity is often depressed by pesticides and major soil disturbance such as mining. These organisms have been targeted for more frequent studies of environmental impact of xenobiotics, toxic metals, and land management practices than the general heterotrophic microbes; and nitrifier populations may be an exception to the general statement that microbial changes are a relatively insensitive measure of soil quality.[18]

Geologic and Topographic Factors

These criteria include physical and chemical characteristics of sub-strata and parent materials which have been shown to greatly influence plant biomass production. For example, soil only a few centimeters thick over bedrock may give a high score based on nutrient content, organic matter, microbial attributes, and have a high biotic diversity. However, in practice it is a functionally poor soil because of inadequate rooting depth and water retention capacity. In this case, poor quality is an irremediable property of the soil.

Chemical Properties

Soil pH has a major influence on both plant diversity and biomass production. High soluble salt content is a primary limitation to plant growth for most species. Toxic natural and man-made organic compounds added to soil can profoundly affect floral and faunal composition. Soils with naturally high concentrations of toxic metals have a distinct and limited flora, as do soils which are contaminated with metals or organic compounds as a result of agricultural or industrial activities. In any of these cases, the toxic content of the soil is a primary restriction on soil quality, irrespective of high values for other measured properties.

Physical Properties

A substantial list of pertinent soil physical properties could be given, but aggregate stucture is suggested as a easily assessed indication of soil quality. In a high-quality soil, the mineral and organic components are organized into porous particles referred to as soil aggregates. Aggregate structure and stability affect soil porosity, water and gas movement within the soil, water retention, microbial activity, and plant growth.[19,20] Loss of aggregate structure increases the potential for erosion by wind and water because displacement of smaller soil particles formed by disintegration of aggregates requires less energy than displacement of larger aggregates. Soils with good aggregate structure diminish surface runoff and erosion when compared to soils with poor aggregate properties.

Biological Activity and Nutrient Cycling

Several proposed criteria address the question of overall carbon and nitrogen budgets of the soil. Natural systems which are not disturbed by man's activities typically have high and stable reserves of organic carbon and nitrogen in the surface horizons. Specific levels of these elements depend on a variety of factors such as parent material, climate, and vegetation.[21] Since natural grassland and forest systems persist for centuries or millennia, they are by definition sustainable and their behavior can serve as a model for sustainable practices to use with managed lands.[22]

Several criteria are used as components to establish net annual changes in organic matter and nitrogen status of the soil. These criteria address the fundamental issue of whether the soil system is maintaining, increasing, or decreasing its organic matter and nitrogen content over time. The question is not simply whether the soil has some arbitrarily chosen amount of some microbial activity or chemical element, but whether the relationship between input and output of a specific element is neutral, positive, or negative over time.

Substantial losses of soil organic carbon and nitrogen occur when native vegetation is removed and annual crops are produced.[4,23] There is an initial rapid loss of soil quality, during which time good crop yields are obtained using only soil reserves of critical plant nutrients. This stage is followed by a period where soil organic matter and nitrogen content remain relatively constant, but plant nutrient availability is low. If external fertilizer is not supplied, these soils are unproductive and are typically abandoned and allowed to revert to native vegetation. Abandonment is followed in a few years by recolonization by perennial forbs, grasses, and woody plants, and soil reserves of organic matter and nitrogen increase until precultivation values are attained. Soil quality increases during this phase. This pattern is seen in tropical slash-and-burn systems[24] and was the historic method of agriculture in eastern North America.[25] Current agronomic data for U.S. soils indicate that one half to two thirds of the yield of nonleguminous crops such as corn and cotton can be attributed to exogenous fertilizer addition (primarily nitrogen); in the absence of fertilizer, corn grain yields were only 30–50 bushels per acre. In contrast, pre-1920 results from several midwestern agricultural experiment stations demonstrated that crop yields were not increased by nitrogen fertilizer addition, i.e., the soil was able to supply adequate nutrients to support grain yields as high as 100 bushels per acre. The inescapable conclusion from these data is that the quality of midwestern soils has declined substantially within 70 years.

Measurement of nutrient and carbon dynamics is relatively expensive and time-consuming. However, there is a substantial body of literature which could be used to generate generic estimates of soil processes. Output from computer-based models of ecosystem processes could also be used in several cases, particularly for criteria involving carbon and nitrogen cycling in soil. A valid scheme of this type should give high values to high-quality soils, irrespective of current use.

Off-site Impacts Associated with Soil Quality Changes

Improving or maintaining soil quality is considered by some scientists as a rather esoteric concept, particularly by those who regard plant biomass production or crop yield as a satisfactory single criterion of soil quality. Changes in soil quality also have a substantial impact on overall off-site environmental quality. The examples given in Table 2 indicate that many land-use practices which diminish on-site soil quality also have an adverse impact on off-site environmental quality. Conversely, improving soil quality can also reduce off-site impacts of land use. It should be evident from the comparisons in Table 2 that the larger incentives to improve soil quality occur at the societal level than at the local (on-site) level, since a substantially larger number of people are affected by adverse off-site impacts than are harmed by local impact. For example, estimates of crop productivity losses due to limited soil erosion (i.e., a few tons of soil per hectare per year) indicate that there is little on-farm economic loss as a result of such erosion. In contrast, the off-site impacts of "tolerable" erosion such as reductions in aquatic biota and water quality are substantial, both by ecological and economic criteria.

POTENTIAL TACTICS FOR DIVIDING SOILS INTO LOW- OR HIGH-QUALITY CATEGORIES

Ecotoxicological concepts that large, persistent, and widespread reductions in key ecosystem properties are of concern can be applied to development of quantitative standards for ecosystem health in general[26] and more specifically to assess soil quality. For example, Nielsen[27] applied the World Health Organization concept of human health to the overall environment as follows:

Environmental health is the extent to which a given ecosystem and its component parts are able to support human activities on the one hand and to sustain populations of other organisms on the other hand. Environmental health therefore is the ability of an ecosystem to cope with human-induced change; it is not the sole objective of conservation, but is the concept emphasizing the balance among social, ethical, aesthetic and biological goals.

A preliminary evaluation of selected soils was made using the criteria given in Table 1. Application of these criteria to managed ecosystems, particularly those used for intensive production of annual crops, indicates that substantial degradation of physical, chemical, and biotic attributes of cultivated soils has occurred in many cases. These decreases in

Table 2. Relationships Between Soil Quality Criteria and On-Site and Off-Site Impacts

Quality criterion	On-site impact	Off-site impact
High proportion of short-season annual plants	Soil organic matter and biotic diversity decreases	Increased erosion, small increase in atmospheric CO_2
Extensive use of exogenous nitrogen sources	Increased growth of crop plant species	Increased nitrogen release into surface and subsurface waters
Frequent pesticide use at approved rates	Decreased floral and/or faunal diversity	Increased pesticide release into surrounding water and air
Poor soil aggregate structure	Decreased aeration, water penetration, and possibly reduced plant growth	Increased soil erosion, possible increased denitrification

soil quality can be attributed to the inevitable reduction in biodiversity accompanying conversion from the highly diverse fauna and flora of natural areas compared to agricultural fields; to the loss of soil structure, organic matter, and nutrients associated with long-term annual crop production; to the adverse impact of nitrogenous fertilizers on some invertebrates;[28] and to the destruction of nontarget invertebrate species by insecticides and herbicides.[29,30]

Some of the "alternative agriculture" methods[31] provide higher soil quality scores than conventional cropping systems[32] when overall farm operations are compared. Higher quality scores for alternative systems are primarily the result of higher crop diversity, higher percentage of land devoted to perennials, lower use of pesticides and inorganic fertilizers in some cases, and intellectual commitment to improving the soil. The economic aspects of these practices were summarized by van Mansvelt.[33]

The question is: what action—if any—should be taken if particular land management systems do not rank as highly as others from a soil quality perspective? Should soil degradation be allowed to continue, should it be arrested at its present status, or should remediation and rehabilitation measures be considered? From a regulatory standpoint, taking an action requires firm evidence of harm resulting from a human activity. If the state of knowledge of a process, an organism, or a system is incomplete, it becomes practically impossible to implement and defend a regulation because gaps in existing scientific knowledge introduce uncertainty when legal challenges of the regulations are made. The greatest deficiencies in available data and potential assessment methods are in the categories of off-site impacts; microbial diversity and activity measurements; and biotic diversity, particularly with respect to placing an economic or environmental value on changing practices to modify off-site impacts, to regulate microbial activity such that less soil degradation takes place, and to enhance biotic diversity.

There is an acute lack of "companion site" studies of different land management systems. Research on biological control capabilities of native forest and grassland soils compared to soils in which annual or perennial crops are grown could provide valuable insights into natural mechanisms of pest control with applications to managed soils.[34,35] Comparative studies would also reduce the uncertainties associated with regional and local variability of soils and climatic conditions and would greatly improve our ability to distinquish between superior and inferior land-use practices.

CONCLUSIONS

Some components of pre-1950 U.S. farming practices could be revived as a means of improving soil quality if the criteria in Table 1 are used. The older systems employed more diverse crop rotations than currently seen, incorporated perennial grasses and legumes into the cropping program, used lower quantities and kinds of pesticides, used less inorganic nitrogen fertilizer, and often included ruminant animals and their manures in a diversified cropping system. There was also a tendency to have a farm woodlot and tree-lined wind barriers between fields. Taken together, these practices increased plant and habitat diversity and indirectly affected avian, mammalian, and invertebrate diversity by providing a broader range of food supply and shelter than current rotations in which short-season annual grains are the predominant vegetation. Less use of pesticides and inorganic fertilizers resulted in less impact on nontarget organisms. Manure additions contributed supplemental organic material to the soil as well as providing a stimulus for microbial and invertebrate populations. Some of the above practices have been revived by practioners of sustainable or alternative agriculture, and a large research effort directed toward contemporary evaluation of the benefits of these alternative practices would have national value.

Some agricultural scientists have objected to the idea of reverting to older practices because these practices may provide lower crop yields and cash income to farmers.[36] If criteria other than cash income alone are used, the older systems are not so disadvantageous as imagined. If Swaminathan's formula (see "Introduction") were used to assess overall performance of various land management options, contemporary systems of annual grain production would have less merit than older systems because of negative impacts such as nitrogen, phosphorus, and pesticide contamination of surface water; higher rates of soil erosion; and dramatic reductions in biodiversity and habitat for noncrop flora and fauna.

Current European and North American trends to include a pollution component into management practices for pesticide use,[37,38] nitrate pollution,[39] and animal waste production and use[40] could be incorporated into North American practices at relatively little cost to the agricultural sector and with substantial benefits to all people. Contemporary concerns in the European community are in sharp contrast to earlier U.S. analyses[41] in which erosion, but not pesticide use, was projected to have an adverse impact on water quality. Water quality reductions associated with fertilizer use in some regions were projected, a conclusion consistent with European experiences.

The few million dollars per year which the U.S. Department of Agriculture has allocated to study alternative farming systems is a trivial fraction of the approximately $800 million research budget of this agency. Diversion of additional funds from studies of conventional cropping and forestry systems to studies of alternative systems would not be an onerous financial burden and would have substantial societal benefit. As described in a previous section, there is an association between soil quality aspects of local land-use practices and off-site impacts; and therefore changes in farming and forestry practices which would improve local soil quality would have significant and positive off-site benefits as well.

REFERENCES

1. van Lier, H. N. "Land-use planning on its way to environmental planning," in *Land-Use and the European Environment*. M. Whitby and J. Ollerenshaw, Eds. (London: Belhaven Press, 1988), pp. 89–107.
2. Swaminathan, M. S. "Chairman's remarks" in *The Biodiversity of Microorganisms and Invertebrates*. D. L. Hawksworth, Ed. (Wallingford, UK: C.A.B. International, 1991), p. 275.
3. *Proceedings of the Soil Quality Standards Symposium*. Document WO-WSA-2, (Washington, DC: U.S. Government Printing Office, 1992), 80 pp.
4. Stevenson, F. J. *Cycles of Soil*. (New York: John Wiley & Sons, 1986), pp. 55–60.
5. Bormann, F. H. and G. E. Likens. *Pattern and Process in a Forested Ecosystem*. (New York: Springer-Verlag, 1979), 253 pp.
6. Jenny, H. *The Soil Resource*. (New York: Springer-Verlag, 1980), 377 pp.
7. Rovira, A. D. and C. B. Davey. "Biology of the rhizosphere," in *The Plant Root and Its Environment*. E. W. Carson, Ed. (Charlottesville, VA: University Press of Virginia, 1975), pp. 153–204.

8. Lochhead, A. G. "Qualitative studies of soil microorganisms. III. Influence of plant growth on the character of the bacterial flora," *Can. J. Res. Sect. C* 18:42–53 (1975).

9. Burges, A. "The decomposition of organic matter in soil," in *Soil Biology*. A. Burges and F. Raw, Eds. (New York: Academic Press, 1967), pp. 479–492.

10. Weaver, J. E. and F. W. Albertson. *Grasslands of the Great Plains*. (Lincoln, NE: Johnsen Publishing Co., 1956), pp. 75–162.

11. Sanchez, P. A. *Properties and Management of Soils in the Tropics*. (New York: John Wiley & Sons, 1976), pp. 478–532.

12. Altieri, M. A. "Increasing biodiversity to improve insect pest management in agroecosystems," in *The Biodiversity of Microorganisms and Invertebrates*. D.L. Hawksworth, Ed. (Wallingford, UK: C.A.B. International, 1991), pp. 165–182.

13. Hole, F. D. "Effects of animals on soil," *Geoderma* 25:75–112 (1981).

14. Zlotin, R. I. and K. S. Khodashova. *The Role of Animals in Biological Cycling of Forest-Steppe Ecosystems*. Translated by Lewus, W. and W. E. Grant. (Dowden, Hutchinson & Ross, Inc., 1980), 221 pp.

15. Bull, A. T. "Biotechnology and biodiversity," in *The Biodiversity of Microorganisms and Invertebrates*, D. L. Hawksworth, Ed. (Wallingford, UK: C.A.B. International, 1991), pp. 203–219.

16. Alexander, M. *Introduction to Soil Microbiology*. (New York: John Wiley & Sons, 1977), pp. 148–202.

17. Kreutzer, W. A. "Selective toxicity of chemicals to soil microorganisms," *Ann. Rev. Phytopathol.* 1:101–126 (1963).

18. Miller, K. W., M. A. Cole, and W. L. Banwart. "Microbial populations in an agronomically managed mollisol treated with simulated acid rain," *J. Environ. Qual.* 20:845–849 (1991).

19. Hillel, D. *Fundamentals of Soil Physics*. (Orlando, FL: Academic Press, Inc., 1980), pp. 265–284.

20. Olson, K. R. "Soil physical properties as a measure of cropland productivity," in *Proceedings of the Soil Quality Standards Symposium*. Document WO-WSA-2, (Washington, DC: U.S. Government Printing Office, 1992), pp. 41–51.

21. Jenny, H. The Soil Resource. (New York: Springer-Verlag, 1980), pp. 246–336.

22. Soule, J. D. and J. K. Piper. *Farming in Nature's Image: An Ecological Approach to Agriculture*. (Washington, DC: Island Press, 1992), 286 pp.

23. Balesdent, J., G. H. Wagner, and A. Mariotti. "Soil organic matter turnover in long-term field experiments as revealed by carbon-13 natural abundance," *Soil Sci. Soc. Am. J.* 52:118–124 (1988).

24. Sanchez, P. A. *Properties and Management of Soils in the Tropics*. P. A. Sanchez, Ed. (New York: John Wiley & Sons, 1976), pp. 347–411.

25. Hopkins, C. G. *Soil Fertility and Permanent Agriculture*. (Boston: Ginn & Company, 1910), 653 pp.

26. Schaeffer, D. J., E. Herricks, and H. Kester. "Ecosystem health. I. Measuring ecosystem health," *Environ. Manage.* 12:445–455 (1988).

27. Nielsen, N. O. "Ecosystem health and sustainable agriculture," unpublished report, Ontario Veterinary College, University of Guelph, Guelph, Ontario, Canada, August 1991.

28. Potter, D. A., B. L. Bridges, and F. C. Gordon. "Effect of N fertilization on earthworm and microarthropod populations in Kentucky bluegrass turf," *Agron. J.* 77:367–372 (1985).

29. Popovici, I., G. Stan, V. Stefan, R. Tomescu, A. Dumea, A. Tarta, and F. Dan. "The influence of atrazine on soil fauna," *Pedobiologia* 17:209–215 (1977).

30. Hoy, J. B. "Effects of lindane, carbaryl, and chlorpyrifos on non-target soil arthropod communities," in *Soil Biology as Related to Land Use Practices*, D. L. Dindal, Ed. (Washington, DC: United States Environmental Protection Agency, Document EPA-560/13-80-038), pp. 71–81.

31. *Alternative Agriculture.* (Washington, DC: National Academy Press, 1989), 448 pp.

32. Drury, C. F., J. A. Stone, and W. I. Findlay. "Microbial biomass and soil structure associated with corn, grasses, and legumes," *Soil Sci. Soc. Am. J.* 55:805–811 (1991).

33. van Mansvelt, J. "The role of lower-input technologies in the future," in *Land-Use and the European Environment*. M. Whitby and J. Ollerenshaw, Eds. (London: Belhaven Press, 1988), pp. 42–53.

34. Bull, A. T. "Biotechnology and biodiversity," in *The Biodiversity of Microorganisms and Invertebrates*. D. L. Hawksworth, Ed. (Wallingford, UK: C.A.B. International, 1991), pp. 203–219.

35. Perfect, T. J. "Biodiversity and tropical pest management" in *The Biodiversity of Microorganisms and Invertebrates*. D. L. Hawksworth, Ed. (Wallingford, UK: C.A.B. International, 1991), pp. 145–148.

36. Black, C. A. "The Alternative Agriculture Report," in *Alternative Agriculture: Scientists' Review*. (Council for Agricultural Science and Technology Special Publication No. 16, Ames, IA, 1990), pp. 68–78.

37. Hoag, D. L. and A. G. Hornsby. "Coupling groundwater contamination with economic returns when applying farm pesticides," *J. Environ. Qual.* 21:579–586 (1992).

38. Kahn, J. R. "Atrazine pollution and Chesapeake fisheries," in *Farming and the Countryside: An Economic Analysis of External Costs and Benefits*. N. Hanley, Ed. (Wallingford, UK: C.A.B. International, 1991), pp. 137–158.

39. Silvander, U. and L. Drake. "Nitrate pollution and fisheries protection in Sweden," in *Farming and the Countryside: An Economic Analysis of External Costs and Benefits*. N. Hanley, Ed. (Wallingford, UK: C.A.B. International, 1991), pp. 159–176.

40. Tamminga, G. and J. Wijnands. "Animal waste problems in the Netherlands," in *Farming and the Countryside: An Economic Analysis of External Costs and Benefits*. N. Hanley, Ed. (Wallingford, UK: C.A.B. International, 1991), pp. 117–136.

41. Crosson, P. R. and S. Brubaker. *Resource and Environmental Effects of U.S. Agriculture*. (Washington, DC: Resources for the Future, 1982), pp. 104–155.

SECTION IV

Future Environmental Management

Joseph M. Abe

INTRODUCTION

Decision makers at all levels of public and private organizations are increasingly aware of the shortcomings of fragmented policies and crisis management. Rapid change and uncertainties, either real or perceived, are driving many people toward an alternative approach to thinking, planning, and event management. Foresight, a key function supporting this new approach, enables organizations to anticipate and prevent problems, plan for uncertainties, and create opportunities. This introduction describes why environmental decision making and planning needs to change, examines what the array of available foresight tools are, and describes when and how these tools can b⌐ implemented effectively.

Over the past 20 or so years, the regulatory approach to environmental management has appeared to achieve significant, while costly, environmental improvements. However, for a number of reasons the current command and control approach cannot ensure a sustainable environment, a heathy society, and a prosperous economy in the long term. There are many indicators that the current regulatory system has outlived its useful life (Table 1A). Rearward looking, fragmented policies have been directed at the symptoms of problems and not the root causes. In many cases, costly regulatory activities have simply transferred pollution within the environment (i.e., pollution shell games). Furthermore, the detection-response intervals for many problems (e.g., chlorofluorocarbons [CFCs] and stratospheric ozone depletion) have often been years or decades, thus making solutions more difficult and costly. Decaying and ineffective infrastructure, budget constraints, barriers to innovative technological and socioeconomic solutions, compliance costs to business, and insufficient state and local regulatory capacity further demonstrate the need to replace the current approach.

239

Table 1. Comparison of Current Regulatory Approach with Forward-looking Systems Approach

A. Current risk-based regulatory approach

Too negative (i.e., emphasis is on problems rather than on opportunities)

Rearward looking (what did we do last year?)

Based on linear and "reductionist" thinking (cannot accommodate the complexity and dynamics of the real world [e.g., multiple stresses to ecosystems, additive and synergistic effects])

Often blind to emerging problems and strategic opportunities

Fragmented (does not allow comprehensive problem/solution assessment)

Restrictive (tends to stifle innovation because it is too prescriptive and inflexible)

Slow to respond to changing conditions

B. Forward-looking systems approach

More positive (e.g., what kind of future do you want?)

Forward looking (e.g., vision creation, trend assessment, issues scanning, and scenario planning)

Based on systems and "holistic" thinking (can accommodate interconnectiveness, complexities, and transient conditions of natural and human systems

Capable of providing early warning signals to allow prevention of early mitigation

Strategic, innovative, and timely (highlights, strategic opportunities, barriers to positive change and provides insight into tomorrow's effects brought about by today's decisions)

Empowering (enables decision makers to break free of "crisis management" by expanding the "thinking" horizon and fosters better management of key resources [e.g., people environmental resources, time, money, technology])

Supportive of sustainable development (sustainable development requires finding harmonious, mutually supportive relationships between human society and the environment; sector-based scenario planning provides insight into the relationships between sector policies [e.g., energy, agriculture, and transportation] and environmental policies; this allows optimal [or win-win], sustainable solutions to be identified and implemented)

Better suited to multilevel policy implementation and coordination (e.g., global, national, and regional policies need to be better coordinated and connected with grassroots implementation at the state and local level)

More responsive (feedback loops to decision makers at all levels, an important characteristic of the systems approach, provide better and more timely information on strategy effectiveness and efficiency; coupled with foresight tools, the systems approach can shrink the detection-response intervals and can even help prevent problems from occurring)

A forward-looking systems approach to planning and decision making appears to be an effective replacement to the current rearward looking regulatory approach (Table 1B). This approach also embodies many concepts and tools found in future studies, strategic planning and management, and organizational learning literature. A systems approach to policy-making would foster cooperation rather than conflict among different parts of the government (e.g., coordination of environmental, energy, transportation, and agricultural policies) and between business and government (e.g., pollution prevention strategies would be developed and encouraged through partnerships and pricing and tax reform). Ultimately, society would factor all consequences (i.e., environmental and social as well as economic) of a product or service or *before* it is produced. A systems approach coupled with foresight tools can enable decision makers at all levels of society to anticipate problems and redirect efforts to more environmental intelligent, sustainable solutions. This approach also encourages optimal (win-win outcomes) solutions to replace suboptimal solutions (win-loose outcomes) associated with the current regulatory approach. That is, policies can be established that *simultaneously* ensure a sustainable environment, a healthy society, and a prosperous economy.

Given the promise of a forward-looking systems approach, what are some useful foresight tools that can enable planners and decision makers to expand their thinking horizon—both in time and breadth of analysis? Table 2 provides a summary of foresight tool categories and the questions that they are intended to answer.

Vision creation involves a normative process used to identify preferred future outcomes. It generally involves both the participation of potential stakeholders as well as expert judgment and opinion; and because of its positive focus, it is well suited for goal setting and consensus building.

Trend analysis and monitoring consists of variety of methods for assessing, tracking, and projecting spatial and temporal distributions of individual variables of interest frequently using econometric, graphic, and other statistical modeling techniques discussed in great detail in other chapters. These techniques require the fitting of historic data with mathematical models, that are then extrapolated into the future (i.e., assumes relationships determined with historic data hold true in the future). In other cases, when single variables are not easily separated or where relationships are not easily described mathematically, expert judgment using Delphi group process techniques is often employed to qualitatively describe important trends.

Issues analysis and scanning examine the emergence of new trends or problems as well as the interaction of multiple trends. This category helps decision makers rank problems and target responses, and identifies major discontinuities and surprises for them. For example: will trends

Table 2. Overview of Foresight Tools

Tools	Applications	Key questions
Vision creation	Goal setting	What do we want to accomplish? By when?
	Strategic planning	How can we achieve results most effectively? Who are the stakeholders? What are dominant external forces?
Trend analysis and monitoring	Track trends	What are past, current, and future and trends?
	Evaluate progress	Will goals be met?
Issue analysis	Opinion surveys	What trends do we see?
Scanning		Do we like what we see?
	Expert workshops	What are emerging issues? How will issues interact?
Scenario planning	Alternative futures	What outcomes could occur?
	Policy options	How will policies change? Which policies acheive desired results?

affecting an issue be mutually supportive or in opposition and what issues will have the greatest impact on my decision?

Finally, scenario planning allows decision makers to assess the future outcomes, both pleasant and unpleasant, resulting from their decisions. It is particularly important for the scenarios to be linked to perceived conditions today affecting a decision. Alternative scenarios, which are plausible stories of the future if particular conditions prevail, provide insight to decision makers about the impacts of their decisions as well as forces beyond their control. Scenario planning provides flexibility to decision makers by allowing the comparison of alternative strategies in advance (hopefully to select the best choice) and by hedging uncertainty by helping them to develop contingency plans.

Foresight tools are particularly useful when they are used in an integrated strategy to support a real-world decision (e.g., should a new power plant be sited in a community?). Well-orchestrated application of tools can enable foresight information to be mutually supportive and complimentary. Successful tool application can be gauged by the extent to which a decision is affected (i.e., is a decision altered as a result of new information or perspectives derived from improved foresight?). It is equally important that the tools be used in an interactive, participatory process within an organization. This helps build consensus and ensures diversity of opinion. The following chapters discuss selected foresight tools and concepts that are relevant to environmental planning and management. Much of the discussion focuses on when and how foresight tools can be effectively used by organizations and decision makers.

REFERENCES

1. Schwartz, P., *The Art of the Long View*, Doubleday, New York, 1991.
2. Taylor, C., *Creating Strategic Visions*, U.S. Army War College, Strategic Studies Institute, Carlisle Barracks, PA, 1991.
3. Senge, P., *The Fifth Discipline, The Art and Practice of the Learning Organization*, Doubleday, New York, 1990.
4. Webler, T., Levine, D., Rakel, H., and Renn, O., A Novel Approach to Reducing Uncertainty, the Group Delphi, *Technol. Forecasting Social Change*, 39(3): 253–263, 1991.

CHAPTER 13

Gauging the Future Challenges for Environmental Management: Some Lessons from Organizations with Effective Outlook Capabilities

Mark A. Boroush

INTRODUCTION

For many organizations these days, estimating the prospective twists and turns of the future operating setting *and using such insights in strategic planning* are increasingly essential. Much of this reflects the more challenging setting now unfolding—an era of intensifying change and uncertainty, a sharp escalation in the competition for resources, and the need to satisfy increasingly diverse and demanding constituencies. This places a premium on sharp concepts of purpose and direction and a roster of action steps which well consider the evolving character of the opportunities and challenges in the external environment of the organization. An effective outlook process—in essence, the analytical effort to anticipate evolving trends and possible "discontinuities"—is an essential foundation for planning.

Many of the best current examples of outlook systems lie in the business sector, where the ebb and flow of the operating setting is often rapid and the sizable risks and opportunity costs of action heighten the value of insights about future trends. Much of the thrust of my remarks draws from this realm.

Such a "business" focus may appear far afield from the complex issues of human health risks and ecosystem impacts that are of prime concern for environmental policymakers. Clearly, the substantive issues and indicators differ. Nevertheless, I think there are some useful lessons on process from these applications that may be of interest to an environmental policy-making community that is, as I understand, increasingly inter-

0–87371–936–0/94/$0.00 + $.50
© 1994 by Lewis Publishers

ested in strengthening its capabilities for anticipatory planning. My remarks below are offered in this spirit.

STRATEGIC MANAGEMENT AND FORECASTING

There is a good deal of attention in management circles now to the concepts of organizational vision, strategic planning, and steps to proactive behavior. The specifics vary from sector to sector (private businesses and corporations, government agencies, and others), but the underlying story is similar. Successful organizations have found a way to advantageously align their internal capabilities and available resources with the opportunities and risks existing in their external operating environment.[1] Identifying, implementing, and sustaining such a balance is the chief task of strategic management.

Very few organizations are sufficiently flexible and adaptive to remain successful through instantaneous adjustment to changing conditions. In practice, some horizon of time is required for planning, action, and goal achievement—all often subject to some uncertainties and factors beyond control. Decision makers must recognize the capabilities of their organization; estimate what challenges and opportunities the external environment could pose as the future evolves; and set their goals, missions, and activities accordingly to maximize prospects for success.

Forecasting plays a critical role in the strategic management process by providing a basis for assembling insights about the future shape key planning variables could take.[2] What matters, of course, varies widely from organization to organization. Businesses in competitive markets are usually quite interested in what the future holds for the economic fortunes of their customers, how fast their markets could grow, or what the early commercialization of new technology could mean for their competitive advantage. Public policy-making organizations often face important uncertainties about factors such as the priorities the public will perceive, the nature of legislation to emerge, and the resources available for program implementation. These are differing questions to be sure, with contrasting technical requirements for forecasting, but similar in importance as critical considerations in planning.

The outlook function provides several tangible products for organization planning. The most immediately obvious role is that it is the source of information and analysis to chart assumptions about the future state of variables critical for planning. However, almost as important, the outlook function provides an internal opportunity for the continuing education of organization managers.[3] When implemented in a continuing and open way, the outlook function promotes discussion around such important issues as: what really drives the business and its prospects for success, what changes in these forces are or could be taking place and what are their implications, and what new actions may be needed to respond effectively in the future? Finally, the thinking and conclusions it stimulates provide a basis for considering and adjusting (if need be) the organization's prevailing strategic vision.

In practical terms, any outlook function must be closely tailored to the characteristics and prime needs of the organization in which it resides. However, several generic activities span the universe from which most are drawn:

- **Trend scanning and monitoring** — Here the effort is to track the progress of developments and the potential for emerging changes and shifts in domains of particular interest (such as new technology, public policy, social and economic trends). In some cases this is done to gain an understanding of where current trends of potential importance are headed. In others, the chief purpose is to look — as an aid for uncertainty hedging — for early indications of which of several possible operating settings appears more likely to emerge.
- **Issues analysis** — This secondary and more analytically focused stage of trend scanning is oriented chiefly toward evaluating prospective developments with high potential to engender issues requiring the prime attention of the organization (e.g., the rise of consumers' interest in environmentally acceptable products or the implications of an aging population for the future pace of economic growth).
- **Scenario forecasting** — Scenarios provide a means to consider the possible paths along which the operating environment could evolve in the future, taking into account prospective developments highlighted by the trend scanning and issues analysis activities.[4] Scenario building often provides a valuable foundation for utilizing complex simulation modeling in long-range forecasting. When developed systematically, scenarios can also provide a basis for considering the implications of uncertainties in key planning variables.

  "Scenario planning" is now a major thrust in many business sectors where uncertainties in key variables are high, stemming from the long duration of the planning horizon and/or an intrinsically turbulent operating environment.[5]

These are general concepts. Their practical application will, perhaps, be clearer in considering a concrete case.

IMPLEMENTATION: A CASE STUDY

Efforts by both private and public sector organizations to probe what the future might hold have expanded considerably since the early 1970s. My experience over the last dozen years suggests that most sizable organizations with significant planning or policy-making requirements have expended at least some effort on trying to divine the longer term future of their operating environment. The depth of these efforts and their ultimate influence on decisions vary widely—from one time, high visibility seminars drawing on a few outside speakers to much more extensive internal forecasting efforts. However, a rapidly growing case literature is emerging. What was chiefly the province of futurists and other seers only two decades ago is now a familiar commodity in planning and policy-making circles.

Much less typical, however, is an organization that has evolved a strong, ongoing outlook process with sustained internal support and close links to the strategic planning cycle. Here the Royal Dutch/Shell Group (one of the world's largest corporations, with numerous operating companies and extensive participation in the global energy industry) is an often cited case. The corporation is widely recognized as an early adopter of alternative scenarios in planning and, more broadly, as a long-range planning innovator. A few comments on the evolution of its planning philosophy and current operational features provide some useful lessons.

Royal Dutch/Shell began to rethink the foundations and content of its planning process in the late 1960s.[6] Heretofore, a single "6-year out" forecast—the first year in considerable detail, the next 5 in only broad terms—served as the principal guide. Yet, the process was widely regarded by the organization's managers as prone to inaccuracies and tending to reinforce a "future as more of the same" perspective. These limitations which became increasingly problematic as the company's emerging requirements for expanded production capacity involved investments with longer lead times and greater exposure to financial risks.

Key people in the corporation, accordingly, concluded that improved planning procedures were needed—ones that promoted looking further out on the planning horizon and provided richer insights on the prospects for emerging trends and changes capable of shifting or disrupting the accustomed shape of the organization's business environment. A series of business "horizon" studies served as an initial thrust in this direction, through which the Royal/Dutch Shell corporate planning department began to search 15 years or more ahead, with much greater attention on identifying the most formative business forces, the potential for significant changes, and the planning variables subject to the greatest uncertainties.

Pierre Wack (head of the business environment division of the Royal Dutch/Shell planning group throughout the 1970s and into the 1980s), in commenting on this part of company history, notes that Herman Kahn's scenario methodology was familiar at the time and perceived as a potentially valuable tool for corporate planning. Wack and others in his group began to draw on the scenario approach, both as an analytical framework to help systematically map out the future possibilities their analyses foreshadowed and as a communications medium to help the organization's managers understand the challenges and opportunities of the future business setting.

These tools proved particularly able to help Royal Dutch/Shell anticipate and plan for the major changes of the 1970s in the global energy industry. Projections of future energy supply and demand patterns prepared by the company's business environment group in the early 1970s suggested the considerable prospect—if not the likelihood—that major discontinuities could soon make their way into the characteristics of international oil markets. Conditions appeared increasingly ripe for a sharp turn to a seller's market, with the Middle East (particularly the oil-rich Arabian Gulf nations) increasingly in the position of the balancing source of supply. Only a global economic depression of major and sustained proportion seemed capable of reducing the demand for Middle Eastern oil to levels that could foreclose the prospect of a future with much higher prices for international oil.

Some 20 years later in the early 1990s, to a world experiencing no less than four major world oil price disruptions, these perceptions may not sound at all prophetic. However, at the time, the business environment group's expectations were well outside the mainstream of company managers thinking and that of the industry generally.

In 1972, and again in 1973, the business environment group prepared a more refined series of scenarios outlining possible future settings for the global oil industry throughout the 1970s. These efforts focused increasingly on the "looming discontinuity" theme; and worked to stimulate the Royal Dutch/Shell managers to rethink their fundamental assumptions about how the oil markets worked, what the new business environment might look like, and what could be done differently to steer the corporation through an unexpected set of business conditions if the sharp turn should come to pass. Not everyone was convinced that the oil industry was soon to undergo a sea change of epochal proportion. Nevertheless, the business environment group did largely succeed in getting managers to examine their assumptions and begin to think more carefully about what contingency plans might be needed to cope with significant changes in the business setting. When the international oil crunch and subsequent global economic shock did begin in late 1973—in reality, sooner than the business environment group's scenarios projected—Royal Dutch/Shell managers were a good deal better prepared to react than the rest of the startled industry.

Not surprisingly, the business environment group visibility and credibility in the organization grew substantially in 1974. The reshaping of global energy demand and supply patterns underway throughout the 1970s provided ample need for Royal Dutch/Shell to rethink its business concept and planning assumptions. Much of the business environment group's outlook and scenario building efforts in this period were targeted on helping organization management understand what new terms and conditions the transformed energy business could impose and what mid-term track important planning variables might take—all to help Royal Dutch/Shell successfully navigate the business "rapids" (as Wack termed the turbulences of the company changing operating environment) surging around them.

Today, the outlook and scenario planning processes Royal Dutch/Shell established during this challenging period for planning continue to function—and enjoy strong internal support as an integral part of the ongoing strategic planning cycle of the organization. Indeed, the company has expended considerable effort since the early 1980s to increase the value of the scenario process for corporation line managers. Insights derived from futures thinking and scenario building are called on in a regular fashion to aid in updating the corporation strategic vision and chart business actions to achieve its goals.

In practical terms, Royal Dutch/Shell's current efforts at these tasks are quite substantial.[7]

A new/revised set of "global" scenarios are prepared about every 2 years by the (corporate) Group Planning Division. These comprehensive projections typically address a planning horizon about 20 years out (further, in some cases) and provide an opportunity for the organization to update its thinking on the potential play of key forces and developments capable of influencing the longer term shape of the operating environment. Periodically, scenarios with a specific "regional" focus are also developed to gain more detailed insights on the economic and energy market forces operating in particular geographic regions of the globe (e.g., Europe, East Asia, the Middle East).

Once compiled, the scenarios are disseminated to Royal Dutch/Shell managers in a visually polished form—with numerous supporting background papers—should amplification of particular issues be desired. Managers are asked to consider the range of scenarios suggested, evaluate their implications, and consider how the principal thrusts of company business strategies would need to differ.

The scenario building activities represent an analytic effort to identify and gauge the longer term "macro" forces shaping the future operating environment of the corporation. The infeasibility of exhaustively considering the full range of developments and outcomes possible over a long future horizon is clearly recognized. However, the broad goals of the process are to identify principal variations in future challenges and opportunities and to address most of the planning issues currently judged important by senior management. The analytical output is deliberately "multiple scenario" in form, explicitly acknowledging the intrinsic uncertainties at play.

The process is chiefly an internal function, involving repeated cycles of review and comment from Royal Dutch/Shell managers. Nevertheless, there is also an active effort to solicit the opinions of leading outside experts in relevant fields (through workshops and brainstorming sessions), to enrich company understanding of key planning issues, and to gain a range of perspectives on the potential for future discontinuities in critical dimensions of the business environment. Methodologically, the forecasting techniques utilized range over a variety of methods, including the informed judgments of the scenario research team, the opinions of outside experts, and quantitative forecasting techniques (both econometric and system dynamic methods).

Importantly, the scenario exercise is well integrated into the Royal Dutch/Shell normal strategic planning steps (at both corporate and operating company levels).[8] The global/regional scenario building is conducted in parallel with "micro" assessments of the operating companies' competitive positions. The results of these separate activities interact through several rounds of discussion, eventually providing a basis to examine—and reshape, if warranted—the organization's prevailing strategic vision (i.e., perceived goals, expected opportunities, and anticipated avenues through which the corporation capabilities can be marshaled to best advantage). These insights in turn provide a basis for considering and comparing the various courses of action available to achieve desired goals.

As a concluding note, it is worth observing that institutionalizing this comprehensive outlook and scenario planning process at Royal Dutch/Shell did not happen overnight. As Wack observes, even with the business environment group's forecasting success in the early 1970s at a time of major change in corporation operating environment, some 5–6 years were required for the mainstream of company managers to become comfortable with the concepts and elements of the process.[9] Along the way, the business environment group had to work hard to learn how to shape the focus of the scenario process in ways that could engage company managers' highest concerns and to modify some deeply entrenched managerial attitudes (particularly, with regard to enhancing managers' willingness to seriously consider the prospect that tomorrow may well not be the same as today and to increase the explicit consideration of planning uncertainties).

THE TOOLBOX FOR LONG-RANGE FORECASTING

Looking ahead over the longer term invariably involves a variety of forecasting tasks—such as estimating the future pace of new technology commercialization, the market penetration of new products and processes, the level and composition of future economic activity, the demographic and other socioeconomic trends, or the outcome of new government legislation and public policy. Naturally, the mix of analytical attention among such forecasting goals depends greatly on the concerns users perceive as important. A good portion of what the long-range forecaster takes on involves shaping and integrating a diverse set of forecasting methods into an analytical framework capable of addressing outlook questions of prime concern.

In an interesting and still useful paper, Georgoff and Murdick[10] several years ago reviewed the variety of forecasting methodologies available to managers in addressing planning problems. These authors identified four major categories: *judgmental methods* (such as naive extrapolation or opinions of expert juries), *counting methods* (including the wide variety of market surveys now frequently commissioned by business), *time series methods* (moving averages, adaptive filtering, Box-Jenkins statistical models), and, finally *association/causal methods* (regression, econometric models, and the like). Depending on the forecasting task at hand, all of these have potential for long-range outlook applications — and nearly all have been so employed at one time or another. Today, long-range forecasting practitioners tend to be eclectic synthesizers across this wide spectrum of available tools.

The specific features required for any given outlook system makes generalization difficult. However, one can point generically to several domains of tools that find frequent application:

- **Scenario building** — The nearly universal "media" now is for considering and discussing prospective long-range trends and events. Not a forecasting methodology *per se*, the principal contribution of scenario building is in providing a framework for systematically exploring the influence of uncertainties in important driving variables.[11]
- **Expert opinion** — This is still an essential resource for the long-range forecasting process. Several mechanisms continue to be used to gather this kind of information: review and synthesis of secondary literatures (professional, trade, commentary, and the like); solicited judgments from a selected group of knowledgeable individuals; and facilitated expert working groups or juries, e.g., through a Delphi survey process.[12]
- **Statistical forecasting models** — Over the last several decades, a great deal of theoretical and empirical work has gone into devising and refining complex, statistical models for forecasting future trends in domains such as economic growth, population demographics, technology trends, and product innovation. The *forte* of these tools generally lies in shorter term forecasting applications, where the case for parameter stability is usually more defensible. However, such methods can often bring a strong causal framework to bear on thinking about the longer term sway of trends and events. Many are routinely utilized and/or adapted for long-range forecasting purposes.

- **Structural simulation modeling** — Increasingly powerful software (from ever more capable spreadsheets to dedicated simulation programs such as Systems Dynamics), in conjunction with the rapidly expanding computational capacities of desktop computers, now provide forecasters with considerable power and flexibility to develop explicit simulation models of the dynamics of the planning environment under consideration.[13] Such models can be designed to link and integrate a variety of intermediate inputs — from judgmental estimates of future developments to outputs of other quantitative forecasting models. Whether chiefly heuristic or more detailed and realistic in the representation of the behavioral features of the system, this kind of model building provides particular value in promoting lucid discussions about which variables and interrelationships matter and in providing a practical foundation for scenario building and uncertainty analyses.

In my experience, the larger and better outlook efforts today actively seek to draw on and integrate findings from methods across all these categories.[14]

It must be recognized, however, that no forecasting method yet available provides a perfect crystal ball. Assumptions and conditionals are legion throughout any effort to foresee the longer term future.

Part of the challenge stems from the power of cascading uncertainties. That is, in complex systems influenced as they are by many factors, even modest uncertainties about the future state of just a few key variables can quickly ratchet up the range and number of possible outcomes to consider. The reality that present scientific understanding of change and adaptation across a wide spectrum of societal phenomena remains finite is also an important constraint. This blunts the ability of forecasters to confidently decipher the twists and turns of history lying ahead. The high priority accorded by the outlook process to the search for "change in the making" imposes a high standard for excellence. Discontinuities are inherently potential developments and forces with low visibility in the present and about which the knowledge base is invariably least well equipped to comment: the unexpected technology breakthrough, the slower or faster than anticipated market penetration of a new process, the unforeseen political coalition that yields watershed changes in public policy, and the like. Admittedly, all of these limitations come into play in shorter term forecasting. Nevertheless, their effects are greatly magnified as the forecasting horizon stretches further and further out into the future.

The range and quality of tools available to forecasters continue to improve. Better methods have elevated the level of debate and enhanced the understanding that can be brought to bear. Even so, uncertainties and imponderables remain a big part of the long-range forecasting trade. The growing interest now in the scenario planning approach, with its emphasis on charting and examining the implications of a wide variety of plausible "what ifs," directly reflects this reality.

THE OUTLOOK PROCESS AND ENVIRONMENTAL PLANNING

Recent conversations with EPA staff and others have made me aware of the growing interest of the environmental management/policy community in improved capabilities for anticipatory planning. No doubt, the many prospects for change posed by the deepening integration of global economic and environmental systems now unfolding, the importance of better envisioning the potential of emerging technologies for cost-effective pollution reduction and prevention, and the need to more sharply define environmental management priorities in an era of tightening public sector resources all contribute to this perception. Clearly, questions about what developments will/could prevail in the future are involved in all of these considerations and make the kind of longer term thinking discussed earlier of immediate relevance.

Defining the elements of an outlook capability for this purpose is a task that deserves careful thought — and lies well beyond what can be accomplished in this brief chapter. However, in the spirit of my earlier remarks, let me conclude with a few observations and recommendations toward this end.

1. The objectives and operations context for contemporary environmental management and policy-making are broad and complex. Even from the vantage point of my only modest experience in this arena, it is apparent that a rich and varied set of considerations have and will continue to impact planning, including:

   - how the workings of the economy and associated social systems may be changing the profile of environmental health risks and ecosystem impacts
   - what priorities for action will emerge from public discourse on the environment and what level of resources will be made available for intervention programs
   - the implications of new technology — both with respect to new impacts and risks created and to improved options for remediation and risk reduction
   - the willingness of consumers and industries to modify their behaviors in light of emerging environmental knowledge and management goals
   - the influence of future developments in the international relations realm, as the policy focus on environmental issues of a global scale expands in the years ahead

   Clearly, this myriad of concerns poses a "tall list" of topics to monitor. As a practical matter, one of the important early issues in establishing an outlook capability will be to chart the substantive priorities for analytic attention. In outlook efforts, just as in other research and inquiry activities, better information and insights tend to result when questions are better defined. Indeed, the question of focus never really goes away, as it needs to be revisited periodically as the process evolves and successive rounds of findings provide a more refined view of what the future may hold.

2. Experience to date suggests that the process components of an outlook effort — including nature of staff and outside expert involvement, form and channel of communication for outputs, linkages to internal planning and decision making cycles, and concept of internal educational objectives — are as important to the overall impact of the function as the substance of the forecasts and analyses that arise.

   The Royal Dutch/Shell case summarized earlier is not the only current example of a well-established outlook function. However, it is one with a long history and numerous experience-based perceptions on how to implement an effective effort. Any organization seeking to establish its own outlook function would be well advised to closely consider the whats and whys behind the approach of this corporation.

3. Plan to utilize a variety of forecasting methods in making the outlook function operate — from simple judgments all the way through to complex quantitative models. No one tool can be expected to meet all needs. The specifics of the forecasting question being asked, as well as the kind of information required for planning, need to dictate the appropriate approach.

   Obviously, the more scientifically credible the methods employed, the better. Nevertheless, long-range forecasting often crosses quickly into the realm of the unknown. It needs to be recognized — particularly, by scientists and technical specialists (whose continuing inputs will be critical) — that judgmental forecasts can be an important resource, as long as they are transparently stated and an effort is made to consider contrasting opinions.

4. The more open and discursive the process, the better. The scanning, issues analysis, and scenario activities making up the outlook function provide a framework for the organization to identify and discuss the forces and developments shaping its future. Ultimately, the process must stimulate critical thinking on the part of company decision makers and managers — enabling, if need be, a reshaping of the concept of what matters and what program priorities should prevail. The chances for this kind of

outcome are enhanced to the degree that the process is open-
ended and debative in form, with a high tolerance for differences of opinion and a willingness to acknowledge uncertainties.

5. Finally, as you embark on an outlook and planning effort, it is
worth recognizing that strategic plans are best viewed as milestones along the way, and not goals to be inflexibly pursued.
The process improves management insight and helps the organization become more responsive and adaptive to change. The
plan helps chart and refine a strategic vision—but is not intended to yield a rigid "masterplan."

REFERENCES

1. Numerous authors now discuss this theme. An older but still useful reference
is D. Schendel and C. Hofer. *Strategic Management: A New View of Business Policy and Planning* (Boston: Little, Brown & Company, 1979).

2. A useful and widely quoted recent discussion is P. Schwartz, *The Art of the
Long View* (New York: Doubleday/Currency, 1991).

3. The recent organizational literature notes the considerable importance of
managers' assumptions about how their business area works, derived chiefly
from their informal, mental models of experience. Getting managers to focus
on and perhaps rethink these assumptions appears to be one of the more
effective avenues of influence for long-range outlook information. See P. M.
Senge, "Mental Models," *Plann. Rev.* 20(2):4–11 (1992).

4. The elements of this approach are discussed in detail in H. S. Becker, "Sce-
narios: A Tool of Growing Importance to Policy Analysts in Government
and Industry," *Technol. Forecasting Social Change* 23:95–120 (1983); M.
Porter, "Industry Scenarios and Competitive Strategy under Uncertainty,"
Comparative Advantage: Creating and Sustaining Superior Performance
(New York: The Free Press, 1985); and I. Wilson, "Teaching Decision
Makers to Learn from Scenarios: A Blueprint for Implementation," *Plann.
Rev.*, 20(3):18–22 (1992).

5. Several excellent examples of the scenario planning approach can be found
in: P. R. Stokke, W. K. Ralston, T. A. Boyce, and I. H. Wilson. "Scenario
Planning for Norwegian Oil and Gas," *Long Range Plann.* 23(2):17–26
(1990) and Systems Planning and Research, Southern California Edison Co.
"Planning for Uncertainty: A Case Study," *Technol. Forecasting Social
Change* 33:119–148 (1988).

6. This section summarizes P. Wack's more detailed discussion of the Royal
Dutch/Shell experience in P. Wack, "Scenarios: Uncharted Waters Ahead,"
Har. Bus. Rev. Sept-Oct:73–89 (1985); and P. Wack, "Scenarios: Shooting
the Rapids," *Har. Bus. Rev.* Nov-Dec:139–150 (1985).

7. S. Partner, "Shell Pioneers Use of Scenarios to Enhance Its Long-Range
Planning," *Bus. Int.* November 12:379 (1990).

8. P. J. H. Schoemaker, and C. A. J. M. van de Heijden. "Integrating Scenarios into Strategic Planning at Royal Dutch/Shell," *Plann. Rev.* 20(3):41–46 (1992).
9. P. Wack, ibid.
10. D. M. Georgoff, and R. G. Murdick. "Manager's Guide to Forecasting," *Har. Bus. Rev.* Jan-Feb:110–120 (1986).
11. Further discussion on the steps involved can be found in D. G. Simpson, "Key Lessons for Adopting Scenario Planning in Diversified Companies," *Plann. Rev.* 20(3):10–17 (1992); M. Boroush and C. Thomas, "Alternative Scenarios for the Defense Industry after 1995," *Plann. Rev.* 20(3):24–29 (1992).
12. Several recent discussions of the Delphi process include G. Rowe, G. Wright, and F. Bolger. "Delphi: A Reevaluation of Research and Theory," *Technol. Forecasting Social Change* 39(3):235–251 (1991); T. Webler, D. Levine, H. Rakel, and O. Renn. "A Novel Approach to Reducing Uncertainty: The Group Delphi," *Technol. Forecasting Social Change* 39(3):253–263 (1991).
13. A wide range of quantitative forecasting methods and their applications in planning are usefully discussed in S. Makridakis, S. Wheelwright, and V. McGee. *Forecasting: Methods and Applications* (New York: John Wiley & Sons, 1983).
14. The general nature of this integration is illustrated in P. Wack, ibid; Stokke et al., ibid; and Southern California Edison, ibid.

CHAPTER 14

Creating Strategic Visions*

Charles W. Taylor

INTRODUCTION

Since the close of World War II, an increasing concern for the world's physical environment has developed gradually in industrialized countries. Interests tended to involve large industries that spewed smoke and automobiles that spewed spent fuel. As scientists, academicians, and bird watchers over the years began to see patterns of pollution to the natural environment and death to plant and animal life, interests broadened to include a global view.

This diverse group began to question what the future might hold for the planet and its people if destruction and contamination of earth environment were to continue unabated for the next 25 to 50 years or more. They also began to ponder how damage to the natural environment can be halted. Various means of predicting, forecasting, and scientific measuring have been used in the past several decades to estimate the damage and the harm yet to come. The U.S. Environmental Protection Agency (EPA) has considered techniques such as surveys, expert forecasts, and other estimative means of viewing the future. This chapter offers a different approach to forecasting large, holistic policy issues facing decision makers that just might be worth trying.

The approach I am presenting is a forecasting and planning model that is suitable as a standard for creating strategic visions for government, business, industry, or academia. Any subject matter—whether it is science; technology; economics; social, political, or religious ideologies; or pollution of the world's natural environment—can be accommodated by the model. The model has three attributes that essentially assure its success. First, the process is highly acceptable to chief executive officers

---

*Portions of this chapter may be quoted or represented without permission, provided that a standard source credit line is included.

(CEOs) and top managers. Second, the product of the process is plausible, that is, believable; and third, because of the assurances built into the process, the model and the final product are marketable.

ACCEPTABILITY

The process achieves acceptability because of the logical ways it brings a variety of players together to create visions of the future. These players are futurists, scenario writers, experts, and planners.

Futurists and scenario writers work together. They meet and discuss past, present, and future problem subject areas with the other players in the process. Futurists, external to the organization, provide broad and relevant alternative forecasts. These are in the form of narrative scenarios for planning at the strategic level. Scenario writers, also external to the organization, provide consistency within each scenario and a harmony between alternative scenarios.

Experts, internal to the organization, assure data and contextual accuracy—as well as plausibility—during the process. An interdisciplinary or special mix of experts is best because of the holistic nature of the world environment. Intraorganizational planners (i.e., also internal to the organization but from different branches or departments) provide relevant trends and milestones for scenario development. Additionally, the planner reveals the real problem areas associated with the projection of trends into the future. What builds the acceptability into the process is that CEOs and top managers can be brought into the process at any step to observe its logic, ask pertinent questions, offer suggestions, and in general develop a sense of belonging as well as an ownership of the forecasts and scenarios. Under the leadership and direction of a futurist, this diverse group brings the power of visioning and creativity together to construct strategic visions, step-by-step, through the process into the future.

PLAUSIBILITY

The process and the strategic scenarios produced by it assume plausibility through the use of a theoretical cone, called *the cone of plausibility*.[1,2] The object of the cone in the process is to serve as an enclosure that circumscribes the thought processes of the players. Scenarios projected within the cone are considered plausible if they adhere to a logical progression of trends, events, and consequences from today to a predetermined time in the future.

Within the cone, experts, planners, and futurists track pertinent trends from the past to the present and into the future in a systematic and logical progression. This maintains plausibility and further increases the acceptance of the scenarios. The players can produce one planning scenario or, preferably, alternative scenarios within the cone simultaneously and incrementally.

The evolving scenarios become increasingly believable or plausible to CEOs and top managers who are invited, periodically, to observe the process of the cone in operation. Their participation in this manner keeps them in touch with future realities. Since the players can display a snapshot of their point of progress in the future, called a planning or forecast focus plane, the finished scenarios bring no surprises or future shock to the CEOs or managers.

MARKETABILITY

The cone of plausibility provides end products, or scenarios, that are marketable because they have been generated through the systematic and logical processes of the cone. As a process in itself, the cone is highly marketable to CEOs because of the enthusiasm it creates in their players, the experts, and the planners.

The cone and its processes offer a means to standardize the methodologies of visioning the future. There is also the possibility that because of the rigors of the processes within the cone, even if not as exacting as those of mathematics, operations research, or systems analysis, the cone brings long-range forecasting and planning closer to a scientific approach. The cone also tends to stimulate goal setting, solution finding, and creativity in the players, as well as to uncover new challenges. Of equal importance, the cone creates a two-way communication between the players and management. These attributes increase the confidence CEOs and managers have in the cone, the process, the players, and the end product scenarios.

CREATING THE STRAGEGIC VISIONS OR SCENARIOS

The players must take several decision steps as they begin to create their strategic visions. For the purposes of this chapter, strategic visions are intuitive, holistic views of plausible realities and futures used for planning.

First, the players must decide how many scenarios they are going to create. There are galaxies of scenarios that can be created for almost any subject matter. The human brain, however, cannot analyze or process the vast amount of data generated by large numbers of scenarios. Computers and the appropriate software are the tools for handling these situations, provided all data can be expressed quantitatively and software programs are available. Much of the data in the social sciences, however, cannot be converted readily to mathematical form; thus the number of scenarios must be only a few, i.e., what the brain can handle.

Experience in generating scenarios has convinced the author of the following: one scenario is predictive. No one can predict the future accurately except by chance. Two scenarios, usually, are best case and worst case futures. Three scenarios almost always provide a middle of the road scenario between the best and the worst. Five scenarios or more tend to become increasingly overwhelming in data and cumbersome to manage. Moreover, their number encourages ranking, e.g., preferred, least likely, or most probable. Any such ranking tends to be predictive of the future. Four scenarios, however, are manageable and allow considerable flexibility in the number of relevant variables. The use of multiple or alternative scenarios tends to improve forecasting accuracy. Four scenarios are the choice number of scenarios to process through the cone of plausibility.

The second and third decision steps for the players are to determine — by consensus — the ten most important elements that influence their forecast or planning subject matter and then determine their rank order of importance to the subject matter. These decisions are used by the futurists to create microscenarios.

Microscenarios are made up of the first four ranked elements determined in the steps above. They consist of four short statements that reflect positive trend attitudes of the leading influencing elements and four that reflect opposing attitudes. After a permutation and sorting of the eight statements, the order of the four final statements in each microscenario set is established at random. The first trend statement of each set becomes the driver trend and the dominant theme for that scenario. The futurists now have prepared four alternative sets of scenarios, each with four trend statements and a dominant theme. These strategic visions can now be processed through the cone of plausibility.

The fourth decision step the players make involves the workshop agenda they will follow as they project the subject matter elements and identify related element problems in the future. Workshops are about 3 days in length and 5 weeks apart. The participants include largely subject matter experts in four groups, each with planners as facilitators and futurists as motivators.

Based on the microscenarios, the experts create visions of the future for each of the scenarios. The planners lead each group of experts, maintain the peace, and record the group progress. After each workshop, the planners in session create responses to the experts' visions. The planners then set new or modified goals to the experts' scenario projections. The futurists constantly urge the experts and planners to project their thoughts into the future.

After each expert workshop and planners' response session, the futurists analyze the projections and responses and compare all relevant data with their notions of the future and those found in futurist literature. Along with the scenario writers, the futurists expand each microscenario to a miniscenario of about 500 words in length. The expanded scenarios provide the basis for the next workshops. The author's experience has shown that often scenarios based on the future expectations of experts contain surprises that the experts themselves did not anticipate. These surprises may be included in the scenarios by the futurists. This procedure continues for the remainder of the workshops. By the completion of the workshops, each scenario may have expanded to fivefold or more to become a macroscenario. These scenarios describe the strategic visions of a selected group of experts, planners, and futurists.

THE ANATOMY OF THE CONE

The cone of plausibility is a theoretical process that can be used by one person or a group of people to project trends and events and their consequences holistically into the future. It is especially suitable for generating alternative scenarios at predetermined points in time. The generic cone is representative of the thought processes used to create strategic visions of the future and is depicted in Figure 1.

The generic cone of plausibility encompasses theoretical projections of four strategic visions or planning scenarios; they are scenarios A, B, C, and D. Each example scenario has a dominant or driver identifier; they are technological, political, economic, and sociological. These identifiers also represent the dominant theme that characterizes each scenario. Each of these themes will command a different vision or scenario of the future.

The trends within each theme are not straight-line projections. There are interactions among trends where dominant trends alter the attitude of less dominant trends or result in discontinuities of others. The probability of the strength of a trend or its continuing influence in its scenario can be determined, as can similar trends in the other scenarios.

Outside of the cone are wild card scenarios, which—if they occur—overwhelm most other visions or scenarios. These disruptive, aberrant, anomalous, or catastrophic scenarios can dominate almost all other trends and events for a period of time—a time long enough to redirect or destroy any interaction or mutual support existing within and among the planning scenarios. The examples shown with the generic cone—a worldwide depression, a major natural disaster, a major war, and a democracy in the U.S.S.R. (now referred to as the former Soviet Union)—are repre-

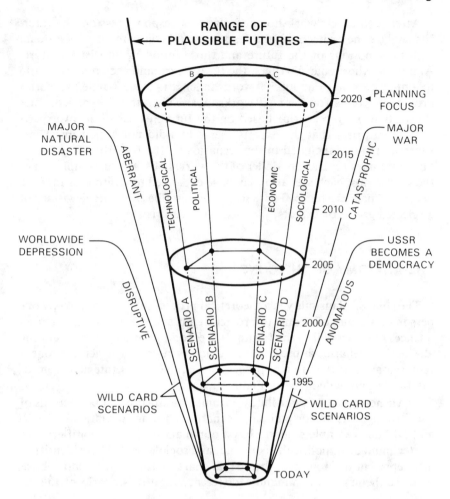

FIGURE 1. The generic cone of plausibility.

sentative of wild card events. The influence such events can exert can be observed by studying the past.

As mentioned earlier, experts, planners, and futurists can track trends from the past to the present and into the future in a systematic and logical way through the use of the cone of plausibility. The influence trends have in the natural or societal environment in which they exist can be observed along a continuum from their origin, or from some point in time in the past to the present. Theoretically, within the logic of the cone, the responses to and the consequences of trends and events at any selected focus plane—even if in the past—can be reconstructed to create plausible scenarios or visions of that environment.

This perspective of creating historic visions is depicted in the drawing of a double cone in Figure 2. Use of the cone for historic research and analysis is an additional feature of this visioning tool. Moreover, this historic linkage adds to the acceptability, plausibility, and marketability of the cone of plausibility. Continued use of the cone builds the mental discipline and conditioning needed to pursue its logic backward or forward in time.

CONCLUSIONS

I have presented **the cone of plausibility** in this chapter as a new method of creating strategic visions or scenarios. The cone is a new technique that can be used by long-range planners and futurists in government, business, and industry to achieve a uniformity of process and interrelated forecasting of most any subject matter. Standardizing the

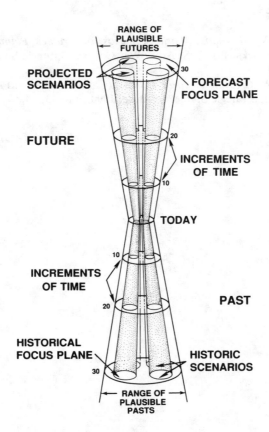

FIGURE 2. The cone of plausibility: past and future.

process of planning and forecasting would be the first step toward developing a standard set of long-range alternative strategic scenarios or visions for organizational planning. The creation and the use of a standardized set of alternative scenarios would improve the accuracy of institutional foresight in the long term, which likely would result in better decisions and policies.

DISCLAIMER

The views expressed in this chapter are those of the author and do not reflect the official policy or position of the Department of the Army, Department of Defense, or the U.S. government.

REFERENCES

1. Taylor, C. W. *Alternative World Scenarios for Strategic Planning.* Rev. ed. (Carlisle Barracks, PA: U.S. Army War College, Strategic Studies Institute, 1990), pp. 1–14.
2. Taylor, C. W. *Creating Strategic Visions.* (Carlisle Barracks, PA: U.S. Army War College, Strategic Studies Institute, 1990), pp. 13, 14.

CHAPTER 15

Exploring Future Environmental Risks

David W. Rejeski

INTRODUCTION

With the exception of a few primitive cultures without time perspectives, a curiosity about the future has infected all major civilizations. Each culture has had its seers, prophets, shamans, and oracles, and virtually all cultures have developed images of the future (both utopian and dystopian) as mechanisms of self-renewal and self-correction. Plagued by uncertainty and driven by curiosity, human beings have been particular inventive in their search for predictive tools, using such approaches as hepatoscopy (the inspection of animal livers), cleromancy (divination by lot), lithology (divination by throwing stones), astrology, and visiting oracles.[1] Mesopotamian priests, long before the time of Christ, regularly evaluated the possible impacts of proposed technological projects. The Judeo-Christian tradition of prophecy was built on the conditional if-then exploration of future actions. Finally, the development and explorations of alternative futures are within the broad tradition of scientific objectivity described by philosopher Israel Scheffler as requiring the "possibility of intelligible debate over the merits of rival paradigms."[2]

Given the historically demonstrable demand for foresight, one can question where our present-day search for oracles will lead. Has science taken over the social purpose of religion and myth in providing predictive guidance? If so, few scientists would readily admit to their new role or do they desire it—fearing a blurring of the distinction between Newton and Nostradamus. Yet science is inherently predictive, seeking to help mankind avoid mistakes by guiding actions through simulation and anticipation.

Though the science of probabilistic risk assessment implicitly deals with the future by assigning probabilities to some future event (a nuclear power plant failure, the chance of contracting cancer, etc.), it is often

267

poorly understood by the public or policymakers who have difficulties in evaluating low-chance events. This may be due, in part, to crucial differences in the type of rationality applied by scientists and laypeople in evaluating risky situations or problems in communicating risks between different societal subcultures.[3]

Over the past decade, a semiquantitative approach to ranking risks — often referred to as comparative risk — has gained popularity and credence. The attraction of comparative risk may well be that it mirrors the lay logic of relativizing a number of potential threats as a guide for action. Though rankings by scientists and the public may diverge, they do provide a basis for a dialogue concerning threats and policy alternatives.

One of the earliest attempts to compare risks was undertaken in 1980 by Dr. Gordon Goodman of the Stockholm Environmental Institute. Goodman evaluated a total of 29 environmental risks using six different criteria. This work was followed by an ambitious study by the Center for Environment, Technology and Development (CENTED) at Clark University which focused on over 90 technological hazards as diverse as automobile crashes, noise from supersonic aircraft, and pesticides.

In 1987, the Environmental Protection Agency (EPA) undertook an internal study to rank 31 major environmental problems for which the EPA had statutory responsibility.[4] This ranking was subsequently reviewed and evaluated by an external group of scientists who further developed the methodology for comparing environmental problems in terms of the human health, ecological, and welfare risks facing the EPA and other government agencies.[5] The methodological approaches have advanced significantly and been applied in virtually all regional areas of the United States, some states, and overseas.[6] In addition to these studies, there are more qualitative, anecdotal risk rankings which have gained credence. One thinks, for instance, of Wilson's[7] four horsemen of the environmental apocalypse: toxics, stratospheric ozone depletion, global warming, and finally death — the unprecedented extinction of species.

In examining these comparative risk approaches one finds that considerations of the future are conspicuously absent. Most emphasize a "snapshot" approach where risks are evaluated at a particular point in time, leaving future projections up to the imagination of the reader. An exception has been work by Norberg-Bohm and colleagues[8] at Harvard which takes an explicit, cross-national look at future risks along the dimensions of future human health consequences, future ecosystem consequences, and magnitude of future consequences.

The work done to date on comparative risk has surmounted enormous conceptual and methodological barriers and truly contributed to increasing the utility of risk-based approaches in policy-making. The work, however, does not fully capitalize on the potential of social imagination.

If such studies are to impact strategic planning and long-term policy-making, they must deal far more explicitly with the future. This means going beyond a simple extrapolation of past risks to incorporate explicit, normative visions of the future. There are a number of reasons why vision will become increasingly demanded and must be included in comparative risk exercises.

WHY VISION?

Risk, Risk Aversion, and the Circle of Blame

It is well known that most people are more risk averse when expected outcomes are negative, yet our entire risk assessment apparatus continually focuses human attention on deleterious impacts to human health and our environment. Barraged daily by media reports of morbidity, mortality, loss of valuable habitats, extinction of species, and newly discovered toxic threats—it is easy to envision a growing "culture of risk aversion," where social imagination is replaced by litigious obsessions with blame and reparation or apathetic withdrawal. Unfortunately, our society is openly adversarial in its dealings with human misfortune and often seeks out an "enemy" for punishment and compensation.[9] Environmental policymakers pay a price for contributing to, and intensifying, a risk-focus in our culture. A risk-averse populace is far less likely to grant public institutions the flexibility and experimental, creative freedom to explore nonregulatory or voluntary approaches to environmental management. The harbinger of the message becomes the message itself.

What if, instead, we estimated the risks associated with positive outcomes? Can we alter the risk analysis process to more fully capitalize on human imagination—creating visions of the worlds we could inhabit and then examining the environmental risks *avoided* by moving down certain paths of social, technological, and economic development? In taking this approach, we are forced to address fundamental questions of human existence such as the way we will live, grow our food, produce our energy, move about, work, process our wastes, etc. We would also provide a fundamental framework for effective strategic planning, using visions to explore and develop long-range objectives, goals, measures, and appropriate strategies and policies.

Philosopher Sagoff[10] recently commented, "Environmentalism has traveled a long road since the 1970s . . . in moving from a preoccupation with technological threats to personal safety and health to a larger concern with the sustainability of ways of life." The recent United Nations Conference on Environment and Development in Rio has underscored the belief, held by growing numbers of people worldwide, that environmental policy debates in the future will be built on the vocabulary and framework of sustainable development. Such debates must go beyond risk to address fundamental decisions concerning global, national, regional, and local paths for social and economic development — in short — "ways of life" questions.

Eroding Scientific Legitimacy and the Decline of Deference

Since the end of World War II, we have witnessed a significant decline in the absolute legitimacy of science. Brooks[11] has commented, "Scientists today are listened to much more but believed much less than they were in those heady days . . . the image of objective, 'value-free' science and scholarship is severely tarnished." The suspicion among policymakers of scientists was echoed even earlier by Winston Churchill, who once remarked, "Scientists should be on tap, not on top." Increasingly, the scientific community itself is raising the fundamental issue of whether we are asking questions of science which cannot be answered by science.[12]

Viewed historically over the past four decades, the unquestioned legitimacy of science has given way to civil legitimacy based on negotiated agreements. The intrusion of the public into scientific debates has eroded the very meaning of science as a consensual, peer-oriented activity within a closed subculture and has surfaced the contradictions between science and democracy. This trend has been accelerated by what some social scientists have termed a "decline in deference," a decreased willingness on the part of the public to defer decisions to scientists and public institutions.[13] Nowhere has this phenomenon been clearer than in the area of risk assessment, where increases in scientific understanding have been greeted by growing skepticism and resistance on the part of the public. Significant investments in risk communication, though helpful and needed, have not (and will not) overcome what is in essence a cultural struggle over the power, legitimacy, authority, trust, and rights of individuals.

Recognizing this dilemma, some countries like Canada have adapted more participative, consultory processes (round tables) aimed at addressing issues and questions which transcend science and require the reconciliation of individual, community, and societal goals. In the United States, the *Globescope Americas* project has a similar participative focus — to bring together a wide range of stakeholders and make sustainable development the foundation of decision making in the United States by the end of the decade.[14]

In future policy-making forums, scientists will be one part of a wider decision making community where legitimacy will be constantly questioned and vision demanded. At a global scale, the crumbling of nation states, the retreating convictions and institutions of the cold war era, and the blurring of distinctions between sacrosanct policy areas will all increase the demand for greater democratization in decision making; these will throw into question the future role of science. Former Czech President, Havel,[15] has noted that we all may be forced to ". . . abandon the arrogant belief that the world is merely a puzzle to be solved, a machine with instructions for use waiting to be discovered, a body of information to be fed into a computer in the hope that, sooner or later, it will spit out a universal solution." In the final analysis, organizations which are heavily dependent on science and science-based analysis for their legitimacy may have to reexamine their methods and mission in the context of existing and emerging democratic structures.

The Language of Community vs the Language of Anonymity

The growing popularity of the risk paradigm has been linked to the evolution of a global society and specifically with the ability of globalization to draw people out of small local communities into larger regional, national, and international spheres. For this to occur, we require a context-free and anonymous symbol system which can support the growth and spread of undifferentiated, large-scale, culturally homogeneous communities.[16] Risk as an abstract, reductionist, scientific symbol set may be as close as we can come to realizing the dream of a universal world language. Risk combined with economic quantification (risk/benefit analysis) may be the ultimate tool for the abstract transcendence of local culture.

The use of the risk language becomes tenuous, however, if the *culture state* takes precedence over the *nation state* or if the local community is viewed as an important locus for the mobilization and implementation of environmental change. If the role of environmental organizations becomes one of *improving the lot of real people living in real places*, then the language of environmental policy can no longer be one of anonymity, homogeneity, and amnesia.[17] We will need a language which is rich in history, context, uniqueness, and place — a new language of environment and community focused on common aspirations and social cohesion. The inability of risk to adequately address issues of environmental justice raised by local, grassroots groups is an acute reminder of our increasing language deficiencies.

Growing Demand for Vision

A recent poll taken by the *Washington Post* and ABC News produced some interesting results. Tied with concerns about pollution and environmental problems was the perception that visions were missing and the long-term needs of the country were not being addressed.[18] Some people have maintained that we need in "fresh face," a new set of strategies and visions for an emerging global era.[19]

There are a number of possible sociocultural explanations for this lack of long-term vision, including recent observations that we are ending a crucial phase in history; are crossing over to another; or are captives of a contented culture which values short-term public inaction (regardless of the seriousness of the long-term consequences) over vision, anticipation, and prevention.[20-22] This is not the first time that symptoms of cultural myopia have been diagnosed, or is it the first time explanations have been rendered. The Dutch social historian, Polack,[23] noted in a classic text that, ". . . for the first time in 3000 years of Western civilization there has been a massive loss in capacity, or even will, for renewal of images of the future." Galtung,[24] in his 10-nation comparative study on future images, concluded ". . . that the tendency to think, or at least to express thoughts, about the future does not seem well developed, it *is mainly located in the direction of technological futures and war/peace problems, not in the direction of social futures.*"

Cultural observers such as Polack and Galtung also discovered something potentially more significant than the decline of future consciousness—the virtual disappearance of utopian thinking as a valid and valued social enterprise. History tells us this may be a profound loss. Again and again, we have been reminded of the role of utopian mentality in preparing blueprints for the future and the importance of utopias as tools of social critique and analysis. This point was clearly made by Mumford[25] when he noted, "A map of the world that does not include Utopia is not even worth glancing at." By avoiding the development of future-oriented, reality-transcending ideas we fundamentally diminish our capacity for effective strategy formation and societal transformation.

All this would not be as disconcerting or alarming if it were not for the truth in Boulding's[44] observation that images of the future are the keys to choice-oriented behavior. Our inability, or unwillingness, to construct such images fundamentally undermines our capacity as an evolving species to anticipate and adapt to change. The Club of Rome has again challenged policy-making organizations to ". . . visualize the sort of world we would like to live in, to evaluate the resources—material, human and moral—to make our vision realistic and sustainable, and then to mobilize the human energy and political will to forge the new global society."[26] A recent study by the World Resources Institute and U.S. Environmental Protection Agency[27] has highlighted the urgent need to ". . . create a national vision of an environmentally sustainable future." Will we respond to these challenges?

WHAT KIND OF VISION?

To meet such challenges we must begin to create positive visions of the future, not just new and increasingly sophisticated risk analysis methods. Science may support this quest for new images of the future, but it will not supply the answers. Central to this challenge is understanding what we mean by vision. To explore this concept, we will use a very broad definition of the term adapted by Sowell[28] to refer to those ideas or beliefs which "shape thought or action."

Preanalytical Cognitive Act

The first, and possibly the most valuable, form of vision is what has been termed a "preanalytical cognitive act."[29] It is precisely this type of vision which can jump the boundaries of the present, the constraints of the past, and the shackles of our cultural and social dogmas. This vision eschews analysis and revels in the gestalt switch. Swift once described this type of vision as, "the ability to see things invisible." Kuhn speaks of the "flashes of intuition through which a new paradigm is born." Psychological research on the nature of insight points to a variety of precursors and antecedents for this type of vision including: tolerance for ambiguity, ability to think across disciplines, playfulness, curiosity, and intrinsic and extrinsic motivation.[30] Unfortunately, most of these qualities are undervalued or repressed in large organizations, including our schools and universities. The facilitation of such vision is unlikely to occur without a fundamental reexamination of our ideas concerning what constitutes intellectual, organizational, and social achievement.

Unarticulated Consensual Pact

There is another type of vision, far more prevalent and often insidious, which impacts the everyday behavior of society. This vision is often disguised by covert rules which govern organizational behavior and is embedded in the unvoiced cultural agreements about the nature of causation, social progress, social decline, and constraints. Such belief systems affect human behavior, scientific inquiry, and problem solving without being articulated and without decision makers being aware of their existence.[31] They function precisely because they go unchallenged and unvoiced, seemingly immune to criticism, shaping our thoughts and actions. It is often only through insight and renunciation that people, organizations, or cultures free themselves from the grip of such visions, as these examples illustrate:

> *My party education had equipped my mind with such elaborate shock-absorbing buffers and elastic defenses that everything seen*

and heard became automatically transformed to fit a preconceived pattern (Arthur Koestler).[32]

> *The system of theories which Freud has gradually developed is so consistent that when one is once entrenched in them it is difficult to make observations unbiased by his way of thinking* (Karen Horney).[33]

In the final analysis, it may be that such grand political or intellectual visions are the most flawed, the most prone to eventual collapse, and the most internally destructive. The inability of institutions to extricate themselves from the grips of such visions is often due to an overdependence on an institutional learning paradigm which some have described as "single-loop learning" where organizational members focus energy on the detection and corrections of errors so as *to maintain the central features* of the organizational belief system.[34]

Constructed Consensus

The third and final form of vision, which we can call a "constructed consensus" or "intentional shared vision," involves a socially mediated attempt to challenge the underlying assumptions driving our everyday thinking and create viable alternatives which are truly embraced by all members of a group. Work done with imaging techniques indicates that individuals can create visions of a world which are radically different than the one they know — worlds which are more harmonious, more egalitarian, less stressful, environmentally sound, and peaceful.[35]

Shared visions emerge from individual visions, from which they derive their power and capacity to attract commitment. Organizations or societies which value vision will encourage members to develop and share personal views of the future. Such organizations will also encourage decision makers to expose, communicate, test, and modify their mental models of the present and visions of the future — allowing an ongoing examination of strategies, objectives, and policies. It is important to point out that this type of vision rarely emerges from the boardroom, the management retreat, or the annual planning ritual which inevitably use the past as a starting point for incremental, reactive thinking about the future.

For shared visions to succeed, people *must believe* that they can shape the future. Unfortunately, such beliefs are seldom reinforced in large institutions where a present-centered, crisis mentality combines with nonsystemic thinking to limit both the existence and breadth of vision. As organizational learning expert Senge[36] has pointed out, ". . . the discipline of building shared vision lacks a critical underpinning if practiced without systems thinking. Vision paints the picture of what we want to create. Systems thinking reveals how we have created what we currently have."

The differences between these three types of vision can be illustrated using a parable told by Comenius[37] concerning the pilgrimage of a priest. At the beginning of his journey, the priest is fitted with glasses which make all relations and events appear as they are purported to be by the representatives of the age (Vision 2). Later, as the journey progresses, the priest manages to peer around the glasses and realizes that other interpretations of the world are possible and desirable (Vision 3). Finally, at the end of the journey, he is fitted with a new set of glasses which allow him to see other worlds, utopian worlds, completely different from anything he had imagined (Vision 1).

RISK AND VISION

The task before us is much like that of the priest, to peer around the glasses or don new glasses — to use vision as a way of (re)organizing our knowledge, values, and social policies. Let us then imagine a comparative risk project driven by vision. The process outlined is tentative and is presented as a challenge rather than a solution. As a sketch, it can only begin to draw on the experiences and literature at our disposal. Four stages are envisioned, which are briefly described below and in Figures 1–4.

Create a Number of Images/Visions of the Future

When one begins this exercise, it may be advisable to avoid a single vision. There are political and intellectual reasons for avoiding a normative approach with a single endpoint: (1) reaching consensus on one desirable future among many stakeholders in a pluralistic society is often impossible, or at least, extremely time-consuming; (2) the existence of a single endpoint makes the exercise vulnerable to attack; and (3) single endpoint definition is difficult — tightly defined endpoints often limit creative thinking concerning alternative development paths, while endpoints which are too open are likely to generate too many alternatives.[38]

Realizing that there is no single "right" vision, an attempt should be made to develop, balance, and clearly articulate visions along various dimensions such as: optimistic, pessimistic, business-as-usual, and romantic/utopian. Achieving such a balance requires considerable attention to the structure of the visioning exercise and its participants. A primary requirement is the dedication of a true multidisciplinary group to the project. Securing broad public participation is another key and

CREATE IMAGES/VISIONS OF THE FUTURE (SCENARIOS)

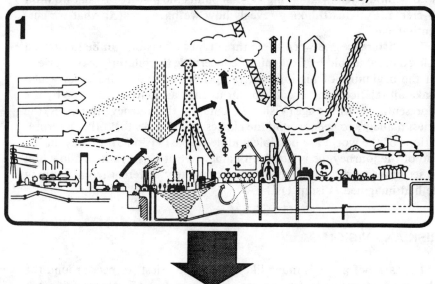

RELATE IMAGES TO RISKS

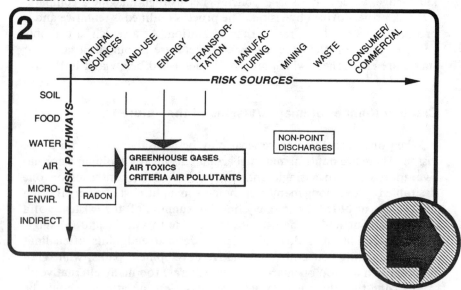

FIGURE 1. Create images/visions of the future (scenarios).
FIGURE 2. Relate images to risks.

CHARACTERIZE SCENARIOS USING RISK SOURCES

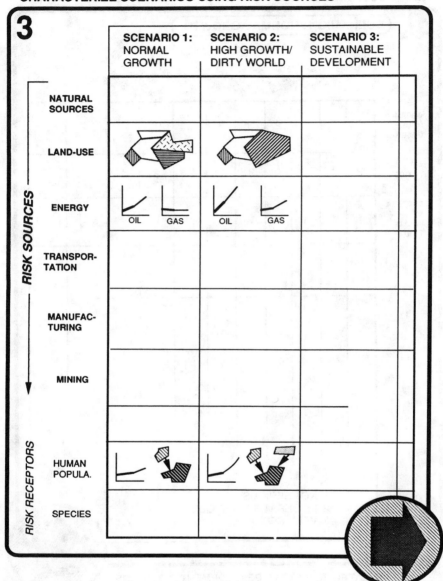

FIGURE 3. Characterize scenarios using risk sources.

RANK SCENARIOS

4a ⟨ **SCENARIO 1: NORMAL GROWTH** ⟩

	INTENSITY	EXTENT	EARLY WARNING	UNCERTAINTY	AVAILABLE TECHNOLOGIES	COSTS	SOCIAL ACCEPTABILITY	ENVIRONMENTAL RESPONSE TIME	SOCIAL ADAPTABILITY		RANK
GREEN-HOUSE GASES	3	3	2	2	3	3	2	3	2		
AIR TOXICS	2	1	1	2	3	3	2	3	3		
CRITERIA AIR POLLU-TANTS	1	1	1	1							

1 NOT SERIOUS
2 MODERATELY SERIOUS
3 VERY SERIOUS

SIMILAR MATRICES FOR ALL DEVELOPMENT PATHS ARE CONSTRUCTED AND COMPARED

FIGURE 4. Rank scenarios.

may require going beyond face-to-face meetings and exploiting information technologies to help support the creation and acceptance of specific visions.[39]

The beginning foundations for such studies may be past visioning exercises which can be reexamined, updated, and expanded. At this point in time, 29 states within the United States have completed such studies, along with a number of cities and regions.[40] The usefulness of such past studies may vary significantly depending on the extent of participation in their creation, relevance to existing social and political conditions, and degree of balance. Some states — such as Hawaii, Minnesota, Washington, Colorado, and Maine — have completed long-term visioning exercises *and* are presently implementing comparative risk projects, offering some unique opportunities to blend risk and vision.

It is crucial that these visions must be carefully crafted and presented. The ability of stories and parables to infect the human imagination and bring about environmental action has been clearly illustrated by such works as Rachel Carson's *Silent Spring* or Garrett Hardin's "The Tragedy of the Commons." We must become storytellers to produce and communicate socially contagious visions.[41] As storytellers, we need a language which is rich in metaphor and analogy (and devoid of jargon) to communicate intelligible visions which invite comment, participation, and political commitment. The stories which we link to these visions should clearly articulate the crucial events and decisions which occurred during their unfolding, becoming histories of the future which reach back and connect to our shared past.

Relate Images to Risks

The next step, which is conceptual rather than operational, is realizing that each vision is linked to future risks through a causal network of risk sources and exposure pathways. The source-pathway-problem relationships can be shown in a two-dimensional matrix (see Figure 2).[42] The vertical axis shows routes of exposure. The horizontal axis lists the sources of chemical and other pollutants which are transported by various processes to biological receptors. Most environmental problems, the typical targets of environmental concern and policy, can be found at the intersection of these two axes.

Interestingly, the risk sources shown on the horizontal axis correspond to many of our primary economic activities. By changing socioeconomic development paths (i.e., adapting new energy policies, transportation approaches, industrial processes, etc.), we will alter the nature of the risks emanating from these sources. By understanding what futures are likely or desirable, the characteristics of the risk sources can be explored. More importantly, this exploration can take place using a language which addresses important questions concerning needs for food, energy, transportation, shelter, commercial products, natural resources, and land.

Characterize Visions Using Risk Sources

Understanding the potential causal linkages between alternative development paths and risk allows a more precise characterization of the risk sources to take place. This characterization may be conjectural (developed by informed groups and educated guesses), or it may draw on quantitative projections which already exist concerning potential energy consumption, land-use patterns, solid waste generation, resource extraction levels, probabilities of accidental releases, etc. In some cases, past visioning exercises may have defined critical junctures and shifts in socioeconomic development without necessarily examining the associated environmental implications ("hard" vs "soft" energy paths, low- vs high-impact agriculture, autobased vs mass transit, etc.).

During this step, one can also add information on receptor populations. For instance, how are human populations expected to grow, stratify, age, and migrate during the time period under consideration or how might the numbers and distribution of vulnerable species change? Throughout this stage, the goal is not quantitative precision, but the creation of context and common understanding for further comparison of the alternative endpoints. This includes the articulation of assumptions underlying development paths, and a clarification of uncertainties and cross-impacts between different future events and decisions.

Rank Alternatives

The next step involves a ranking of the alternative development paths and utilizes a methodology developed at the International Institute for Applied Systems Analysis (IIASA) in Austria for the European Community.[43] During this step, the impacts of our hypothesized futures on various environmental problem areas are compared using an ordinal ranking system.

The ranking factors used are: problem intensity (severity per unit of human or sensitive biotic population), extent (geographic scale), early warning (can the problem be detected with enough lead time for social responses to be effective?), uncertainty, available technologies (can existing technologies solve or ameliorate the problem?), cost (how expensive is it to manage the risk to tolerable levels?), social/political acceptability (are there social or political barriers to dealing with the problem?), environmental response time (if we act, how long will it take the environment to recover?), and social adaptability (if the problem cannot be forestalled or cured, are we willing to adapt?). Many of these factors, such as the availability of early warning indicators and technologies, are directly related to the preventability of future problems and can be used to identify the pollution prevention potential implicit in different development paths.

By making avoided risks and potential trade-offs explicit, this exercise can lead participants to converge on a desired endpoint (if this has not already occurred). A number of other useful exercises might then be undertaken, including an analysis of barriers (what keeps us from reaching specific endpoints?), an examination of levers (how can we facilitate and accelerate change in certain directions?), a discussion of institutional and social challenges (what can we do?), and an examination of the social values which underlie a particular development path. This exercise can also underpin the development of an implementation plan (with cost estimates) focused on necessary short-term steps down the longer path. Again, the goal is to encourage participation, visionary speculation, and reflection within a framework which can conceptually link future choices with risks.

POSTSCRIPT

The process described in this chapter is not science but a blending of art and science, commitment, and conjecture. The French poet Jean Girardoux once made the point that the elite often watch catastrophes from their balconies. Risk analysis has often operated from the balcony of erudition, engaging in vacuous monologues with little understood publics who are surrounded by real and imagined threats. Futurists, for their part, have often been as prone to the lapses of irrelevance and the tyranny of pundits. It may now be time to abandon lofty positions and engage in direct discourse over possibilities for the future—to seek and create new visions.

REFERENCES:

1. David, F. N. *Games, Gods and Gambling* (New York: Hafner, 1962); and Lewinsohn, R. *Science, Prophecy and Prediction* (New York: Harper, 1961).
2. Scheffler, I. "Discussion: Vision and Revolution: A Postscript on Kuhn," *Philos. Sci.* 39(3) (1972).
3. One can differentiate between the types of rationality which guide our assessment of risks: the "technical rationality" of the scientists, or the "cultural rationality" displayed by the public which draws on traditional peer groups and shared social perceptions rather than expert opinion. See Krimsky, S. and Plough, A. *Environmental Hazards: Communicating Risks as a Social Process* (Dover, MA: Auburn House, 1988).

4. "Unfinished Business: A Comparative Assessment of Environmental Problems," U.S. EPA Office of Policy, Planning and Evaluation (February 1987).
5. "Reducing Risk: Setting Strategies for Environmental Protection," U.S. EPA Science Advisory Board Report SAB-EC-90–021 (September 1990).
6. All of the EPA 10 regional offices have completed comparative risk projects, as well as the states of: Washington, Colorado, Louisiana, Vermont, and Michigan. In addition, comparative risk methodologies are finding increased use outside of the United States. See, for instance, "Ranking Environmental Health Risks in Bangkok, Thailand," U.S. EPA and U.S. AID Working Paper (December 1990); or Sebastian, Iona et al. "Urban Industrial Pollution Management with Community Participation: The Case of Eastern Europe," Proposal for the World Bank (April 1991).
7. Wilson, E. O. "Biodiversity, Prosperity, and Value," in *Ecology, Economics, Ethics: The Broken Circle,* Bormann, F. and Kellert S., Eds. (New Haven, CT; Yale University Press, 1991).
8. Norberg-Bohm, V., Clark, W., et. al. "International Comparisons of Environmental Hazards: Development and Evaluation of a Method for Linking Environmental Data with the Strategic Debate on Management Priorities (Draft 3.3)," J. F. Kennedy School of Government, Harvard University, unpublished report (1992).
9. An overview of cultural blaming patterns can be found in: Douglas, Mary "Risk and Blame," Talk at the London School of Economics, February 21, 1990.
10. Sagoff, M. "The Great Environmental Awakening," *Am. Prospect* 9:39–47 (Spring, 1992).
11. Brooks, H. "Expertise and Politics: Problems and Tensions," in *Proc. Am. Philos. Soc.* 119:257–261 (1975).
12. Alvin Weinberg has coined the term "trans-scientific" to describe questions which are questions of fact, i.e., they can be stated in the language of science, but are unanswerable by science; they transcend science. See Weinberg, A. M. "Science and Trans-Science," *Minerva* 10:209–222 (1972).
13. Laird, F. "The Decline of Deference: The Political Context of Risk Communication," *Risk Anal.* 9:543–550 (1989).
14. Global Tomorrow Coalition *Globescope Americas: Charting a Sustainable Future* (Washington, DC: Global Tomorrow Coalition, 1992).
15. Havel, V. "The End of the Modern Era" *New York Times*, March 1, 1992.
16. This argument has been made by Mary Douglas, building largely on Ernest Gellner's work on nationalism and the symbolic needs of nation states. See Gellner, E. "Nationalism and the Two Forms of Cohesion in Complex Societies," in *The Proceedings of the British Academy*, Vol. LXVII (London: Oxford University Press, 1982).
17. It has been suggested that the domain of "quality of life" provides the proper strategic focus for the Environmental Protection Agency. If that is the case, then integrated environmental policy in, and for, the community will become increasingly important. See Landy, M. et. al. *The Environmental Protection Agency: Asking the Wrong Questions* (NY: Oxford University Press, 1990).
18. *Washington Post*, November 4, 1991, p. A12.
19. Gelb, L. "Fresh Face," *New York Times Mag.* December 8, 1991.

20. Fukuyama, F. *The End of History and the Last Man* (New York: Free Press, 1992).
21. Borgmann, A. *Crossing the Postmodern Divide* (Chicago, IL: The University of Chicago Press, 1992).
22. Galbraith, J. K. *The Culture of Contentment* (Boston, MA: Houghton Mifflin, Co,. 1992).
23. Polak, F. *The Image of the Future* (Amsterdam, Holland: Elsevier Scientific Publishing Co., 1973).
24. Galtung, J. et al. *Images of the World in the Year 2000: A Comparative Ten Nation Study* (Atlantic Highlands, NJ: Humanities Press).
25. Mumford, L. *The Story of Utopias* (NY: Boni & Liverlight, 1922).
26. King, A. and Schneider, B. *The First Global Revolution: A Report by the Council of the Club of Rome* (NY: Pantheon Books, 1991).
27. "Challenges Ahead for the U.S. Environmental Protection Agency in the 21st Century," World Resources Institute & U.S. Environmental Protection Agency, unpublished report, December 1, 1992.
28. Sowell, T. *A Conflict of Visions: Ideological Origins of Political Struggles* (NY: Quill, 1987).
29. Russell, B. *Skeptical Essays* (New York: W.W. Norton & Co. 1938).
30. Hein, P. "Creative Thinking in Science and in Human Relations," in: Ganelius, T., Ed., *Progress in Science and Its Social Conditions—Nobel Symposium 58* (New York: Pergamon Press, 1986).
31. The mechanisms used by the scientific community to suppress attacks on its dominant visions have been explored by Polanyi, M. in: *Personal Knowledge: Towards a Post Critical Philosophy* (NY: Harper and Row, 1964). Polanyi cites three strategies:
    1. *Circularity:* If a belief system is called into question, this doubt is countered by reliance on other belief systems which are not under attack at the time.
    2. *Self-expansion:* The system can be expanded to account for any piece of evidence which seems to refute it.
    3. *Suppressed nucleation:* The system of beliefs stabilizes itself against criticism by suppressing the germination of any alternative systems.
32. Koestler, A. *The God That Failed* (London, 1950).
33. Horney, K. *New Ways of Psychoanalysis* (London: Routledge, 1939).
34. Argyris, C. and Schoen, D. *Organizational Learning: A Theory of Action Perspective* (Reading, MA: Addison-Wesley, 1978).
35. Boulding, E. "Image and Action in Peace Building," *J. Soc. Issues* 44(2):17–37 (1988).
36. Senge, P. *The Fifth Discipline: The Art and Practice of the Learning Organization* (New York: Doubleday, 1990).
37. Comenius, J. A. *The Labyrinth of the World and the Paradise of the Heart* (Chicago: The National Union of Czechoslovak Protestants in America, 1942).
38. Ingelstam, L. "Forecasting for Political Decisions," in: *Policy Analysis and Policy Innovation*, Baehr, P. R. and Wittrick, B., Eds. (London and Beverly Hills: Sage Publications, 1981).

39. There have been some interesting uses of information technologies to widen democratic participation in creating common futures. The most publicized was New Zealand's Televote Project which was conducted in mid-1981 by a permanent government agency, The Commission for the Future, created to involve the public in long-range future studies. See "Teledemocracy" *Futurist* 15(6)6-9 (December 1981).

40. For a review of state visioning exercises see: Keon, C. "Foresight Activities in State Government," *Futures Res. Q.* 7(4)49-60 (Winter, 1991); interesting examples include: *Sustainable Design for Two Maine Islands*, Boston, MA: The Boston Architectural Center, (1985); *Knowing Home: Studies for a Possible Portland*, Portland, OR: RAIN (1981); Bezold, C. and Olson, R. "The Future of Florida: Four Scenarios for the Sunshine State," *Futurist* (October 1983); and Gappert, G., Ed. *The Future of the Urban Environment* (Sage Publications, forthcoming).

41. See Michael, D. "The Futurist Tells Stories," in *What I Have Learned Thinking about the Future Then and Now*, Marien, M. and Jennings, L., Eds. (New York: Greenwood Press, 1987).

42. This matrix was adapted from "Relative Risk Reduction Project: The Report of the Human Health Subcommittee (Appendix B)," U.S. EPA Science Advisory Report SAB-EC-90-021B (September 1990).

43. Stigliani, W. M. et al. "Future Environments for Europe: An IIASA Study of Some Implications of Alternative Development Paths," *Sci. Total Environ.* 80 (1989).

44. Boulding, K., *The Image: Knowledge, Life and Society*, (Ann Arbor: University of Michigan Press, 1956).

SECTION V

Geographic Information Systems (GIS)

Alvin M. Pesachowitz

INTRODUCTION

Once environmental statistics data and information have been collected, compiled, analyzed, and modeled they only have value if they can be communicated. A clear, accurate, and effective presentation is essential to make the most use of the data and information we do have. An integral part of communicating environmental statistics is the use of geographic information systems (GIS). GIS is an integrated technology of which the central part is a database management system structured to handle spatial data so that they can be presented in a common framework. The components of a GIS system include: entry, storage, manipulation, analysis, and display. The key purpose is to organize and analyze data to determine the relationships that can only be seen in a visual presentation. One important consequence of presenting data in this way is a cost saving through the reduction of duplication, focusing on relationships not clear from other methods. This approach thus can point to the most efficient ways to protect public health and the environment. The U.S. EPA spends over $500 million annually on monitoring environmental parameters; and by integrating the results in a presentation method such as GIS, these resources can be used more effectively. The use of GIS at the U.S. Environmental Protection Agency (EPA) is described in Chapter 16. Also discussed is the use of advanced computer technology to manage, analyze, and display large volumes of spatially related data and to provide rapid information retrieval and display of multiple data layers to help analysts identify spatial relationships. GIS also provides a more effective vehicle for communicating with the public.

There are uncertainties in the data and information concerning the environment, where they are and how to get them. A commonly asked question is how can you find and retrieve the data and information you need when you do not know what are available, where they are, or how to get them? One of the problems in accessing data is the lack of definitions and standards in the form of protocol specification. Such standards would provide a bridge between library science and the information service community so that data and information can be more completely and efficiently intercommunicated. Chapter 17 describes such a system called the wide area information servers (WAIS) which would provide a single interface mechanism. The important factors in this system are language, feedback of information to further refine a search, consistency, format, search methodology, publicity and access, and hardware and software requirements.

CHAPTER 16

Geographic Information Systems (GIS) for Environmental Decision Making

Alvin M. Pesachowitz

ABSTRACT

Protecting the environment is a job that is inherently geographic in nature. Understanding the spatial relationships of natural resources is critical to successfully accomplishing the EPA mission. Geographic Information Systems (GIS) provide for the first time a set of tools which allow us to integrate and analyze our rich base of existing environmental data in a spatial context. With GIS, we can dramatically combine data about air, water, and soil to better visualize and understand the natural interactions between these media and highlight areas of environmental interest and concern.

Specifically, GIS is an advanced computer technology that allows users to collect, manage, analyze, and display large volumes of spatially related data, e.g., geographic, cultural, political, environmental, and statistical. Rapid information retrieval and display of these multiple data layers help analysts identify spatial relationships that might have otherwise have gone undetected.

This chapter will discuss the importance of the application of GIS technology to the environmental decision-making process.

INTRODUCTION

The U.S. Environmental Protection Agency (EPA) is responsible for implementing the U.S. federal laws designed to protect the environment. The EPA endeavors to accomplish its mission systematically through integration of a variety of research, monitoring, standard-setting, and enforcement activities. As a complement to its other activities, the EPA coordinates and supports research and antipollution activities of state and local governments, private and public groups, individuals, and educational institutions. The EPA monitors the operations of other federal agencies with respect to their impact on the environment. With the growing emphasis on global environmental issues, the EPA is also becoming more actively involved in environmental activities with its neighboring countries in North America as well as abroad.

Today, the EPA is in the process of institutionalizing the use of geographic information systems (GIS) technology in nearly all aspects of the agency's mission. This chapter will relate how GIS was introduced into the EPA and provide examples of GIS applications to illustrate the use and power of this technology in environmental protection. After nearly a decade since GIS was first utilized at the EPA, resource managers are now able to document some of the benefits realized from this technology. However, many of the most significant benefits are not at all what one would expect, or at least, not what the EPA expected when we began.

GIS AND GEOGRAPHIC INITIATIVES AT THE EPA

GIS is an integrated technology the heart of which is a database management system tailored to handle locational data that can be analyzed and presented in a common spatial framework. GIS technology includes components for the entry, storage, manipulation, analysis, and display or presentation of spatial data. The key purpose of GIS is to place data in a common spatial framework and examine the relationships that become evident only where data are integrated in this way.

In environmental protection, the EPA can use GIS to manage and analyze data about water flow and chemistry, soil characteristics, meteorology and demography, etc. and then manipulate and present these data and their relationships. These visual displays have proved invaluable in supporting decision makers and in discussing pending decisions with the public, legislators, and others who would otherwise either be outside the decision process or who would participate in the process without a firm grasp of the factors under consideration.

The introduction of GIS technology at the EPA has coincided with the development of a new long-term environmental strategy that emphasizes "geographic initiatives." These initiatives focus on areas delineated by distinctive geographic features such as the Great Lakes, the Chesapeake Bay, and the Gulf of Mexico. This new geographic perspective acknowledges the natural interaction between the air, water, and soil in the environment and the influence that natural boundaries have on this interaction. The EPA will rely on the integrating capabilities of GIS to analyze and portray these often complex relationships as it addresses the substantial environmental challenges of today and the future.

GIS IMPLEMENTATION IN THE EPA

The Environmental Protection Agency did not automatically embrace GIS technology, instead a thorough evaluation was performed by two Research and Development laboratories located in Corvallis, OR and Las Vegas, NV. Both groups of scientists were working on complex studies that required analysis of spatial information, including the long-term response of surface water to acidic deposition and the movement of groundwater contaminants through complex subsurface geology. Each group decided quite independently that they needed a special tool to manage these complex data sets and to perform and display the analysis required by policy-makers. In the case of Las Vegas, they were also interested in finding a way to fully utilize remote sensing data in environmental analyses.

In 1983, both laboratories acquired early versions of GIS technology. After about 2 years, GIS was introduced on an experimental basis in the regional office in Atlanta, GA to evaluate how GIS could be applied to routine environmental protection work—permitting facilities, compliance and enforcement work, and managing hazardous waste sites, etc. In the United States, most environmental protection activities, although authorized or mandated by federal laws, are delegated to and carried out by state governments. The state of Georgia's environmental protection agency was invited to participate in the Atlanta regional GIS program. Specifically, a GIS tool was developed to assist state permit writers in evaluating landfill permit applications by identifying suitable sites for landfills.

A great deal was learned from the Atlanta experience about what would be entailed in delivering GIS technology to the EPA regions, including:

- how to purchase, install, and support GIS computer technology
- what it costs to acquire and operate a GIS installation
- what is entailed in acquiring and assuring the quality of spatial data of many different types, and where such data can be found in the United States

Since the Atlanta pilot, GIS technology has been installed at the EPA headquarters, in all 10 regional offices, and in many of the more than two dozen laboratories. The agency-wide implementation strategy occurred as follows:

- First, to limit the initial capital investment in what was still a very new and rapidly evolving technology, GIS software was installed on mini-computers that we already owned and that were nearing the end of their useful life in their original role of supporting administrative functions. It was far more important to assure the success of this new technology by investing agency resources in training and generating useful applications than it was to have the best available hardware. This approach allowed the installation of GIS in all 10 EPA regional offices within 3 years.

- Second, before a region could install a GIS, two conditions were required: (1) the region had to assemble a multidisciplinary team of up to five people to operate the GIS as a service to the region programs; and (2) the GIS team had to be placed in a crosscutting organization within the region, and not in one of the individual line organizations that addresses only one "media"—air, water, etc.

- Third, having gained experience in using GIS throughout the EPA, and having developed a large body of strong adherents to this technology, the agency received the capital to modernize its GIS technology. In this modernization program, the EPA capitalized on newly available local and wide-area network technology that allows direct access to GIS throughout the professional workforce. Subsequently, the EPA completed the purchase of new GIS systems for all regional offices— expecting this technology to be in all or most EPA laboratories and analytic facilities and at the headquarters policy staffs, which had historically been limited to hard copy products.

The EPA GIS expenditures to date approach $15 million since the technology was first introduced in the mid-1980s. This investment has been primarily focused on the acquisition of GIS technology and associated expertise. The EPA established agency-wide contracts for the acquisition of GIS software and hardware, including training and support. It is expected that the explosive growth of GIS utilization currently being experienced within the EPA will undoubtedly lead to increasing expenditures in the future that will be applied in greater portions to spatial data acquisition and management activities.

GIS requires a nontraditional organization structure that unlike the traditional ADP organization where it may not be used effectively, if at all. Computer experts know hardware and software, but do not necessarily know data. The real costs, problems, and benefits of GIS are tied to spatial data, not technology.

In the EPA case, GIS offered a unique opportunity to change the way people work together. GIS provides a new and extremely powerful way to visualize complex relationships and solve problems in environmental protection, in a way so attractive that people have been willing to surrender organizational primacy in order to gain access to this technology. The agency has attempted to capitalize on this opportunity by trading technology for organizational reform.

This last point is crucial to understanding the benefits of GIS to the EPA. Environmental protection in the United States and in most nations has evolved from earlier programs to manage water, land, and other natural resources; from public health programs focused on sewage treatment and drinking water supplies; and more recently from efforts to reduce harmful air pollution. Each of these programs grew up separately and were later combined into something called "environmental protection."

Common sense and science have shown that the environment is, in fact, an integrated whole; and efforts to understand and protect the environment on the basis of separate "media" — air, water, land, etc. — are doomed to fail. Unfortunately, environmental protection agencies have had great difficulty in integrating their inherited "single media" traditions and predecessor organizations into a single, coherent and effective program. Thus, early pollution control often produced pollution transfers — sludge incineration generates harmful air pollution, "tall stack" air pollution strategies produce acid rain, etc.

In addition to "soft" obstacles such as custom, organizational politics, scientific disciplines, etc., integration of environmental programs and bureaucracies have been blocked by one very concrete obstacle — the near impossibility of integrating data across the separate media to support holistic environmental assessments and strategies. GIS, in combination with major advances in computing power and data resources, is proving to be a key force in overcoming this obstacle in the United States. Perhaps the revolution in the information technology world will produce a new offering able to eliminate organizational politics.

GIS APPLICATIONS IN THE EPA

Before exploring the economic and other benefits produced by GIS, it will help to briefly describe some examples of GIS applications within the EPA.

San Gabriel Basin Superfund Site

The San Gabriel basin in southern California includes portions of Orange County and the City of Los Angeles, and its aquifer provides drinking water to an estimated 1.2 million people. In 1984, the 170 mi^2 (440 km^2) basin was placed on the EPA National Priorities List (NPL) due to extensive groundwater contamination from four hazardous waste sites. In 1985, the EPA Environmental Monitoring Systems Laboratory in Las Vegas, NV (EMSL-LV) implemented the agency's first Superfund site GIS application at the request of the Region 9 Superfund program. (Superfund is the U.S. mechanism for identifying and remediating abandoned hazardous waste sites which have significant public health implications.)

The San Gabriel basin investigation focused on the evaluation of contamination extent, prevention of exposure to contaminated groundwater, and development of a basin-wide remediation strategy. GIS was used throughout the remedial investigation/feasibility study to effectively manage, analyze, and query the vast amount of spatially referenced data that were used to characterize the site. There are approximately 1500 wells at the four San Gabriel sites where monitoring of volatile organic compounds (VOC) still occurs. The application included the analysis and display of three-dimensional groundwater monitoring information, e.g., well log data interpolated as contaminant plumes. These plumes were used to identify changes in contamination movement in relation to groundwater flow patterns, temporal changes in groundwater quality and contaminant migration, water management scenarios (pumping impacts), and sources of contamination. Utilizing this information, the remediation plan at San Gabriel calls for managing groundwater withdrawals in order to slow and eventually reverse the spread of contamination.

The findings of a GIS cost/benefit analysis performed on this application have documented significant cost savings and both tangible and intangible benefits from using a GIS when compared to traditional cartographic and computer-aided design (CAD)-based methods. Perhaps the most dramatic cost savings was reflected in terms of the reduction in labor hours required when a GIS was used for analyzing and mapping the contaminated groundwater plumes from data gathered at the 1500 monitoring wells. This represents a ninefold decrease in the labor hours required to analyze and plot the well data, interpolate the plume boundaries, and prepare the maps.

In addition to improvements in the remediation planning process, numerous other benefits of GIS were realized during the project:

- GIS maintained a central, convenient access point for all the data necessary for plume analysis, greatly facilitating routine access, including all historic data.
- GIS allowed rapid selection and subsetting of the data for analysis and mapping.
- GIS facilitated the use of very sophisticated groundwater flow models, and much of the analysis would not have been completed in a timely manner if they were required to perform the same work by hand.
- Because some of the base data for the project (including census demographic data and the road networks used to accurately locate many of the monitoring wells) were available digitally, GIS actually facilitated the production and acquisition of new data that would have been required regardless of the method of analysis and integration.

- The scientists on the project emphasized that the enhanced communication of their scientific results to the public was a primary benefit of the GIS technology.

The economic impact of GIS on this project continues today. Because all of the data were collected and stored in digital format, those data are now available for use by other agencies. The regional water authorities are now using the GIS database for natural resource and human health risk assessments. The EPA has implemented a new recycling program — not paper or plastic or used oil, but data.

Shaver's Farm, Georgia

Shaver's Farm is located in central Georgia. In the early 1970s portions of this land were leased to chemical companies, where they buried hundreds of barrels of toxic chemicals at the site. The companies concealed their activities, and no one knew for certain what they were disposing. In 1987, a report came to the EPA that possible uncontrolled hazardous waste disposal had been occurring, and a groundwater monitoring assessment revealed the presence of a toxic herbicide.

Following a round of legal paperwork with the chemical companies believed to be responsible for the contamination, the EPA was not satisfied with the answers although the company legal obligations had been fulfilled. Consequently, the EPA performed a geophysical assessment of the site with three different surface-based sensors: electron magnetometer, proton magnetometer, and WADI — a low-frequency radio-wave detection device.

This approach demonstrated the unique value of the GIS. Taking the point measurements from the three devices, GIS produced contours which showed the anomalies each device detected. GIS then allowed the rapid overlaying of this data for a visual merging for presentation. Within 2 weeks of the time the field crew delivered its data from the geophysical recorders, the GIS analyst had delivered an initial assessment to the site manager. This information allowed the site manager to instruct the GIS team about how best to represent the data.

Because of data which were collected with highly accurate spatial coordination, digging at several sites identified by the geophysical sensors unearthed over 1200 barrels. This procedure has resulted in a 100% predictive accuracy rate with the GIS providing a very rapid turnaround on data analysis, interpretation, and display. As with the San Gabriel site, manual methods would have required several additional weeks worth of work, and each additional combination of the data from the three recorders would have added an equal amount of recalculation and redrawing time. GIS coupled with other advanced technologies, combined to make a real breakthrough in the remediation of a life-threatening abandoned toxic waste dump.

Rocky Mountain Arsenal Area Cancer Study

The Rocky Mountain Arsenal is a large area near Denver, CO, owned by the U.S. Army which has been publicly criticized over the past few years because of the release of toxic chemicals into the environment from the site. Recent studies of the incidence of cancer in the area have revealed higher than normal rates of certain types of cancer. Immediately, people jumped to the conclusion that the arsenal was responsible and demanded retribution from the government for injury. However, the ability of GIS to integrate data from various isolated sources suggests that this is not necessarily the case. When the areas around the arsenal were divided into three zones and other sources of data were applied, a surprising relationship was revealed. While two of the surrounding zones had cancer rates that fell within the bounds of what would be expected, the one zone which displayed the higher rates or incidences of cancer also had a preponderance of hazardous waste sources in the immediate area. This spatial relationship suggests that the arsenal may in fact not be the causal link with the increased cancer rates that everyone assumes.

The ability of GIS to integrate information from diverse data sets provides the potentially monumental economic benefit of precluding the success of a class action lawsuit for damages against the U.S. government. However, of greater importance is the ability to evaluate the potential impacts of those identified hazardous waste sources in an effort to reduce the potential human health risks to the surrounding population. It will be difficult to assess the total impact, but it can be stated conclusively that the analysis which revealed these relationships probably would not have been undertaken without the abilities afforded by GIS. It should be noted that the EPA Region 8 GIS team performed this analysis and displayed the results within a 24-hr turnaround, assisting the epidemiologists who were preparing to release their findings to the public.

Oregon Clean Water Strategy

The Oregon Clean Water Strategy (CWS) GIS provided the state of Oregon environmental managers with a capability to dynamically compare, contrast, and ultimately prioritize cross-programmatically water management activities and resource allocation decisions. The Oregon CWS GIS explicitly relates known pollution impacts to a publicly determined value system for waters (such as drinking water, fishing, swimming, fish spawning areas, etc.) on a geographically specific (i.e., stream-by-stream) basis. The unique benefits of this GIS approach to setting water management priorities are threefold.

- First, Oregon was able to prepare comprehensive maps and lists of all state streams and waterbodies in order of management priority based on selected ranking method.
- Second, Oregon officials were able to use these lists to communicate their priorities to many other federal, state, and regional organizations; and to develop a cross-organizational consensus on where common problems existed.
- Finally, Oregon was able to use the Oregon CWS GIS and its outputs to develop geographically specific agreements with "sister" agencies to jointly manage the affected resources and coordinate cleanup strategies.

The Oregon CWS GIS was effective because of its powerful data integration and communication capabilities, the latter translating sophisticated analysis into intuitive and credible information.

ECONOMIC AND OTHER BENEFITS OF GIS TO THE EPA

The true benefits of information technology are typically found in doing projects that would otherwise be impossible, not from performing current tasks more efficiently. This is the lesson of the past 20–30 years of experience in applying information technology to all fields except perhaps the most routine, labor-intensive functions such as those found in financial institutions.

The previous examples and others illustrate how GIS has helped the EPA do its work more effectively, and in doing so how to reduce the ultimate cost of performing the work at hand. GIS has improved the EPA efforts to communicate with the public and as a result gain acceptance for essential measures to protect human health and natural resources. Since it is often difficult to act without public consent, GIS helps reduce the endless debates and delays that can afflict the decision-making process in a democratic society. There are also a growing body of examples of GIS helping to perform critical management functions, such as integrating field samples from several different contract teams working on different phases of a single hazardous waste site cleanup, thereby helping to obtain value for the fees paid and avoid unnecessary work. There is also strong evidence that GIS is shortening the process of cleaning up hazardous waste sites, resulting in the potential for enormous cost savings.

The real benefits of GIS, however, come in forms that are very difficult to measure. These benefits are the result of being able to think and do things that are impossible to imagine without the spatial integration and visualization made possible by GIS. Nonetheless, economic and other equally important benefits can be imputed, as shown by the following examples:

- The EPA estimated expenditures for environmental monitoring now exceed $500 million annually. Unfortunately, much of monitoring data have been essentially inaccessible to anyone other than the initial investigator/user. GIS is changing that and in the process is yielding enormous improvements in data quality, which inevitably improves when data are used rather than merely gathered and stored. The total "life cycle" cost of the EPA investment in GIS is about $38 million, or less than 7% of 1 year of spending on monitoring. It will not require much in the way of improved data or use of data for GIS to produce a favorable benefit/cost ratio.

- EPA programs are evolving into integrated, multimedia efforts keyed to environmental results, rather than intermediate and often misleading measures such as the reduction of emissions without regard for cost or ultimate environmental benefit. GIS has been an important factor in bringing about this shift, by providing the analytic underpinning for program integration and by leading/forcing changes in organization and policy. There is a growing body of evidence that an integrated approach to environmental protection does avoid enormous costs to the U.S. economy and yields far superior protection of human health and natural resources.

- The EPA is moving to a long-term strategy that emphasizes "geographic initiatives" and organizes around major geographic features, such as the Great Lakes and the Chesapeake Bay. This strategy is based on the belief that "people protect that which they love." To protect these resources, people will take actions voluntarily that are beyond the reach of legal coercion.

- GIS has opened the door to vast amounts of data that had been inaccessible to the EPA for all practical purposes. Examples include soils and geologic data and land-use information. Although the EPA is understandably pleased by the integration of internal data and programs, the benefits of integration with data and programs outside the agency may prove to have far greater benefits.

- GIS allows for the protection of the environment in ways that would seldom have even been considered in the past. To illustrate, state environmental agencies have used GIS technology to influence transportation and land-use planning — an area with great impact on long-term environmental trends — but previously beyond the effective reach of case-by-case regulation.

- The EPA is now implementing a nationwide Ecological Monitoring and Assessment Program (EMAP) that would be impractical without GIS as its basic data management strategy. This program promises to provide the first statistically reliable information about the status and trends of ecosystems, and insight into the conditions that are affecting their overall health. Initial results suggest that this approach will allow for targeted programs to mitigate the effects of pollution on ecosystems; by doing so this will help avoid broadbrush regulatory strategies that impose enormous costs on American industry.

CONCLUSIONS

While the potential for GIS to strengthen environmental protection in the long term is enormous, the application of GIS technology to environmental protection presently offers only very modest, near-term cost savings. However, GIS technology appears to offer very large-scale, long-term economic and policy benefits by enabling new program strategies; by permitting the use of previously inaccessible and extremely valuable data; by supporting new, more valid monitoring methods; and by driving fundamental changes in traditional institutional roles and structure. In view of this, the strategy for implementing GIS in an environmental protection agency or ministry should be designed to both foster and capitalize on these long-term changes.

The EPA mission is clearly geographic in nature. Geography will serve as the common integrating factor that enables the combination of environmental and natural resource data in ways that will allow more effective protection of the environment. Substantial benefits of applying GIS to solving environmental problems are already being realized throughout the EPA, and its use will most certainly play a critical role in meeting the increasingly complex environmental challenges of the future.

CHAPTER 17

Mechanisms to Access Information About Spatial Data

Eliot J. Christian and Timothy L. Gauslin

ABSTRACT

Most existing database products and information services either are targeted to specific communities or require the user to know details of the database. Consequently, users who need to access multiple sources have to go through several steps and learn a variety of search and retrieval approaches. The library and information services community has published an open standard for information search and retrieval, known as Z39.50. The standard is being implemented very widely, and there is a public domain implementation of Z39.50 called Wide Area Information Servers (WAIS). WAIS is targeted to users with limited computer skills to access a wide range of information from a single, intuitive interface. Versions of WAIS are available for most computers, and there are already hundreds of information sources accessible. WAIS is being used not only for bibliographic information, but also for providing online access to directories of data. As an integrating technology, WAIS promises to vastly simplify the problems of access to scientific and technical data and information.

BACKGROUND

Researchers face a fundamental problem — *How can you find and retrieve the data and information you need when you do not know what are available, where they are, or how to get them?* Many information services and databases are available to help users find and retrieve information, but most are targeted to specialists or require the user to already know how the data are organized. Since most researchers gather information from a variety of sources, their frustrations in dealing with complex and divergent approaches may be inhibiting growth in the information services market. Dow Jones News Service and the accounting firm Peat Marwick teamed with Apple Computers and Thinking Machines to develop a facility that allows users with limited computer skills to access personal, corporate, and published information from one interface—Wide Area Information Servers (WAIS, pronounced "ways").[1]

WAIS development is occurring in the context of an explosive growth of digital information, especially as the definition of information expands beyond alphanumeric to graphic and multimedia (e.g., audio, music, video). Yet network connectivity through utilities such as the Internet is also growing explosively,[2] as is the availability of computing power ranging from ubiquitous desktop workstations to massively parallel supercomputers. Although conventional brute-force methods can search sources containing several trillion characters of text and other data types, better techniques are needed to make such searching quick and affordable.[3]

AN OPEN STANDARD

In the late 1980s, the information services community was completing a crucial piece of work—an open standard known as the Information Retrieval Service Definition and Protocol Specification (Z39.50). Open standards such as Z39.50 are crucial in addressing the problem of information access. By employing a commonly accepted computer-to-computer protocol, information can be stored on systems having very different hardware and software, yet a single interface can retrieve information from all of them.

The Z39.50 standard is a product of the National Information Standards Organization (NISO), accredited to the American National Standards Institute (ANSI). NISO focuses on standards for libraries, information services, and publishing; and Z39.50 is fully compatible with the NISO standard for library catalogs (Z39.2) originally promulgated by the Library of Congress and known as machine readable cataloging (MARC).

Both Z39.50 and Z39.2 have corresponding International Standards Organization (ISO) standards. Z39.50 is an applications service within the family of ISO standards comprising the Open Systems Interconnection (OSI) model.[4] It is also fully compliant with transmission control protocol/internetwork protocol (TCP/IP) as implemented on the international network of networks known as the Internet.

Adoption of the Z39.50 standard protocol throughout the international library community fits well with the WAIS goal of making information search coherent across different services. The developers of WAIS worked very closely with NISO and proposed enhancements to the

1988 version of the Z39.50 standard. Most of those proposed enhancements are reflected in the 1992 version of Z39.50.

The 1992 version of Z39.50 has a built-in mechanism to assure consistent operation across various implementations. This is achieved by having all computer-to-computer interchanges precisely represented in a standard computer language known as Abstract Syntax Notation (ASN.1). The standard also establishes a process for the formal registration of additional objects and structures expressed in ASN.1, so that the information interchange can rapidly and reliably expand into new areas. Further enhancements to the Z39.50 standard are being actively pursued, and a Z39.50 Implementor's Group provides a forum for raising issues and finding consensus.[5]

CLIENT/SERVER INTERACTION

WAIS implements Z39.50 in a client/server mode of computer interaction.[6] In a typical search for textual information, the client software prompts the user to select which information sources to include in the search and to enter a search request. Once the search request is entered by the user, the client software converts the search words to the standard information retrieval protocol (Z39.50) and presents the search request in turn to each server associated with a selected source. The server software matches the search request words to the contents of all documents in each selected source. The client software receives search results from all of the servers and presents to the user a list of all document titles found. (See Figure 1.) When the user selects a title from the list, the client software requests the server to pass the document content, and the client then presents the document to the user.

DOCUMENT SCORING

Up to this point, the Z39.50 client/server interaction is fairly conventional, and many information service providers will fit their current products into this model. One very useful feature of Z39.50 is that the user sees documents listed in a ranked order based on relative scores assigned to documents by the servers.[6] The algorithm employed for the scoring of documents is likely to be a fascinating area of development. In a sense, the scoring algorithm is where judgment is applied about the likelihood that a particular document will be seen as relevant by the user. The public domain version of WAIS has a fairly simplistic scoring algorithm that stresses the frequency of occurrence, with different weights depending on where the words occur and whether the words occur so often as to be nonspecific.

WAIS Query

File   Setup   Aids   Help

Tell me about:
toxic metals

Similar to:

Search
Add Doc
Delete Doc

Resulting Documents

Score	Size	Src	Title
▓▓▓▓▓	1.8K	ESDD	OFFICE OF HAZARDOUS MATERIALS DATA MGMT. TOXIC CHEMICAL
▓▓▓▓▓	1.7K	DOE	effects of sewage sludge and toxic metals upon vesicular-
▓▓▓▓▓	6.9K	NOAA	Toxicants in the Chesapeake Bay water column—TOXIC.WATE
▓▓▓▓▓	3.9K	NOAA	Toxicants in the Chesapeake Bay finfish and shellfish—T
▓▓▓▓▓	4.0K	NOAA	Heavy metals and pesticides in stream bottom sediments i
▓▓▓▓▓	6.8K	NOAA	Toxicants in the Chesapeake Bay sediment—TOXIC.SEDIMENT.
▓▓▓▓▓	2.7K	NOAA	Trace metals in sediments of Raritan Bay.
▓▓▓▓▓	.9K	ESDD	REVIEW OF POTENTIAL TOXIC PITS CLEANUP ACT FACILITIES
▓▓▓▓▓	.9K	ESDD	SOUTHERN CALIFORNIA DHS TOXIC SUBSTANCES SECTION FILES
▓▓▓▓▓	1.0K	ESDD	TOXIC CHEMICALS IN GROUNDWATER PROGRAM FILES
▓▓▓▓▓	3.4K	NOAA	Distribution of five metals in sediments from the New Yor
▓▓▓▓▓	.9K	ESDD	NORTH COAST CALIFORNIA DHS TOXIC SUBSTANCES SECTION FILE
▓▓▓▓▓	1.2K	ESDD	OFFICE OF HAZARDOUS MATERIALS DATA MANAGEMENT GUIDE TO T
▓▓▓▓▓	1.0K	ESDD	NORTHERN CALIFORNIA DHS TOXIC SUBSTANCES SECTION FILES
▓▓▓▓▓	3.1K	NOAA	Trace metals quality control analysis (Mississippi, Alab
▓▓▓▓▓	2.8K	NOAA	Trace metals in marine biota and sediments collected fro
▓▓▓▓▓	3.6K	NOAA	Metals in interstitial waters of the New York Bight dredg
▓▓▓▓▓	3.4K	NOAA	Distribution of metals in Elizabeth River sediments, Vir
▓▓▓▓▓	2.1K	NOAA	Distribution of metals in Baltimore Harbor sediments.

Status: Found 20 references
Querying /usr/opt/wais/db/nedies at 130.11.48.107

FIGURE 1.   Results of a search are displayed as a list of all document titles found.

REQUESTS IN ENGLISH

A critical issue in searching for information is that the results of the search can be compromised if the user lacks the skills needed to use the client software to specify the search. Because the WAIS goal is to avoid nonintuitive query languages, users simply enter a search request in WAIS in English. The WAIS software does not analyze grammatical structures or meanings of the words, but depends on the scoring of documents to emphasize words — primarily nouns — that are more specific. Conversational embellishments will tend to get low scores since they are unlikely to be of specific value in the target documents. While the simplicity of the WAIS search request has a certain appeal, real users cannot be expected to bring with them all of the key words needed to find everything that may be relevant. WAIS therefore provides an elegant mechanism to refine the search so that the session converges on meaningful results — an approach known as "relevance feedback."

RELEVANCE FEEDBACK

A WAIS information search and retrieval session can be thought of as similar to the experience of using a library.[1] A library user may begin by consulting a card catalog or index, or by asking a reference librarian for help. At this point, the user is searching for documents based on a few key words (e.g., subject, title) or names (e.g., author). As the user reviews the documents found, he or she may note other key words or names that could lead to additional relevant documents. A feedback situation develops as the user modifies subsequent searches based on results found in prior searches. Ideally, the user stops searching when all of the most relevant documents are found.

In WAIS, a user may highlight any retrieved document or part of a document that seems to be closer to the user's interests. When the user designates that highlighted portion of a document as relevance feedback, WAIS handles the words in that portion as another search request. Some WAIS developers see relevance feedback as crucial to achieving the goal of eliminating complicated query languages.

WAIS SERVERS

WAIS information servers are registered to the Directory of Servers currently maintained on the Internet by Thinking Machines. Registration requires a commitment to keep the server in a reliable operational mode. The registration entry includes a narrative text description of the contents of the sources reachable through the server; this description is itself indexed for searching. Also listed is information that will be used by the client software to contact the server (e.g., TCP/IP node name), as well as information on charges and billing for use of the server if not free.[6] (See Figure 2.)

Any server capable of responding to Z39.50 information retrieval requests can be an information server. Information servers can be local (on the workstation or local area network) or remote (accessible via TCP/IP, X.25 networks or asynchronous dial-up). WAIS does not require central coordination unless the server is to be advertised through the Directory of Servers. In fact, an information server registered to the Directory of Servers can itself act as a subordinate directory of servers administered locally. By describing sources under various directories of servers, the sources may be organized in whatever relationships make sense and yet allow users to search as many sources as desired.

One feature of WAIS allowed but not required by the Z39.50 protocol is that the client/server interaction is "stateless."[6] At the application level, each request from the client to the server is a separate process that is not associated to any previous request. The server does not maintain information about the client between requests. (For efficiency, the communications link itself may be retained across requests.) This feature is significant for situations where a user may want to search hundreds of sources on dozens of servers at a sitting.

INFORMATION SOURCES

Information servers exist to provide access to the information sources placed on them. These sources are compilations that may include a variety of formats. Such formats are known as "document types," although information need not be textual. While all Z39.50 clients and servers support search and retrieval of textual information, support for other document types that may have been registered in Z39.50 (e.g., MARC bibliographic format, graphics, hypertext, video) is negotiated when the client initiates its relationship with a server.

When sources are created, defining the document types allows the server to use the appropriate translation between the specific query format of the source and the Z39.50 protocol. The public domain WAIS package includes assistance in creating information sources. Indexing software is provided in the WAIS package for several common document types consisting of text, graphics, and bibliographic references in MARC. Source code in the "C" programming language is provided for adding other document types. If access to other data structures is required, the server interface routines are also designed to be customized.

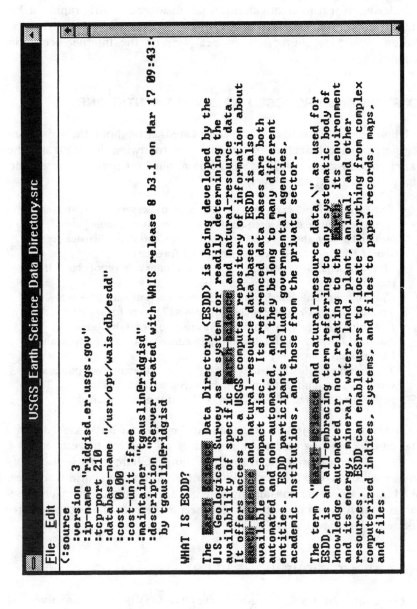

```
USGS_Earth_Science_Data_Directory.src

File Edit

(:source
 :version  3
 :ip-name "ridgisd.er.usgs.gov"
 :tcp-port 210
 :database-name "/usr/opt/wais/db/esdd"
 :cost 0.00
 :cost-unit :free
 :maintainer "tgauslin@ridgisd"
 :description "Server created with WAIS release 8 b3.1 on Mar 17 09:43:
  by tgauslin@ridgisd

WHAT IS ESDD?

The ████ ████ Data Directory (ESDD) is being developed by the
U.S. Geological Survey as a system for readily determining the
availability of specific ███████████ and natural-resource data.
It offers access to a USGS computer repository of information about
████ ██████ and natural-resource data bases.  ESDD is also
available on compact disc.  Its referenced data bases are both
automated and non-automated, and they belong to many different
entities.  ESDD participants include governmental agencies,
academic institutions, and those from the private sector.

The term \"████ ██████ and natural-resource data,\" as used for
ESDD, is an all-embracing term referring to any systematic body of
knowledge, automated or not, relating to the █████, its environment
and its energy, mineral, water, land, plant, animal, and other
resources.  ESDD can enable users to locate everything from complex
computerized indices, systems, and files to paper records, maps,
and files.
```

FIGURE 2. A document can describe an information source.

A typical customization would be to use search requests to access a relational database such as Ingres or Oracle using Structured Query Language (SQL).

Indexing of text to create an information source is fairly rapid—a 30-megabyte file was indexed in about 20 minutes on a Data General Aviion. Searching sources either locally or on the Internet occurs in seconds.

EXISTING AND PROPOSED WAIS IMPLEMENTATIONS

Hundreds of information servers located throughout the world are registered in the Directory of Servers. The following list illustrates the variety of uses to which WAIS is being applied (at present, there is no charge for using these servers):

- *Thinking Machines* maintains the Directory of Servers, WAIS documentation and source code, Frequently Asked Questions, some patents, molecular biology abstracts, a cookbook, the Central Intelligence Agency *World Factbook*, and weather maps and forecasts.
- *Massachusetts Institute of Technology* maintains a server of classical and modern poetry.
- *Cosmic* maintains descriptions of government software packages.
- The *Library of Congress* is creating a WAIS server for its catalog.
- *Columbia Law Library* maintains abstracts of legal decisions.
- *National Institutes of Health* publishes announcements of opportunities for research grants.
- The Internet facility known as *"archie"* creates directories of public software, data, and information from hosts connected to the Internet. These archie directories are available as WAIS information sources.[2]

The range of subject matter that can be covered by WAIS could be viewed as a blessing or a curse depending on whether the user needs a broad or narrow search. As in any publishing medium, there is also likely to be a wide range in the credibility of information from various sources. WAIS developers are considering whether sources should carry endorsements that they would earn through peer review, in the manner of respected scientific journals.

WAIS APPLICATIONS IN THE USGS

The United States Geological Survey (USGS) began its involvement with WAIS through its initiative to enhance the Earth Science Data Directory (ESDD). ESDD is maintained by USGS as a useful source for references to earth science data (including many at the state level) and a comprehensive list of data holdings relevant for arctic research. WAIS is especially appropriate for ESDD because the ESDD user community ranges from local citizens to international global change researchers. The ability of WAIS to place the ESDD in the context of other information sources is especially powerful for these users.

A subordinate directory of servers focused on earth science data and information will be published in the WAIS Directory of Servers. That directory will be maintained by the USGS, but will list sources from other organizations as well. Among the sources already implemented on a demonstration basis are the interagency Global Change Master Directory, the National Oceanic and Atmospheric Administration (NOAA) National Environmental Data Referral Service, the Department of Energy Climate Change Directory, the USGS ESDD and its subset known as the Arctic Environment Data Directory, the USGS Water Data Abstracts, the USGS Library catalog, and the USGS Spatial Data Clearinghouse. The Spatial Data Clearinghouse is itself a directory to several subordinate sources such as the Distributed Spatial Data Library, the Geographic Names Information System, the Map Chart Information System, the Cartographic Catalog, and the Aerial Photo Summary Records Systems.

The USGS is adding to WAIS features required for ESDD such as phrase searching, key word searching within fields, and location searching. (The use of a map to assist in location searching is illustrated in Figure 3.) These capabilities are being introduced within the constraints of the Z39.50 standard. With this approach, users of the USGS/WAIS client software are able to access any Z39.50 server, but have additional capabilities when accessing one of the USGS servers. USGS is also including the ability for a user of the client software to drop from a WAIS session into an automated log in to existing data systems such as the USGS Global Land Information System. WAIS then acts in a complementary role, simplifying navigation among disparate systems and allowing the data systems to be seen in a broader context with the rest of the world of information resources.

In any large organization, data and information are already being made available in a digital form in many instances. From a resources viewpoint, to put that information onto a WAIS server is a small step. The following are examples of some potential applications of WAIS in organizations such as the USGS:

- *Announcements*—These include news releases, schedules of upcoming events, calls for papers, training opportunity announcements, computer virus alerts, requests for proposals.
- *Catalogs and publications*—These include library catalogs, electronic copies of publications, conference proceedings, abstracts of publications,

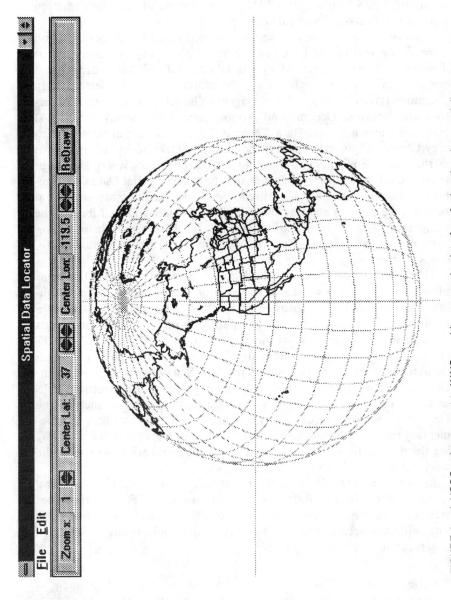

FIGURE 3. A USGS enhancement to WAIS provides a mapping interface for locating data spatially.

newsletters, technology assessment reports, and information product descriptions and ordering procedures.

- *Data accessed as information* — The fully qualified names of data fields can be searchable so that a user can have the equivalent of an SQL query without having to learn the query language.
- *Education* — The simplicity of the WAIS user interface may make it a good tool for teaching students how to navigate the world of scientific and technical information.
- *Electronic filing* — Personal or office-wide files of correspondence can be maintained using WAIS.
- *Full-text search interface for CD-ROMs* — WAIS may provide a good base for a standard user interface for text searching (such a standard is currently being pursued by CD-ROM developers and users).
- *Graphics* — These include scientific visualization product demos; architectural drawings such as building and floor layouts; and forms distribution for local printing, with or without fill-in software.
- *Information access from data* — Textual and graphic information can be linked to specific data fields in a database (e.g., station history linked from the data values for that station).
- *Management information, manuals, and handbooks* — These include strategic plans; all-employee letters; security guidelines; user's manuals; procedures manuals; models for commonly used documents such as position descriptions; vacancy announcements; personnel regulations; and personnel locator information such as phone directory, organizational directory, and electronic mail directory.
- *Software* — These include directories of public domain software, including searching program descriptions and source code, distribution of software.

As WAIS becomes more widely used by researchers, the opportunity arises to simplify the process by which data contributors interact with their data and information repositories. For example, Apple Computers has created a dynamic program that periodically retrieves requested information in an unattended mode and presents to the user a "personalized newspaper" showing the latest information as it is published.[1] Another idea is to have available to the user an option to contribute data and information. The software would help the contributor to document the product appropriately and would then route the product through an electronic review process. The process would culminate in the contribution coming under long-term stewardship and being included in an appropriate information dissemination management system.

While this list only scratches the surface of possible applications, WAIS may not always be the most appropriate solution. For instance, until user authentication is included, information servers published via WAIS should not include restricted access information. There are also issues to be considered in establishing a formal review process for releasing digital data and information via WAIS.

WAIS AND GLOBAL CHANGE

The ESDD interfaces with the interagency Global Change Master Directory, a source for references to key global change data. The global change data management community is using WAIS as an adjunct to the Global Change Master Directory. As a data directory tool, it is possible to rapidly correlate the Global Change Master Directory to other already existing data directories relevant to global change research. For example, NOAA has a directory with about 25,000 data set references and the Interuniversity Consortium for Political and Social Research has another directory referencing about 28,000 data sets. This approach would be especially useful for the very broad community of users intended to be served by the Consortium for International Earth Science Information Network (CIESIN). WAIS has also been suggested for the "Directory of Directories" effort taking place under the International Council of Scientific Unions' Committee on Data, and for providing a generalized "gateway to government information."

One of the traditional barriers to information access has been that the custodians have been unable to agree on common formats. This is hardly surprising in view of the widely disparate needs of the different user communities, and the fact that there is a noticeable lack of consensus even in fairly homogeneous user groups. In the WAIS approach, there is no requirement to recast the data or information into a different format and there is no prerequisite to seek agreement on a single style of presentation. WAIS sources will hold information in the formats that make sense to the custodians as they serve their primary customers, yet the information is still accessible to secondary users. A variety of WAIS clients will be built, with each one handling the search and presenting information in the manner most appropriate to the targeted user.

The ability of WAIS to handle different information sources through a single user interface makes it possible for researchers to explore publications and data sets concurrently. The federal research libraries involved in global change research (primarily, NASA, NOAA, USGS, and USDA) are very interested in the potential for WAIS to bridge between the data and information worlds. Also, WAIS is seen as a useful way to connect textual information into a data system. For example, when a user is researching an existing data set, it would be useful to provide immediate access to all of the associated documentation about that data set. Using WAIS, the associated documentation could extend beyond the data set itself to include publications which reference the data set or engineering specifications of the instruments used.

Since WAIS simplifies searching for information across many dimensions, it would be a natural tool for use in the Global Change Research Information Office. This new office is mandated by the Global Change

Research Act of 1990 to provide information internationally on global change research and technologies associated with mitigating or adapting to global change.

SOFTWARE REQUIREMENTS AND SUPPORT

WAIS software is in the public domain and is distributed free of charge. WAIS is written in the "C" computer language and the client software is available for: a range of Unix workstations in X-windows, MS-DOS in either Windows 3.0 using Dynamic Link Libraries or in character mode, the Apple MacIntosh and the NeXT computer. The server software runs on these computers plus DEC VMS, IBM MVS, and the Connection Machine, among others. WAIS software and documentation is available by anonymous File Transfer Protocol (FTP) in the directory "pub/freeware" on the Internet host "ftp.wais.com."

Any WAIS client can access WAIS servers through TCP/IP software. TCP/IP is usually included with Unix workstations, and both MS-DOS versions (Windows and character mode) include public domain TCP/IP software. WAIS clients and servers have been implemented without communications on Unix, MacIntosh, and MS-DOS. This allows WAIS to also be used to access data and information distributed on floppy disks or CD-ROMs. Terminals having only asynchronous dial-up capabilities can access WAIS by telephone connection to a client running on a Unix machine through a facility known as SWAIS (Simple WAIS).

WAIS is undergoing very rapid evolution through the contributions of hundreds of developers worldwide. At present, a major focus is on making a WAIS fully compliant with the NISO Z39.50–1992, to further simplify the development process, and to make it easier for users to customize applications.

WAIS SUPPORT

In pursuing WAIS, the USGS is working closely with the broader WAIS community, which includes about 150 universities and a number of major corporations (Apple, Sun, NeXT, DEC, Microsoft, Dow Jones, Peat Marwick, Mead Data Central, and Thinking Machines, among others). The original leader of the public domain WAIS initiative was Thinking Machines. The Clearinghouse for Networked Information Discovery and Retrieval (CNIDR) received a grant from the National Science Foundation to become the focus for long-term support of WAIS, in a manner similar to the public domain Kermit software support provided by Columbia University. Those heavily involved in WAIS development are working very closely with NISO, as well.

CONCLUSION

WAIS represents more than just another slick new piece of software. Although it is based on the Z39.50 protocol born in the library science and information services community, WAIS bridges to the computer science and data processing services community. Such bridges hold out the promise of revolutionary improvements in information services and in how information is handled in data processing systems.

The capabilities demonstrated by the Z39.50 standard and the WAIS implementation challenge us to examine how we think of access to data and information and how we make public the data and information we possess.

FURTHER INFORMATION

Information on WAIS is available from Barbara Lincoln (barbara @wais.com), as well as Brewster Kahle (brewster@wais.com).

A copy of the NISO Z39.50–1992 standard can be purchased from the National Information Standards Organization, P.O. Box 1056, Bethesda, MD, 20827. The maintenance agency for NISO Z39.50 is the Library of Congress, and the Library of Congress contact is Ray Denenberg.

Within the USGS, the authors are currently a focal point for WAIS development and applications. Timothy Gauslin completed a version of the WAIS software for the MS-DOS Windows environment and is a primary contact for technical help. Eliot Christian has been involved in outreach to the broad community of federal agencies. Tim and Eliot may be contacted at USGS, 802 National Center, Reston, VA, 22092. Eliot's phone number is (703) 648-7245; Tim's is (703) 648-5980 (Tim's TCP/IP address is tgauslin@isdres.er.usgs.gov).

REFERENCES

1. Kahle, B. and Medlar, A. "An Information System for Corporate Users: Wide Area Information Servers," Thinking Machines Corp., April 1991.
2. Kehoe, B. P. *Zen and the Art of the Internet* (Chester, PA: Widener University, 1992).
3. Markoff, J. "For the PC User, Vast Libraries," *New York Times*, July 3, 1991, p. C1.
4. *ANSI/NISO Z39.50–1992: Information Retrieval Service Definition and Protocol Specification* (Bethesda, MD: National Information Standards Organization, 1992).
5. Denenberg, R. Personal communication (1992).
6. Stein, R. M. "Browsing Through Terabytes," *Byte*, 157, May, 1991.

SECTION VI

Environmental Statistics

Barry Nussbaum

INTRODUCTION

In Chapters 18 through 22 the interest and concern about assessment of the environment from several aspects are demonstrated. One can argue about the relative merits of studying and assessing the status of human health, environmental health, and ecological health, but no one will argue that all the indicators are important. Certainly the environment is fragile enough that data and resultant inference will be regarded as some sort of guideline into the ultimate formulation of policy to keep the earth and its inhabitants in a health state. Therefore the accuracy and quality of the data must be top notch. This requirement for first-class, high-quality data is the topic that runs through this section on environmental statistics.

While the statistician may have the news, good or bad, it is imperative that all the parties know what the statistician is talking about. Chapter 18 attempts to provide a uniform list of tools to help planners and decision makers determine the quality of the data they were analyzing. The chapter is designed to actually satisfy a number of needs for the decision maker. Two of them are to provide a set of tools so that a secondary user could have the proper information to determine whether the data is suitable for secondary use, and to educate that secondary user in the proper terminology to use this requesting information concerning the adequacy of a database. This recognizes that a secondary user has to make some decisions concerning the adequacy of the data for his needs; and undoubtedly he recognizes that he wants the data to be "good"; yet he may not know how to measure what is good. The chapter indicates the need for the secondary user to be educated in the nomenclature of the tools that will measure the adequacy of the data. The chapter could have gone one step further and suggested that the primary user, perhaps even the statistician himself, would do well to review the nomenclature and make sure he is using the tool or set of tools he really intends to use.

That such a "reality check" is required among the statistical community is evident from Marker's observation that within the U.S. Environmental Protection Agency (EPA), the term "accuracy" has different meanings depending on which office (air, water, solid waste, toxics, . . .) is using the term. Indeed, the term accuracy even has two different meanings within one office. His recognition of these difficulties is a welcome admission in a field wherein circular conversations are commonplace because of the lack of adherence to common parlance even among professionals in the same field.

Marker presents statistical terms such as standard deviation, standard error of a sample estimate, and measurement bias to name a few. What distinguishes this presentation is that each term is explained at four levels. The first level is designed to present a simple definitional statement of the tool. The second is also definitional, but is a bit more complex by including the mathematical formulas for the quantifiable terms. The third and fourth levels are more theoretical. Most likely the most critical level is the first one. This is probably the level most overlooked, or perhaps more accurately, most "presumed known." In this regard the first level may have been the most difficult to establish, but is also the most important.

Three of the chapters in this section describe applications of and problems involved in the collection and analysis involving environmental statistics methods in the areas of chemical contamination in U.S. coastal waters (Chapter 19), dose estimation (Chapter 20), and site selection for area remediation (Chapter 21).

The section ends with Chapter 22, describing analysis of the need to collect, analyze, and present environmental statistics data and information in a comprehensive and complete way. This will necessitate a center for the activity, and the chapter describes current work in this direction at the U.S. EPA.

CHAPTER 18

The Quality of Environmental Databases

David A. Marker and Svetlana Ryaboy

INTRODUCTION

The quality of environmental databases can be affected by many sources of variability and bias. These include field instruments, laboratory analyses, collection of physical samples, population variability, and survey sampling components of error. This chapter reports on an effort by the Environmental Protection Agency (EPA) to develop a comprehensive list of the factors that impact the quality of environmental databases, and to standardize the set of indicators that should accompany any database. It is hoped that such indicators would allow secondary data users to better understand the utility and limitations of available data, thereby improving the quality of their inferences. This chapter describes a report, "Tools for Determining Data Quality,"[1] prepared by Westat for the Environmental Statistics and Information Division, Office of Policy, Planning, and Evaluation, U.S. Environmental Protection Agency.

It is hoped that this report will achieve five goals. These are to:

- provide a list of tools for determining data quality that are associated with any database produced or used by the EPA
- have producers of all EPA databases provide information using these tools in order to allow secondary data users to determine the suitability of the database for their analytic or decision-making purposes
- similarly, secondary data users will use this terminology when they request information on the quality of existing databases
- serve as a checklist for authors of both the methodology sections of final reports and data quality objectives (DQOs)
- illuminate the trade-offs between cost and quality that are always encountered when collecting and analyzing data

315

DEFINITION OF TERMS

The biggest potential roadblock to successful development of a list of tools which would accomplish the above goals is the language differences between statisticians, chemists, and other environmental scientists. This is exacerbated by the language disagreements within each of these fields. It is therefore necessary to define how a few key words are to be used before actually describing the report. The first term to be defined is data quality.

Data quality cannot be described in the abstract; instead it must be defined relative to a given use for the data. A given set of data may be of excellent quality for the purpose of guiding the design of a future data collection effort. At the same time, this same data set may be completely incapable of assisting in development of regulations for the studied analyte. These different determinations of quality may result from different quantitative requirements such as levels of precision or from qualitative requirements such as differences between the survey population and the populations of inference.

The second term requiring definition is accuracy. Accuracy has been used by other researchers to mean either of two very different concepts. An excellent discussion of this issue is provided by Kirchmer.[2] "It is interesting to note that while statisticians [at the EPA] have clearly recognized the two ways of defining accuracy, the definition that equates accuracy with bias or systematic error has been commonly used in water analyses. [Others define accuracy in terms of] the total error of a result; that is, accuracy represents the combined random and systematic errors of results and is said to improve as the total error decreases."

It is interesting to note that the EPA Office of Air uses yet another definition for accuracy: the error identified when conducting a performance evaluation audit, with precision referring to the same kind of error, but when it is identified by analysis of quality control samples.

The definition of accuracy as the combination of random (precision) and systematic (bias) errors is the one commonly accepted by professional statisticians. In the *Dictionary of Statistical Terms*, Kendall and Buckland[3] define accuracy in terms of this overall combination. They mention that the term has also been used to refer to either unbiasedness or precision, but they state that "neither usage can be recommended."

"With the definition commonly used [at the EPA], methods giving very imprecise results can be characterized as accurate, when individual analytical measurements are clearly not accurate. A definition of accuracy based on individual analytical measurements, which includes the effects of random as well as systematic errors, is clearly more useful."[1] The Water Resources Division of the U.S. Geological Survey (USGS) also defines accuracy to include both types of error.

The EPA Quality Assurance Management Staff (QAMS) defines accuracy to include "a combination of random error (precision) and systematic error (bias) components which are due to sampling and analytical operations . . . [QAMS] recommends that [the term accuracy] not be used, and that precision and bias be used to convey the information usually associated with accuracy." *This report supports the definition of accuracy including a combination of precision and bias. However, to avoid unnecessary confusion, we headed chapters with the words precision, bias, and mean square error* (MSE), where mean square error is defined as the sum of the variance (the inverse of precision) and the square of the bias.

ORGANIZATION

The list of quality measurement tools is organized according to six elements of quality: *precision, bias, mean square error, representativeness, comparability, and completeness.* Examples of the tools corresponding to the first four elements are as follows (comparability and completeness are general elements of quality that are not comprised of a separate list of tools):

- precision —
  standard deviation
  standard error of the mean
  field instrument and laboratory measurement
    variation
  seasonality
  physical support

- bias —
  average percent recovery
  measurement equipment bias
  biases from questionnaires

- mean square error —
  measurement MSE
  survey sampling MSE
  nonresponse (or nonanalyzed) MSE

- representativeness —
  populations and coverage
  time period covered
  detection, quantification, and reporting limits
  minimum reliably detected concentration

Each of the tools are defined at two levels (see Exhibit 1). Level I provides clear, nonmathematical, definitions. Level II is more detailed, providing mathematical or statistical formulas for the quantifiable data quality tools. Readers solely interested in definitions of terms are urged to examine level I, skipping the more theoretical discussions of level II.

(Level I definitions can be easily found by looking for the boldfaced/ italicized areas of the text.) It is important to recognize that some components (e.g., nonresponse) can affect many elements (e.g., nonresponse variation and nonresponse bias). This interrelationship between elements is discussed in Chapter 8 of the Westat report, following the separate chapters for each quality element.

Chapter 9 of this same report provides suggestions for further research to improve the utility of the document. Included in this chapter are level III definitions for selected tools. These level III definitions contain "state-of-the-art" or alternative statistical procedures. It is hoped that in the future such definitions could be developed for all appropriate tools.

EXAMPLE DEFINITIONS OF TOOLS

The list of tools represents a combination of standard statistical terms such as standard deviation, confidence intervals, and populations; and environmental science terms such as physical support, average percent recovery, and reporting limits. Depending on the background of the user, some of these definitions will be obvious, while others may be completely new.

Exhibit 1 shows the definitions for two measures of precision: the standard deviation and standard error of a sample estimate. This demonstrates the differences between level I and II definitions. These definitions are straightforward for statisticians, but not for all environmental scientists. Readers may be surprised by the treatment of the finite population correction (fpc). Many survey sampling statistical texts include the fpc when they define the standard error. Many environmental texts do not bother with the fpc because in the vast majority of environmental studies the sampling fraction is quite small. The fpc is appropriate in some environmental data applications such as when there are only a limited number of point sources for pollution. Thus in this document the fpc is described, but not until after introducing the standard error.

Exhibit 2 shows the definitions for two sources of bias: average percent recovery and measurement (equipment) bias. The first of these is a well-known source of bias in environmental data; the second is common to both statistical and environmental sampling.

The next example is the definition of populations and coverage. This issue of representativeness is not usually mentioned in environmental science and it does not have a uniform terminology in the statistical community. We have used the terminology found in Groves.[4] Exhibit 3 shows definitions for survey population, frame population, target population, frame undercoverage, and population of inference. Level II provides an example of each of these terms for a hypothetical environmental survey.

The last example (Exhibit 4) describes how the completeness of collected data is qualitatively measured against the study objectives. These objectives may be in the form of formal data quality objectives (DQOs) or the planned uses of the data. Two potential users of the study, with different objectives, might find very different levels of completeness achieved by the same set of data. This use of the term completeness is specific to the EPA, but provides a method of evaluation that may have application to a much wider audience.

ACKNOWLEDGMENTS

The list of tools for determining data quality and their definitions were selected and developed by David Marker and Svetlana Ryaboy. They would like to acknowledge the input and suggestions that led to significant improvements by the following Westat staff: Bob Clickner, Steve Dietz, David Morganstein, and John Rogers. The EPA review and guidance were provided by John Warren, Pepi Lacayo, and Bob O'Brien of the Statistical Analysis and Computing Branch of the Office of Policy, Planning, and Evaluation (OPPE); Nancy Wentworth and Dean Neptune of the Quality Assurance Management Staff of OPPE; and Terrence Fitz-Simons of the Monitoring and Reporting Branch of the Office of Research and Development.

DISCLAIMER

This chapter does not reflect official positions of the U.S. Environmental Protection Agency.

EXHIBIT 1. DEFINITIONS OF THE STANDARD ERROR AND THE STANDARD ERROR OF A SAMPLE ESTIMATE

Standard Deviation

Level I

  The standard deviation is a measure of the spread of a set of numerical values (e.g., concentrations) around their overall average. This is the most frequently used measure of spread. It is the measure of spread associated with the individual values of the population under study.

Level II

  The standard deviation is the square root of the average squared deviation of the individual values from their true mean. It may be estimated from a simple random sample of size n as:

$$s = \sqrt{\frac{1}{n-1} \sum_{i=1}^{n} (x_i - \bar{x})^2} = \sqrt{\frac{1}{n-1} \left(\sum_{i=1}^{n} x_i^2 - n\bar{x}^2 \right)}$$

where $\bar{x}$ = the sample average.
  The following useful rule can be applied when the data are assumed to be normally distributed. The chance of a random interval of the following form containing the next individual sampled value is:

Interval	Probability
$\bar{x} \pm s$	0.68
$\bar{x} \pm 1.645\ s$	0.90
$\bar{x} \pm 1.96\ s$	0.95
$\bar{x} \pm 3\ s$	0.997

Example 1. Calculation of Standard Deviation

  Given sample solutions with concentrations:

$$x_1 = 1.5 \text{ ppb}$$
$$x_2 = 2.0 \text{ ppb}$$
$$x_3 = 1.0 \text{ ppb}$$
$$x_4 = 1.5 \text{ ppb}$$
$$x_5 = 2.5 \text{ ppb}$$

then

$$\bar{x} = (1.5 + 2.0 + 1.0 + 1.5 + 2.5)/5 = 1.7$$

$$\sum_{i=1}^{5} x_i^2 = 1.5^2 + 2.0^2 + 1.0^2 + 1.5^2 + 2.5^2 = 15.75$$

and

$$s = \sqrt{\left(\frac{1}{4}\right)(15.75 - 5(1.7)^2)} = \sqrt{\left(\frac{1}{4}\right)(15.75 - 14.45)}$$

$$= 0.57$$

The standard deviation (as well as the sample average) can be highly influenced by extreme values, that is, values that are extremely high or low as compared with the rest of the data. For example, suppose that in Example 1 the concentration of the last sample is $x_5 = 49$ ppb instead of 2.5 ppb; then:

$$\bar{x} = (1.5 + 2.0 + 1.0 + 1.5 + 49)/5 = 11.0$$

$$\sum_{i=1}^{5} x_i^2 = 1.5^2 + 2.0^2 + 1.0^2 + 1.5^2 + 49^2 = 2410.5$$

and

$$s = \sqrt{\left(\frac{1}{4}\right)(2410.5 - 5(11.0)^2)} = \sqrt{\left(\frac{1}{4}\right)(2410.5 - 605)}$$

$$= 21.25$$

This shows that the selected sample should be very carefully examined to ensure the reliability of the actual numbers before making any conclusions based on it.

Standard Error of a Sample Estimate

Level I

Samples are selected in order to make estimates of population parameters. For example, in Example 1 a sample of five solutions is selected, and the sample mean is used as an estimate of the population mean.

Similarly, one could estimate the proportion of values above or below a certain level. The standard error of a sample estimate is a measure of the variability in that estimate that occurs in repeated sampling, when employing the same sampling procedure. Its value depends on the process variability, on the number of observations in the sample, and on how the sample was obtained. Unlike the standard deviation, which refers to the variability of individual measurements, the standard error of the estimate is a function of the sample size.

Level II

When the data are obtained from a simple random sample, the standard error of the sample mean can be estimated by the standard deviation divided by the square root of the sample size. That is, the estimated standard error of the mean = $s.e.(\bar{x}) = s/\sqrt{n}$. In Example 1, the standard error of the mean is $0.57/\sqrt{5} = 0.254$.

When estimating a proportion p (e.g., the proportion of samples containing detectable levels of an analyte) from a simple random sample, the estimated standard error of the proportion may be estimated by:

$$s.e.(\hat{p}) = s_p = \sqrt{\frac{\hat{p}(1-\hat{p})}{n}}$$

where $\hat{p}$ is the sample estimate of the proportion. It is worth noting that in all cases $s_p \leq 0.5/\sqrt{n}$.

Example 2. Calculation of the Standard Error of p

Assume that 10% of 300 samples contained detectable levels of the analyte:

$$\hat{p} = 0.10$$
$$n = 300$$

$$s_p = \sqrt{\frac{0.1(1-0.1)}{300}} = 0.017$$

When the population is finite, the sample is collected using a random sample without replacement, and the sampling fraction exceeds 5–10% of the entire population; then a finite population correction (fpc) should be taken into account when calculating the standard error.[4] The standard error is multiplied by $\sqrt{(1 - n/N)}$, where n is the sampling fraction, and N is the population size. For example if (n =) 300 out of (N =) 800 factories were selected for the sample, the sampling fraction is $n/N = 300/800 = 0.37$. Then for example 2:

$$s_p = \sqrt{\frac{\hat{p}(1 - \hat{p})}{n}} \; \left(1 - \frac{n}{N}\right) = \sqrt{\frac{(0.1)(0.9)}{300} (1 - 0.37)} = 0.013$$

EXHIBIT 2: DEFINITIONS OF AVERAGE PERCENT RECOVERY AND MEASUREMENT (EQUIPMENT) BIAS

Average Percent Recovery

Level I

Average percent recovery is a measure of how well laboratory equipment, protocols, and technicians can detect known concentrations of a contaminant. For example, a given sample may be known to contain 10 ppm of nitrates, but repeated measurement of this sample by a laboratory might on average only detect 8 ppm.

Level II

Average percent recovery is the ratio of the average detected concentration to the average known concentration, in known concentration samples. Ideally, multiple samples are examined over a reasonable range of concentrations at each laboratory. Consistency in percent recovery can then be examined across concentrations and laboratories. Lack of consistency can be used to suggest improvements in quality assurance procedures.

Average percent recovery is almost always less than 100%. If no adjustment is made in the data analyses, it will result in a downward bias in both average concentrations and percent detections. Dividing measured concentrations by the average recovery will adjust for most of this bias in estimating average concentrations. It will not, however, adjust for the underestimate in either average concentration or percent detected resulting from samples measured below the reporting/detection limit with true concentrations that are above the limit (unless chemists report all measurements that are below their reporting/detection limits). Thus, it is important to develop protocols and use laboratory equipment that can achieve an average percent recovery as close to 100% as possible.

Example 6. Calculation of Average Percent Recovery

Five known concentration samples with 50 ppb are sent to the laboratory. The average reported value is 40 ppb.

$$\text{Average percent recovery} = \frac{40 \text{ ppb}}{50 \text{ ppb}} = 0.80 = 80\%$$

If a field sample is measured with 60 ppb, an unbiased measurement is:

$$\frac{60 \text{ ppb}}{0.80} = 75 \text{ ppb}$$

Suppose that 10 ppb is the detection limit at the laboratory, and a field sample is measured (but not confirmed) around 9 ppb. It is therefore reported as nondetected. Given the average percent recovery of 80%, an unbiased (not confirmed) estimate for this field sample is 11 ppb, which is greater than the detection limit. Thus the percent detected and average concentrations will be biased by the combined impact of the average percent recovery and the detection limit.

Measurement (Equipment) Bias

Level I

There are two ways in which measurement bias can result from field instruments and/or laboratory equipment. First, analogous to average percent recovery in a laboratory, field instruments may on average detect an amount not equal to the true amount being measured. This may be adjusted for by recalibration of the equipment and/or the development of mathematical adjustments to the raw data.

Second, estimates are sometimes based on the maximum of a series of field instrument or laboratory equipment measurements. For example, a building may be considered to contain lead-based paint if the maximum of multiple readings from that building exceeds a certain lead concentration. In such cases, lack of precision in the individual equipment measurements combines with sampling variation to increase the variability of the individual readings. This increases the chance that the maximum of multiple readings will exceed the cutoff value, thus upwardly biasing the estimates of the percent of buildings containing lead-based paint.

Level II

All of the calculations in Example 6 can apply to measurement equipment bias as easily as they do for average percent recovery. Five measurements are taken of a substance known to contain 50 ppb, but the average measurement from the field equipment is 40 ppb. The second way in which measurement error can introduce bias is demonstrated by the following example.

Example 7. Bias Introduced by Measurement Error

We would like to estimate the percent of individuals with lead blood levels above $10\mu g/dL$. The following true (unknown) concentrations of lead are found in a sample of 100 individuals.

Lead levels ($\mu g/dL$)	Number of individuals
1–7	60
8–10	20
11–13	10
≥ 13	10

Due to measurement error, the concentrations of 20% of each category are measured in the next smaller category and another 20% are measured in the next larger category. This results in 4 of the 20 individuals with true lead levels of 8–10 $\mu g/dL$ being measured with levels exceeding the cutoff of 10 $\mu g/dL$. Two of those with true lead levels between 11 and 13 $\mu g/dL$ have their levels measured below 10 $\mu g/dL$.

This results in a sample estimate of 22 out of 100 individuals exceeding the cutoff, when only 20 of them had true concentrations above the cutoff. This bias is exacerbated when the maximum of multiple readings is used to determine contamination.

EXHIBIT 3. DEFINITION OF POPULATIONS AND COVERAGE

Populations and Coverage

Level I

The survey population consists of those members of the frame population with a positive probability of being included in the study. This excludes those parts of the frame that consistently refuse to participate, plus those which cannot be analyzed according to the established survey protocols. For example, if water samples are required to be selected from individual wells, then tubular well systems in which water from multiple wells is piped into a common reservoir before being piped to the surface would not be part of the survey population, even if they are listed on the frame.

The frame population is the population from which the sample is drawn. The frame might consist of one or more of the following: lists of units (e.g., houses, individuals, wells, factories), lists of areas (typically as defined by the U.S. Bureau of the Census), or lists of telephone exchanges (for random digit dialing surveys).

The target population is the population of which the frame population is hopefully representative and all inclusive. This population will frequently cover the same time period as the frame population, but may include additional members for whom no frame exists and for which it is not deemed cost efficient to try and sample. For example, a telephone survey is frequently used even when the target population includes all homes in the United States. Those households without telephones may be in the target population, but they are not in the frame population. Similarly a survey of asbestos in buildings may select its sample from a list of buildings built before a certain date which contain boiler rooms and other locales where asbestos is most likely to be found. Frame undercoverage is most typically used to refer to the difference between the target population and the frame population.

The population of inference is the population about which it is desired to draw conclusions based on the data collected by the study.[4] This population may frequently be larger than the target, frame, and survey populations. For example, while a study is typically conducted during a short period of time, it is common to want to draw conclusions about past; and, in particular, future occurrences on the same population of elements. When a study is to be used to support regulatory action, the population of inference is typically the population to whom the regulation would apply. For example, in determining whether to ban or severely limit the application of a certain pesticide, the EPA will survey the current and historical applications of the pesticide. The population of

*applications about which the EPA wants to draw conclusions (infer-
ences) is all future applications. These are the ones that the EPA hopes to
ban or control.*

Level II

Assume the EPA is considering severely restricting the household use
of an herbicide. It is desired to collect data on all recent users of the
herbicide in order to draw inferences about the impact on future users.
To collect this information quickly and cost effectively it is agreed that a
telephone survey of homeowners will be conducted. The population of
inference is future household users of the herbicide. The target population is recent household users. The frame population is those households
with telephones. The survey population is those households with telephones that are able to respond to the survey.

The survey population excludes frame population households that are
nonEnglish speaking (unless the interviewers are multilingual), deaf
households, households whose telephone is primarily used for business
purposes (telephone household surveys typically screen out business
phones), households that are on vacation during the entire data collection period, and households that refuse to participate in any surveys. The
frame population excludes nontelephone households, which are a part of
the target population. Finally, the target population excludes households
that have not recently, but may in the future, use the herbicide.

When discussing frame undercoverage, it is important to try and estimate the size of each of these excluded groups. For groups with members
making up a more than trivial percentage of the population, it is also
important to try and estimate whether the responses from the under-
covered would be different from those who are covered.

EXHIBIT 4. DEFINITION OF COMPLETENESS

Completeness

Level I

*During the design stage of the survey, data quality objectives (DQOs)
are specified. These quantitative statements about the desired quality for
the collected data are predicated on a series of assumptions about the
underlying system being studied. Completeness compares the final
results of the study against DQOs in terms of precision, bias, mean
square error, and representativeness. Completeness can also be measured
by the ability to use final data to conduct the analyses that were planned,*

even if DQOs are not met. Lack of completeness may be a result of poorly executed studies, or of initial assumptions that are found to be invalid once the data are collected. At the most simplistic level, completeness can be measured as the proportion of DQOs which have been met by the survey or the proportion of planned analyses that can be supported by the data. An examination of the actual quality of survey data should try to address the causes for any lack of completeness.

Level II

Lack of completeness can indicate major shortcomings in the utility of a study. Many of the planned analyses are likely to be impossible to conduct from such a study. Sometimes DQOs were established to allow for testing a specific hypothesis with a certain degree of confidence in the results. When these objectives are not met either the hypothesis to be tested may need to be modified or the degree of confidence in the results may be reduced.

When DQOs are met, it does not imply that the planned hypothesis tests will detect significant differences. It may be that there are no real differences between the true values being compared in the test. Meeting DQOs simply allows you to have the desired level of confidence in the results of your test.

The following four situations demonstrate the importance of considering the planned usage of the data when determining the completeness of a study. The purpose of the study is to test the hypothesis that the average concentration of dioxin in surface soil is no more than 1.0 ppb. The established DQO specified that the sample average should estimate the true average concentration to within ±0.30 ppb with 95% confidence.

1. $\bar{x}$ = 1.5 ppb ± 0.28 ppb   satisfies DQO and study purpose
2. $\bar{x}$ = 500 ppb ± 0.28 ppb   satisfies DQO and study purpose
3. $\bar{x}$ = 1.5 ppb ± 0.60 ppb   does not satisfy either
4. $\bar{x}$ = 500 ppb ± 0.60 ppb   fails DQO but meets study purpose

For all but the third situation, the data that were collected completely achieved their purpose. That is, they either met the data quality requirements originally set out, or if they failed the DQO they still achieved the purpose of the study.

When a study is found to lack completeness, the reasons for this shortcoming should be investigated. It may be a result of poor assumptions on which the DQOs were established, poor implementation of the survey design, or the design proved impossible to carry out given the monetary and/or time limitations that were established. Regardless of the reason, this information is vital to the planning of future studies.

Lack of completeness should always be investigated and the lessons learned from conducting the study incorporated into the planning of future studies.

REFERENCES

1. Marker, D. A. and Ryaboy, S. "Tools for Determining Data Quality," report submitted to U.S. EPA Environmental Statistics and Information Division, July 31, 1992.
2. Kirchmer, C. J. "Quality Control in Water Analyses," *Environ. Sci. Technol.*, 17(4), 1983.
3. Kendall, M. G. and Buckland, W. R. *A Dictionary of Statistical Terms*, 4th ed., Longman Inc., New York, 1982.
4. Groves, R. *Survey Errors and Survey Costs*, John Wiley & Sons, New York, 1989.
5. Cochran, W. G. *Sampling Techniques*, John Wiley & Sons, New York, 1977.

CHAPTER 19

The National Oceanic and Atmospheric Administration (NOAA) National Status and Trends Mussel Watch Program: National Monitoring of Chemical Contamination in the Coastal United States

Thomas P. O'Connor

ABSTRACT

Since 1986, the Mussel Watch project within the NOAA National Status and Trends (NS&T) program has been chemically analyzing sediments and molluscan tissues collected at sites throughout the coastal United States. Data from sediment analyses have been used to describe the status, or spatial distribution, of contamination on a national scale. The molluscan data, on the other hand, are used primarily to describe and follow temporal trends in contaminant concentrations. On the basis of accumulating knowledge on both the status and trends of chemical contamination, the NS&T program has changed to some extent in terms of the chemicals it measures, its frequency of collection at individual sites, and its level of replication.

INTRODUCTION

The Mussel Watch project within the National Status and Trends (NS&T) program of the National Oceanic and Atmospheric Administration (NOAA) began in 1986 with the goal of determining the status and trends of chemical contamination in the coastal and estuarine United States. With that goal as a guiding principle, developing a program required making choices. Sites for sample collection had to comprise a set that could represent conditions the United States as a whole. The time scale for trends needed to be selected. Specific chemicals needed to be chosen. The matrices, or types of samples to be analyzed, had to reflect chemical concentrations at that site and time.

While all of these choices are to some extent arbitrary, the NS&T program has annually measured concentrations of chemicals, which are mostly priority pollutants, in triplicate samples of sediment and in bivalve mollusks collected at fixed sites. As the program has evolved and reported on the spatial distribution of contamination and its temporal trends, those initial choices are constantly reexamined. Chemicals have been added and subtracted, sites have been added, annual sampling continues for mollusks but not for sediment, and replicate samples are no longer collected. It is instructive to examine how the NS&T program selected its sites and its chemicals, summarized its data in terms of status and trends, and adjusted its sampling strategy based on conclusions drawn from those analyses.

SITE SELECTION

The National Research Council[1] recently reported a need for large-scale, long-term monitoring in the coastal United States. The council noted that more than $100 million is spent each year on compliance monitoring, i.e., testing wastewaters and other materials prior to discharge, or making measurements near discharge points as prescribed by regulation. Since compliance monitoring covers very small spatial scales, national programs such as the NS&T program are needed to focus on wider scales. It is on this wider scale that national benefits should be derived from expending billions of dollars on regulating discharges of chemical contaminants. Because the Mussel Watch project is designed to describe chemical distributions over these national and regional scales, its sampling sites need to be representative of rather large areas. To this end, no sites were knowingly selected near waste discharge points or poorly flushed industrialized waterways. Similarly, to avoid extremely local contamination, the program avoids collecting mollusks from artificial substrates such as pilings which are often chemically treated to inhibit decay.

A further requirement of the Mussel Watch project is that indigenous mussels or oysters be collected for analysis. Mollusks can be put into cages to monitor chemical concentrations at any particular spot; and caging is done, for example, to monitor contamination near waste discharges[2] or during at-sea disposal of dredged material.[3] For practical purposes, though, a program visiting sites throughout the nation must rely on natural populations being available at the time of annual sampling. Thus Mussel Watch sites can only be located where indigenous populations of mussels or oysters exist.

Last, the NS&T Mussel Watch project is not the first use of mussels on a national scale for chemical monitoring in the coastal United States. In particular, Goldberg et al.[4] sampled about 100 sites each year between 1976 and 1978. To allow comparisons of data with that program, when possible its sites were incorporated into the NS&T program.

Mussel Watch sampling sites are not uniformly distributed along the coast. Within estuaries and embayments, they average about 20 km apart, while along open coastlines the average separation is 70 km. Almost half of the sites were selected in waters near urban areas, within 20 km of population centers in excess of 100,000 people. This choice was based on the assumptions that chemical contamination is higher, more likely to cause biological effects, and more spatially variable in these waters than in rural areas.

In 1986 a total of 145 Mussel Watch sites were sampled. In 1988 a few sites were added on the East Coast to fill in large spatial gaps between sites, and 20 new sites were selected in the Gulf of Mexico for the specific purpose of gathering samples closer to urban centers. Results from the initial sampling showed that the highest chemical concentrations were near urban areas on the East and West Coasts, and that few sites in the Gulf of Mexico could be considered contaminated. Since urban centers along the Gulf are further inland than those on other coasts, an attempt was made to sample further inland. The major limitation on doing that, however, is that oysters are not found at salinities below about 10 ppt. By 1990 a total of 234 sites had been sampled, with further additions made to test the representativeness of earlier sites.

CHEMICAL SELECTION

The elements and compounds measured in the NS&T program are listed in Table 1. Not listed are Si, which is measured in sediment but not mollusks; and Sb and Tl, which were not detected in mollusks during the first 2 years of the program and are no longer measured. Except for Al, Fe, and Mn, the elements in Table 1 are all possible contaminants in the sense that their concentrations in the environment have been altered by human activities.[5] The mere existence of the chlorinated organic compounds and butyltins is due to human actions. Polycyclic aromatic hydrocarbons are similar to metals in the sense that they occur naturally. They are found in fossil fuels such as coal and oil and are produced during combustion of organic matter. Their environmental presence, however, is also attributable to humans because they are released in the use and transportation of petroleum products and a multitude of human activities, from coal and wood burning to waste incineration create polycyclic aromatic hydrocarbons (PAH) compounds in excess of those that exist naturally.

Almost all the chemicals in Table 1 are also on the list of 127 priority pollutants created by the United States Environmental Protection Agency (EPA) in the late 1970s. Priority pollutants are chemicals which commonly appeared in discharged wastewaters and for which there were existing stocks of chemicals with which to make standard solutions.[6] Some of the alkylated PAH compounds and the butyltins in Table 1 are not priority pollutants because standards did not exist or the environmental occurrence of the compound was not recognized in the late 1970s.

The 74 priority pollutants that are not being measured are almost all low molecular weight, highly soluble, volatile organics. While they can be toxic, they have relatively low tendency to leave the aqueous phase and be accumulated in organisms or on surfaces of sediment. Gossett et al.[7] measured concentrations of priority pollutants in effluent from a large sewage treatment plant and in organisms and sediments collected near the end of the discharge pipe. Compounds with low octanol-water coefficients (i.e., those with a relatively strong tendency to remain dissolved in water) were usually not detected or were in very low concentrations in organisms and sediments. In general, organic compounds monitored by NS&T have octanol-water coefficients greater than about 5000.

There are four priority pollutants that qualify for inclusion on the NS&T list of measured chemicals but are not measured: diethylhexylphthalate (DEHP), pentachlorophenol, endosulfan, and toxaphene. The latter two are pesticides; pentachlorophenol is a commonly used biocide; and DEHP is the most widely produced of the phthalates used extensively as plasticizers and, unlike other phthalates, resists degradation.[8] They are all compounds with high octanol-water coefficients and could be expected to be found in organisms and tissues. They have not been included among the NS&T chemicals because each requires special handling. All the organic analytes in Table 1, except butyltins, are extracted together from sediments or tissues. One aliquot of that extract is analyzed by gas chromatography-electron capture detection (GC-ECD) for the chlorinated compounds and gas chromatography-mass spectroscopy (GC-MS) for the polycyclic aromatic hydrocarbons. Measuring toxaphene requires a separate cleanup of that extract to remove the polychlorobiphenyls (PCBs). One form of endosulfan (endosulfan sulfate) and pentachlorophenol require an entirely separate extraction procedure. Phthalate analysis is hindered by the ubiquitous presence of phthalates in all the plastic components of a laboratory. If DEHP were added to the NS&T list, special precautions—including isolation from the main analytical laboratory—would be required.

A third category of possible compounds is contemporary pesticides. Except for endosulfan, all the chlorinated pesticides and PCBs on the

Table 1. Chemicals Measured NOAA NS&T Program

DDT and
its metabolites

2,4'-DDD
4,4'-DDD
2,4'-DDE
4,4'-DDE
2,4'-DDT
4,4'-DDT

Chlorinated pesticides
other than DDT

Aldrin
Cis-chlordane
Trans-nonachlor
Dieldrin
Heptachlor
Heptachlor epoxide
Heptachlor
Heptachlor epoxide
Hexachlorobenzene
Lindane (γ-BHC)
Mirex

Polychlorinated biphenyls

PCB congeners 8, 18, 28,
44, 56, 66, 101, 105, 118,
128, 138, 153, 179, 180,
187, 195, 206, 209

Tri-, di-, and mono-butyltin

Polycyclic aromatic
hydrocarbons

2-Ring
    Biphenyl
    Naphthalene
    1-Methylnaphthalene
    2-Methylnaphthalene
    2,6-Dimethylnaphthalene
    1,6,7-Trimethylnaphthalene

3-Ring
    Fluorene
    Phenanthrene
    1-Methylphenanthrene
    Anthracene
    Acenaphthene
    Acenaphthylene

4-Ring
    Fluoranthene
    Pyrene
    Benz(a)anthracene
    Chrysene

5-Ring
    Benzo(a)pyrene
    Benzo(e)pyrene
    Perylene
    Dibenz(a,h)anthracene
    Benzo(b)fluoranthene
    Benzo(k)fluoranthene

6-Ring
    Benzo(ghi)perylene
    Indeno(1,2,3-cd)pyrene

Major elements

Al Aluminum
Fe Iron
Mn Manganese

Trace elements

As Arsenic
Cd Cadmium
Cr Chromium
Cu Copper
Pb Lead
Hg Mercury
Ni Nickel
Se Selenium
Ag Silver
Sn Tin
Zn Zinc

Priority Pollutant list and all those measured by NS&T are banned for use in the United States. We are monitoring what should be the gradual disappearance of these compounds as they disperse, get buried in sediment, or very slowly decay. There are, however, many compounds that are registered for use on crops. Common herbicides are atrazine, alachlor, 2,4-D, and metalochlor; and common insecticides are carbaryl, carbforan, chlorpyrifos, and methyl parathion.[9] Most contemporary pesticides share two common characteristics, either of which excludes them as possible candidates for annual monitoring. Generally they have low octanol-water coefficients so that when they are flushed from fields into natural waters, they tend to stay dissolved rather than be accumulated by organisms or adsorbed onto sediments. The few reports of the occurrence of contemporary pesticides in the environment,[10,11] and the reports tabulated by Pait et al.[9] all emphasize aqueous phase measurements. Second, the contemporary pesticides have relatively short half-lives. It is of course desirable that toxic chemicals break down rather than accumulate in the environment, but that also makes it difficult to monitor such compounds unless sampling is done near the time of application.

In 1988 the NS&T Mussel Watch project did attempt to measure atrazine, alachlor, propanil, methyl parathion, and carbaryl in oysters from at two sites in the Maryland portion of Chesapeake Bay and one site in Winyah Bay, South Carolina. These pesticides are in use in those regions but none were detected in oysters. The annual NS&T sampling occurs in winter. Conceivably, if the oysters had been collected in spring at the time when pesticides are applied, they might have been found in the oysters. However, just as NS&T sites are chosen to be representative of areas rather than isolated "hot spots," it would not be representative of Chesapeake Bay or Winyah Bay if results depended on collecting in conjunction with pesticide application.

One contemporary insecticide, chlorpyrifos, has been detected in sediments off Central America.[12] This compound is unusual among modern pesticides because of its very high octanol-water coefficient of more than 100,000. It is used in the United States[9] and is a potential analyte for the NS&T program.

Since mollusks are collected annually, there is opportunity to expand the list of analytes at some sites in some years. As explained this was done for three rural sites in an attempt to monitor some contemporary pesticides. It was done as well for butyltins in 1987 and 1988. In that case tributyltin and its breakdown products, dibutyltin and monobutyltin, were found almost at all tested sites; and analyses for butyltins are now regularly done at all sites. In this case we expect to monitor a decrease in butyltin resulting from a 1988 law banning their further use as an antifouling agent on boats less than 25 m in length. A third category of chemicals, radionuclides, have been measured at a few sites on a one-time basis.[13] Here the objective was to gather a set of current values for fallout nuclides (Cs-137, Am-240, and Pu-239,240) to compare with values obtained by Goldberg et al.[4] in the late 1970s.

SPECIES SELECTION

Mussels or oysters are sampled at each Mussel Watch site. They were chosen as sentinel species because they are sessile, hardy, and unlike fish have limited capacity for metabolizing rather than simply accumulating polycyclic aromatic hydrocarbons. For essentially the same reasons, other programs have collected or are collecting mollusks for chemical analysis. Data from these programs are to some extent comparable with those from the NS&T program. An example of an earlier nationwide program is that of Goldberg et al.[4] in the late 1970s. That program itself was preceded by one conducted by Butler,[14] which from 1965 to 1973 measured pesticide concentrations in marine organisms including mussels and oysters. The state of California has been monitoring coastal waters through analyses of mollusks since 1977.[2] Among other countries systematically monitoring through analyses of mollusks are France,[15] Hong Kong,[16] Korea,[17] and New Zealand.[18] The Intergovernmental Oceanographic Commission of the United Nations began an International Mussel Watch program in 1992 with the collection of samples in South and Central America.[19] Cantillo[20] has compiled a bibliography of more than 1200 citations where authors report chemical concentrations in mussels and oysters.

To extend as much as possible the ability to compare among sites, the ideal species would be available in all coastal areas of the United States. There is no such species, but bivalve mollusks are fairly cosmopolitan. In clockwise progression around the continental United States, the mussel *Mytilus edulis* is collected at all sites from Maine to Delaware Bay, the oyster *Crassostrea virginica* is collected from Delaware Bay south and through the Gulf of Mexico, and on the West Coast the mussels *M. edulis* and *M. californianus* are gathered.

At three sites in Long Island Sound both the oyster *C. virginica* and the mussel *M. edulis* have been collected. At one site at the mouth of the Columbia River the mussels *M. edulis* and *M. californianus* were both collected.

Comparisons between species at common sites (Table 2) show that the trace elements Ag, Cu, and Zn are enriched in oysters by more than a factor of 10 relative to mussels. Conversely, Cr and Pb are more than three times higher in mussels. For other elements and for organic compounds there is no strong species effect between mussels and oysters. There are no important differences for elements or organic compounds

Table 2. Ratios of Mean Concentrations of Trace Elements and of Aggregated Groups of Organic Compounds Measured in Two Species at the Same Sites in 1988 and 1989.

Site[a]	Yr	Species[b]	Ag	As	Cd	Cr	Cu
CRSJ	88	mc/me	0.76	1.13	0.65	0.67	0.97
CRSJ	89	mc/me	.[c]	0.95	0.68	1.18	0.76
LIHR	88	cv/me	14.04	0.82	1.93	0.19	31.47
LIHR	89	cv/me	6.15	0.80	1.87	0.21	32.89
LINH	89	cv/me	4.87	0.85	2.21	0.21	17.11
LIPJ	89	cv/me	20.63	0.81	1.71	0.24	25.26

Site[a]	Yr	Species[b]	Hg	Ni	Pb	Se	Zn
CRSJ	88	mc/me	0.52	1.09	0.84	0.92	0.90
CRSJ	89	mc/me	0.66	0.74	2.19	0.82	1.21
LIHR	88	cv/me	0.77	2.43	0.30	0.70	40.00
LIHR	89	cv/me	0.65	1.91	0.24	0.59	38.33
LINH	89	cv/me	0.66	0.89	0.24	.[c]	28.71
LIPJ	89	cv/me	1.07	2.07	0.28	0.84	47.73

Site[a]	Yr	Species[b]	ΣCdane[d]	ΣDDT[d]	ΣPCB[d]	ΣPAH[d]
CRSJ	88	mc/me	1.22	0.27	0.53	1.04
CRSJ	89	mc/me	.[c]	.[c]	.[c]	.[c]
LIHR	88	cv/me	0.87	0.86	0.63	0.63
LIHR	89	cv/me	1.27	1.11	0.96	0.94
LINH	89	cv/me	1.01	0.95	0.98	0.89
LIPJ	89	cv/me	.[c]	.[c]	.[c]	.[c]

Note: Ratios are underlined if they are ≥ 3.0 or ≤ 0.3.

[a]Sites designated as CRSJ, LIHR, LINH, LIPJ are Columbia River South Jetty, Long Island Sound Housatonic River, Long Island Sound New Haven, and Long Island Sound Port Jefferson, respectively.

[b]Species designated as mc, me, and cv are *Mytilus californianus*, *M. edulis* and *Crassostrea virginica*, respectively.

[c]Ratios are missing (.), if the chemical was not detected in one or both species.

[d]The aggregated groups of organic compounds are: ΣCdane = the sum of cis-chlordane and trans-nonachlor and two minor components, heptachlor and heptachlorepoxide. ΣDDT = sum of six compounds (o,p and p,p isomers of dichlorodiphenyldichloroethylene [DDE], dichlorodiphenyldichloroethane [DDD], and DDT). ΣPCB = sum of 18 congeners, ΣPAH = sum of 20 polycyclic aromatic hydrocarbons.

between the two species of mussels. (The threefold difference in ΣDDT at the Columbia River site is ignored because the concentrations were very low in both mussel species.) There is some recent discussion among malacologists over whether the West Coast organism called *M. edulis* is actually *M. galloprovincialis* in California and *M. trossulus* toward the north. In fact, the three mussels may be strains of a single *Mytilus* species.[21] Given this uncertainty, the mussels collected the Columbia River site may have been *M. trossulus* or even *M. galloprovincialis* instead of *M. edulis*. However, the lack of concentration differences between two *Mytilus* species at that site has been taken to validate comparisons among all mussels collected in the program.

Expanding the geographic extent of the Mussel Watch project has required collecting other species. There have been sites in Hawaii since the program began, and the species has been the oyster *Ostrea sandvicensis*. To obtain molluscan data from Puerto Rico the mangrove oyster *Crassostrea rhizophorae* was obtained. Zebra oysters, *Dreissena polymorpha*, were collected at seven sites in the Great Lakes in 1991. At one site in the Florida Keys, the smooth-edged jewel box *Chama sinuosa* was obtained. In all these cases it is probably valid to compare concentrations of organic compounds with those from sites in the continental United States, but the trace element concentrations may be species specific.

SAMPLING STRATEGY

Mollusks are collected annually and in the winter (November through March) with each site occupied within 30 days of an annual target date. This timing is due to several reasons. Annual differences are the temporal scale of the program. Any season, as long as it was consistently used, would serve the objective of being annual. However, since mollusks may change their contaminant concentrations either by losing contaminants or losing biomass during the spawning process,[22] and since spawning usually occurs in spring and summer, it was considered prudent to sample in winter prior to spawning.

Every chemical analysis performed on mollusks is performed on composite samples of whole soft parts of 20 oysters or 30 mussels. Separate composites are collected for inorganic and organic analyses. From 1986 through 1991, three separate composites were collected at each site in each year. Thus, for example, if a site was sampled every year from 1986 through 1990, the NS&T database would contain 15 values for the concentration of each chemical (3 per year for 5 years). Multiple samples yield estimates of the variance of annual means at each site, and it remains the NS&T practice to collect three composites if a site is being sampled for the first time. However, because estimates of annual variance are not used in analyses of trends, in 1992 the NS&T program began collecting a single composite at annually sampled sites.

Sediment collections were made at every Mussel Watch site in 1986 and 1987. Since mussels or oysters are often found on hard bottom substrates such as rocks or shells, it was usually not possible to collect sediment at exactly the same location as mollusks. The protocol in 1986 was to collect three separate samples of the upper 2 cm of sediment at separate sites all within 500 m, of the mollusk collection. Since 1987, the distance between sediment sampling stations has remained at 500 m; but in order to improve the ability to collect muddy rather than sandy samples, the center of the sediment sampling site can be as far as 2 km from the mollusk site. Given the distances between sites and the national scale of the NS&T program, an offset of 2 km can generally be considered negligible

Since 1988, sediment collections have been made only at Mussel Watch sites not already sampled in prior years. This has been due to the fact that in the absence of data on rates of sedimentation and biological mixing of sediment, the NS&T program cannot know the time scale represented by the upper 2 cm of sediment. For example, if mixing and deposition rates at a site are such that the upper 2 cm of sediment is a mixture of particles deposited over the prior 10 years, sediment collections in consecutive years would essentially be annual collections of the same sample.

Since NS&T is designed to detect annual trends, chemical concentrations in 1 year have to be independent of those from other years. This independence does not exist for surface sediment samples, but there are data demonstrating that chemical concentrations in mollusks can reach altered steady-state concentrations over time scales of a few months or less if concentrations change in the surrounding water or the food they filter from that water. Roesijadi et al.[23] showed that when mussels were moved from a clean to a contaminated location in Puget Sound and vice versa, concentrations of trace elements increased in the formerly clean mussels and decreased in the mussels moved to the clean location. In a similar reciprocal transplant experiment, Sericano[24] found PCBs, dichlorodiphenyl-trichloroethane (DDT), and PAH compounds to increase or decrease in response to their surroundings. Pruell et al.[25] are among the many investigators to have shown that, under laboratory conditions, mussels accumulate or depurate contaminants as their exposure to chemicals is altered.

ANALYSIS OF STATUS

The status of chemical contamination is its geographic distribution and trends are its temporal change. To document status, the NS&T program has relied on data from analyses of sediment. All reports on status[26-29] are essentially comparisons among sites and have all emphasized sediment data over that from analyses of organisms. This is simply due to the strong species effect that for some elements precludes valid comparisons of molluscan data between sites with mussels and those with oysters.

While there is no species effect on sediment concentrations, chemical concentrations do vary with sediment grain size; and it is not appropriate to compare concentrations measured in sandy sediments with those found in mud. If comparisons ignored the effect of grain size, sites with sandier sediments would invariably appear the least contaminated. Sites receiving a relatively large amount of chemical contamination could go unrecognized simply because the sediments at that site were sandy and were, therefore, incapable of accumulating contamination.

The NS&T approach to the grain size effect has been to simply divide all concentrations by the fraction of sediment in the sample that was silt or clay (i.e., the fine fraction particles < 63 μm in diameter). This adjustment, in effect, assumes that no chemical contamination is associated with sand and that sand in the sample was only diluting the contaminant concentration. If the fine fraction was less than 0.2, meaning that more than 80% of the sample consisted of particles > 63 μm, the sample was not used for comparative purposes. That sample was not part of the calculated mean for the site. This treatment of the mostly sand samples excluded the mathematical possibility an adjusted concentration could be unrealistically high only because a low raw concentration was divided by a small fraction.

There was a premium, therefore, in finding muddy sediment samples with which to compare among sites. It has already been mentioned that a search for muddy sediments could extend up to 2 km from a mussel watch site. Nevertheless, of the 216 Mussel Watch sites sampled between 1986 and 1989, only 45 of them yielded just sand in every sample. That left data from 171 sites with which to make comparisons. Since there was no information on sediment deposition or mixing rates, it was assumed that all sediment data were temporally equivalent. Thus all nonsandy sample data, regardless of year, from each site were used to determine mean concentrations for that site.

These data were augmented by data from 62 sites sampled by the NS&T Benthic Surveillance project between 1984 and 1986. This project, through 1986, was similar to the Mussel Watch project in the sense that it sampled surface sediment, while fish livers were analyzed rather than whole soft parts of mollusks. The Benthic Surveillance project differs from the Mussel Watch project in that it has a component that seeks evidence of biological responses to contamination among the collected fish. This search for biological effects has, since 1987, forced the Benthic Surveillance project to sample in areas of extreme contamination. These include the small-scale patches of contamination near outfalls or in industrial waterways that are not considered representative of a general area.

The Benthic Surveillance data through 1986 and the Mussel Watch data through 1989 have been combined to form a set of sediment data on chemical contamination throughout the coastal and estuarine United States. The overall concentration distributions for each contaminant are approximately lognormal, and "high" concentrations were defined as those exceeding the mean plus one standard deviation of the lognormal distribution. Those high concentrations are useful for comparisons within the NS&T data set and with other reports on sediment contamination. The high concentrations in units of microgram per gram of dry fine-grained sediment for each contaminant are (in parentheses): Ag (1.2), As (24), Cd (1.2), Cr (230), Cu (84), Hg (0.49), Pb (89), Sn (8.5), Zn (270), ΣPAH (3.9), ΣDDT (0.037), and ΣPCB (0.20).

Most of the high concentrations for any particular contaminant were found at sites near the urban areas of Boston, New York, San Diego, Los Angeles, and Seattle. Reports on the NS&T results[26-29] always recognize that there are innumerable examples of extremely high concentrations that are missed by the program because it samples only "representative" sites. Cantillo and O'Connor[30] compared the NS&T sediment data with data on trace elements in sediments from throughout the world. They found that logarithmic cumulative distribution plots of NS&T and worldwide data had similar means and standard deviations but diverged at the extremes. The worldwide data containing lower low concentrations and higher high values. The offset at the low end was attributed to the worldwide data set containing concentrations in sandy sediments while such sediments were excluded from the NS&T compilation. The high end of the worldwide data set contained concentrations from "hot spots," e.g., wastewater discharge ports or industrialized waterways. Such data are in the worldwide literature because it was usually gathered in the context of documenting particularly egregious environmental impacts. The NS&T Benthic Surveillance project, for example, samples such locations with the specific purpose of testing for effects under extreme conditions.

ANALYSIS OF TRENDS

The effect of species on concentration limits the use of chemical concentrations in mollusks for defining the status of contamination. However, mollusk data are well suited to analysis of trends. Here, the important parameter is the annual direction of change within a site. Species are irrelevant as long as they remain constant at each site.

Temporal trends in molluscan concentrations have been sought through two nonparametric statistical tests on the data for 1986 through 1990.[31] One, the Sign test, is based on the fact that there are many sites with which to examine year-to-year changes. The other, Spearman rank correlation, examines the correlation between the ranks of concentration and year. Ranks rather than actual concentrations are used because one or a few unusually high or low concentrations near the beginning or end of a sequence can yield a significantly high parametric correlation, when in fact there is no consistent pattern to the year-to-year changes. The Spearman test is sensitive only to the direction of change rather than its size, its statistical strength is limited by the number of years for which there are data. With only 5 years of data, the "n" in the test is five; and to be significant at the 95% level of confidence, correlations coefficients must be at least 0.9. This means that only very strong, almost monotonic trends could be detected.

The most common observations from the first 5 years of NS&T Mussel Watch data were that of no statistical change between years on a national scale and that of no strong trends in concentrations at individual sites. However, where the Sign test did find differences between the 2 years (1986 and 1990) they were only decreases. There were significant Spearman correlations between concentration and year in 13% of the site/ chemical combinations; and, among those, decreases were twice as frequent as increases. The overall conclusion was that levels of contamination in the coastal United States are decreasing.

That initial trend analyses was limited to the 141 Mussel Watch sites that were sampled annually over the first 5 years. Over that period, however, 233 sites had been visited. Four of the original sites were not sampled in two of the next 5 years because mussels or oysters were not available. In 1988, 1989, or 1990 55 sites were added to the program, and they will not be examined for trends until 1993 or later. There are 33 sites where mollusks were sampled only once or, perhaps, twice. These are sites where mollusks are no longer found or where sites were added to check on representativeness of unexpectedly low or high levels of contamination at regularly sampled sites.

There is now a grid of about 200 sites at which trends can be monitored through systematic collection and analysis of mollusks. It is not necessary to continue collecting samples at every site in every year. As time passes, it becomes progressively less important to have data for each year over the entire span of years. The bare minimum of years necessary to calculate a meaningful Spearman rank correlation with 95% confidence is four. Tests were run at year five, and trends needed to be very strong to be found at all. If we were willing to wait 9 years to make the first attempt at trend detection and had sampled in alternate years, we would still have an "n" of five, would still have required a correlation coefficient of 0.9, and still only strong trends would have been detected. If data were available for each year over 9 years, weaker trends could be found because the correlation coefficient would only need to be 0.60. However, with 5 years in hand and starting to sample in alternate years, there would be an "n" of seven in the ninth year and a required correlation coefficient of 0.71. Identified trends would need to stronger than if data were available each year, but trends would still be fairly accessible. The difference between required correlation coefficients corresponding to annual and alternate year sampling becomes progressively smaller as time passes. At 13 years, for example, the coefficients for the two sampling frequencies are 0.6 and 0.48; and by year 21, they become 0.37 and 0.41.

There are valid technical and obviously economic reasons to sample in alternate years. On the other hand, there is value in continuing to sample all 200 sites. O'Connor[31] tested 14 chemical concentrations at 141 sites with 5 years of data and found 253 correlation coefficients ≥ 0.9. Since the test was run on 1974 site/chemical combinations (141 $\times$ 14) and since a 95% level of confidence was accepted, there would be 99 (0.05 $\times$ 1974) cases where the correlations were seemingly significant; but they were, in fact, random. Adding chemicals and sites will increase the number of random correlations since, by definition, they will always be 5% of the combinations. Identifying the real trends requires similar behavior at nearby sites. With the first 5 years of data O'Connor[31] was able to distinguish trends in Long Island Sound, Delaware Bay, and Terrebonne Bay as real trends because the same chemicals showed the same trends at nearby sites. The advantage of maintaining a 200-site network is the ability to identify real trends on the basis of geographic proximity among statistically identified trends.

The Sign test says nothing about individual sites; but if the number of sites is large, e.g., 50 or more, it can determine the significance of the direction of change between any 2 years. This method has been applied to the NOAA Mussel Watch data and also used to compare the NOAA Mussel Watch data with that from the previous program[4] of the late 1970s.[32] In the recent data, for example, it was concluded that total butyltin concentrations in the coastal United States are decreasing because its concentration was higher in 1989 at 103 of the 149 sites sampled in both 1989 and 1990. By the Sign test, the random chance of finding 103 out of 149 changes in one direction is $< <0.05$. At 39 of the 50 sites where both the earlier and the NOAA Mussel Watch Programs sampled mollusks, the Pb concentrations were higher in the 1970s. This was taken as evidence that Pb has decreased and was attributed to the fact that between the earlier and more recent measurements use of leaded gasoline decreased dramatically as new cars in the United States required unleaded fuel. Application of the Sign test on a national scale does not require a 200-site network; however, if sites are sampled in alternate years, there will be only 100 sites per year. That is still more than needed for national assessments but provides sufficient samples to look at year-to-year changes separately along East, West, and Gulf of Mexico Coasts.

OTHER U.S. MONITORING PROGRAMS

Through the estuaries component of its Environmental Monitoring and Assessment program (EMAP-EC), the United States Environmental Protection Agency has been conducting coastal and estuarine monitoring since 1990.[33] In that year EMAP-EC sampled in the Virginian province, i.e., Chesapeake Bay northward to the tip of Cape Cod. In 1991 monitoring continued in that province, and began in the Louisianian province and all the U. S. coast along the Gulf of Mexico except southern Florida. In future years other provinces are to be added until EMAP-EC covers the entire coastal United States. The part of EMAP-EC dealing with chemical contamination measures the same chemicals as NS&T in sediments, but EMAP-EC differs from NS&T in a number of ways. Most importantly, EMAP-EC does not analyze mussels or oysters and EMAP-EC does not sample at fixed sites. Site selection is a rigorously random procedure. In 1990, in the Virginian province, 136 sites were occupied. Assuming similar coverage in all six provinces of the coastal and coterminous United States, EMAP-EC when it is national could be annually analyzing sediments from about 800 sites. This is more than three times the number of sites for which the NS&T program has data, and in addition to the previous discussion this is a reason for NS&T to emphasize analyses of mollusks. Joint NOAA/EPA assessments of coastal contamination are being developed using data from both programs.

Since its inception, the NOAA program has included a chemical quality assurance component. Central to it is an annual intercomparison exercise where all laboratories analyze common samples supplied by the U. S. National Institute of Standards and Technology for organic compounds and the Canadian National Research Council for trace elements. Since 1990, the cost of this program has been shared with EMAP-EC; and all labs analyzing samples for both the NOAA and EPA programs are participating.

Beyond that, however, there are myriad monitoring programs in the United States being conducted by state or local agencies. Usually the sampling grids for these programs provide more spatial resolution than NS&T or EMAP-EC, but over small spatial scales. The national programs serve the more local programs in two ways. First, there are points in space where the local and national programs overlap. These provide a basis for extrapolating the local results to larger scales. Second, the annual intercomparison exercises are open to the state and local programs. Thus, while their data remain confidential, programs beyond NS&T and EMAP-EC can document how well they agree with all other participants in the intercomparison exercises.

CONCLUSION

Monitoring is a series of systematic measurements over time or space or both. The temporal component of the NS&T program consists of annually collecting mussels or oysters at about 200 sites and analyzing them for the chemicals listed in Table 1. Results have been used to test whether the imposition of controls on chemical discharges has resulted in decreases in levels of environmental contamination.

While estimates of concentrations that cause biological effects are also essential, determinations of the spatial distribution, or status, of chemical contamination are needed to define the extent of contamination problems in the coastal United States. Molluscan data provide spatial information, but the need to collect mussels at some sites and oysters at others excludes comparing concentrations of trace elements with concentrations that are strongly affected by species. Sediment contamination, on the other hand, is free of biological influence, so that surface sediments were used to determine the status of contamination. However, even here, not all sediments have an equal capacity for accumulating contamination and analytical results needed to be adjusted by the proportion of sediment particles in the fine grain (<63 μm in diameter) size range. Because the time scale represented by the upper 2 cm of a sediment column is unknown without site-specific knowledge on rates of deposition and bioturbation, sediment data cannot be used to monitor temporal trends and annual collections of surface sediment ceased in the second year of the NS&T program.

Other important changes since the program began in 1986 have been the realizations that determining temporal trends does not require replicate analyses in any given year, and that as years go by there is less need for sampling every year rather than less frequently, but nonetheless systematically. Trends are identified not just by correlations between concentrations and time at individual sites but by parallel behavior at nearby sites.

Finally, the NS&T program is not alone in monitoring chemical contamination in the coastal United States. Its emphasis is on the national scale. On large regional scales, the U.S. EPA EMAP-EC program measures the same chemicals in sediments. Together NS&T and EMAP-EC sponsor an annual chemical quality assurance program that is open to organizations monitoring on state and local levels. Combining data from these programs increases the spatial resolution of the large-scale undertakings such as NS&T and expands the perspective of the smaller scale efforts.

REFERENCES

1. National Research Council (NRC). 1990. *Managing Troubled Waters; The Role of Marine Environmental Monitoring.* National Academy Press, Washington, D.C. 125 pp.
2. Martin, M. 1985. State mussel watch: toxics surveillance in California. *Mar. Pollut. Bull.* 16: 140–146.
3. Arimoto, R. and S. Y. Feng. 1983. Changes in the levels of PCBs in *Mytilus edulis* associated with dredged-material disposal. *In* Kester, D. R. et al. (Eds.) *Wastes in the Ocean. Vol. 2. Dredged-Material Disposal in the Ocean.* John Wiley & Sons, New York, pp. 199–212.
4. Goldberg, E.D., M. Koide, V. Hodge, A.R. Flegal, and J. Martin. 1983. U.S. mussel watch: 1977–1978 results on trace metals and radionuclides. *Estuarine, Coastal Shelf Sci.* 16: 69–93.
5. Nriagu, J. O. 1989. A global assessment of natural sources of atmospheric trace metals. *Nature* 338: 47–49.
6. Keith, L. H. and W. A. Teillard. 1979. Priority pollutants. I. A perspective view. *Environ. Sci. Technol.* 13: 416–423.
7. Gossett, R. W., D. A. Brown, and D. R. Young. 1983. Predicting the bioaccumulation of organic compounds in marine organisms using octanol/water partition coefficients. *Mar. Pollut. Bull.* 14: 387–392.
8. Shelton, D. R., S. A. Boyd, and J. M. Tiedje. 1984. Anaerobic biodegradation of phthalic acid esters in sludge. *Environ. Sci. Technol.* 18: 93–97.
9. Pait, A. S., A. E. DiSouza, and D. R. G. Farrow. 1992. Agricultural Pesticides in Coastal Areas: A National Summary. NOAA, National Ocean Service, Office of Ocean Resources Conservation and Assessment, Rockville, MD, 112 pp.
10. Goolsby, D. A., R. C. Coupe, and D. J. Markovchik. 1991. Distribution of Selected Herbicides and Nitrate in the Mississippi River and its Major Tributaries, April through June 1991. Water-Resources Investigations Report 91–4163, U.S. Geological Survey, Denver, CO, 35 pp.
11. Holden, L. R., J. A. Graham, R. W. Whitmore, W. J. Alexander, R. W. Pratt, S. K. Liddle, and L. L. Piper. 1992. Results of the national alachlor well water survey. *Environ. Sci. Technol.* 26: 935–943.
12. Readman, J. W., L. Liong Wee Kong, L. D. Mee, J. Bartocci, G. Nilve., J. A. Rodriguez-Solano, and F. Gonzalez-Farias. 1992. Persistent organophosphorus pesticides in tropical marine environments. *Mar. Pollut. Bull.* 24: 398–402.
13. Vallette-Silver, N. and G. G. Lauenstein. 1992. Radionuclides in bivalves collected along the Coastal U. S.: Preliminary results. MTS '92. Marine Technology Society, Washington D.C.
14. Butler, P. A. 1973. Residues in fish, wildlife, and estuaries. *Pestic. Monit. Bull.* 6: 238–262.

15. Claisse, D. 1989. Chemical contamination of French coasts: the results of a ten-year mussel watch. *Mar. Pollut. Bull.* 20: 523–528.
16. Phillips, D. J. H. and W. W. S. Yim. 1981. A comparative evaluation of oysters, mussels, and sediments as indicators of trace metals in Hong Kong waters. *Mar. Ecol.* 6: 285–293.
17. Korean Ocean Research and Development Institute (KORDI). 1990. A Study on the Coastal Water Pollution and Monitoring—Third Year. BSPG 00112-315-4, Korean Ocean Research and Development Institute, Seoul, Korea, 261 pp.
18. Aukland Regional Water Board (ARWB). 1990. N.Z. Steel Limited Environmental Monitoring Program. Fourth Annual Report. Technical Publication 68. Aukland Regional Water Board, Aukland, New Zealand, 78 pp.
19. Tripp, B. W., J. W. Farrington, E. D. Goldberg, and J. Sericano. 1992. International mussel watch: the initial implementation phase. *Mar. Pollut. Bull.* 24: 371–373.
20. Cantillo, A. Y. 1991. Mussel Watch Worldwide Literature Survey—1991. NOAA Technical Memorandum NOS ORCA 63. 143 pp.
21. Seed, R. 1992. Systematics evolution and distribution of mussels belonging to the genus *Mytilus*: An overview. *Amer. Malac. Bull.* 9: 123–137.
22. Phillips, D. J. H. 1980. *Quantitative Aquatic Biological Indicators*. Applied Science Publishers, London, 488 pp.
23. Roesijadi, G., J. S. Young, A. S. Drum, and J. M. Gurtisen. 1987. Behavior of trace metals in *Mytilus edulis* during a reciprocal transplant field experiment. *Mar. Ecol. Prog. Ser.* 15: 155–170.
24. Sericano, J. 1993. The American oyster (*Crassostrea virginica*) as a bioindicator of trace organic contamination. Ph. D. Thesis. Texas A & M University, College Station, TX. 242 pp.
25. Pruell, R. J., J. G. Quinn, J. L. Lake, and W. R. Davis. 1987. Availability of PCBs and PAHs to *Mytilus edulis* from artificially resuspended sediments. *In* Capuzzo, J. M. and D. R. Kester (Eds.) *Oceanic Processes in Marine Pollution. Vol. 1. Biological Processes and Wastes in the Ocean.* Krieger, Malabar, FL, pp. 97–108.
26. NOAA. 1988. National Status and Trends Program for Marine Environmental Quality Progress Report—A summary of selected data on chemical contaminants in sediments collected during 1984, 1985, 1986, and 1987. NOAA Technical Memorandum NOS OMA 44, Rockville, MD. 15 pp. and appendices.
27. O'Connor, T. P. and C. N. Ehler. 1991. Results from NOAA national status and trends program on distributions and effects of chemical contamination in the coastal and estuarine United States. *Environ. Monit. Assess.* 17: 33–49.
28. O'Connor, T. P. 1990. Coastal Environmental Quality in the United States, 1990. Chemical Contamination in Sediments and Tissues. A Special NOAA 20th Anniversary Report, NOAA Office of Ocean Resources Conservation and Assessment, Rockville, MD, 34 pp.
29. NOAA. 1991. Second Summary of Data on Chemical Contaminants in Sediments from the National Status and Trends Program. NOAA Technical Memorandum NOS OMA 59. 29 pp. and appendices.

30. Cantillo, A. Y. and T. P. O'Connor. 1992. Trace element contaminants in sediments from the NOAA national status and trends program compared to data from throughout the world. *Chem. Ecol.* 7: 31–50.
31. O'Connor, T. P. 1992. Recent Trends in Coastal Environmental Quality: Results from the First Five Years of the Mussel Watch Project NOAA Office of Ocean Resources Conservation and Assessment, Rockville, MD, 29 pp. and appendices.
32. Lauenstein, G. G., A. Robertson, and T. P. O'Connor. 1990. Comparisons of trace metal data in mussels and oysters from a mussel watch program of the 1970s with those from a 1980s program. *Mar. Pollut. Bull.* 21: 440–447.
33. Environmental Protection Agency (EPA). 1990. Environmental Monitoring and Assessment Program. Near Coastal Program Plan for 1990: Estuaries. U. S. Environmental, Protection Agency Report EPA/600/4-90/033, Washington D.C., 217 pp. and appendices.

CHAPTER 20

Some Problems of "Safe Dose" Estimation

Asit P. Basu, Gary F. Krause, Kai Sun, Mark Ellersieck,
and Foster J. Mayer, Jr.

SUMMARY

In this chapter the current status for estimation of "safe doses" in carcinogenic experiments is examined. After considering various definitions of safe doses, a method of standardizing safe doses based on some objective parameters is introduced; and a procedure of estimating standardized safe doses under competing risks is discussed. Examples are given to illustrate how raw data can be analyzed by using the proposed methods.

INTRODUCTION

Environmental risk assessment has statistical issues which are essential to our ability to make scientifically sound statements. Several major statistical problems involved in the quantitative assessment of environmental risks are collection of quality control data, modeling data based on scientifically rigorous ground, efficient estimation of parameters, and standard criteria in chemical regulation. Due to the breadth of this area, this chapter is focused on examining some existing frameworks for carcinogenic risk assessment, and attempts to propose a standard measurement for carcinogenic risk and a procedure of estimating such standard risk under competing risks.

To make an informed decision we need clear and comparable information on the degrees of toxicity or carcinogenicity of chemical substances and mixtures. One hindrance to such decision making is that we often do not use coherent concepts and standard scientifically rigorous approaches to integrate and analyze data. Some concepts and procedures of estimating safe doses are such examples.

In the context of carcinogenicity, quantitative risk assessment is often concerned with "safe levels of exposure" for potential carcinogens. A safe dose for a carcinogen is usually defined as the largest amount of a chemical which produces a specific small level of risk. Mathematically, a safe dose is usually defined in terms of cancer incidence probability. First, a permissible increment over the spontaneous cancer incidence probability is specified; then the safe dose x is defined by the equation:

$$\pi = F(T^*, x; \alpha, \beta) - F(T^*, 0; \alpha, \beta) \qquad (1)$$

where $F(T^*, x; \alpha, \beta,) = P\{T \le T^* | x, \alpha, \beta\}$ is the cancer incidence probability at the dosage level x and by the exposure time T^*. α and β are unknown parameters. The acceptable level is determined by regulatory agencies and is usually very small, e.g., 10^{-5} or 10^{-6}.

Hartley and Sielken[1] introduce another definition which specifies tolerance (TOL) rather than π and defines the safe dose x by the equation:

$$TOL = Q(T^*, x; \alpha, \beta)/Q(T^*, 0; \alpha, \beta) \qquad (2)$$

where $Q(T^*, x; \alpha, \beta) = 1 - F(T^*, x; \alpha, \beta)$ is cancer free probability at the dosage level x and by the exposure time T^*. The tolerance TOL is interpreted as a permissible percentage reduction in the spontaneous cancer free proportion. TOL is usually specified as a number slightly less than 1, e.g., 0.9999 or 0.9999.

Notice that both Equations 1 and 2 depend on T^*. Therefore, a safe dose defined by either Equation 1 or 2 might be safe regarding a particular time, but "unsafe" regarding other times. In other words, at the same permissible level regarding different exposure times, there exist different safe doses for each exposure time. Therefore, one will have a number of safe doses at the same permissible risk level if he or she chooses different exposure times.

We now face a problem: how should we establish safe level of exposure for a certain cancer inducing substance? In other words, how do we set a criterion in choosing a standard safe dose among so many safe doses for a given permissible risk level? If safe doses can be standardized in some meaningful way, we may make an index of standardized safe doses which gives measurements quantifying the differences in degree of carcinogenicity among chemicals and the differences in sensitivity among animal species. This could be used as "standardized information" for chemical regulation and environment policy decision. In the next section, a method of standardizing safe doses is introduced.

Another problem relating to safe doses estimation is the use of competing risks models to adjust for different mortality rates across doses. Safe residual doses for potential cancer-inducing substances are usually very low, and it is often impossible to observe enough tumors to make a meaningful estimate of the time to tumor distribution. Therefore, carcinogenic experiments are conducted in animals at much higher than usual dosages to induce more tumors. If different mortality rates are due to causes other than the tumor of interest, the safe dose estimates must be adjusted for competing risks. The most common different type of mortality between dose groups is (as Kodell et al.[2] point out) a significantly reduced survival for high doses relative to low doses, which causes a higher degree of censoring of the tumor response at higher doses. In the "Competing Risks" section, a procedure of estimating standard safe doses under competing risks is discussed.

A METHOD OF STANDARDIZING SAFE DOSES

Since safe doses depend on the choice of exposure time T*, they appear as subjective. Therefore, it seems of interest to characterize carcinogenic substances by a set of objective parameters. In other words, there is a need in chemical regulation to choose a standard safe dose among so many subjective safe doses.

In most cases of carcinogenicity studies, no human data are available; and one turns to experiments on animals. In this section, attention is focused on the issue of establishing a safe level of exposure for a potential carcinogen to a certain kind of animal. Since different animal species have different life spans, it is not plausible to set a fixed exposure time for all kinds of animals in defining safe doses; instead, scientists commonly investigate carcinogenic risks at the median life (see, e.g., Hartley and Sielken;[1] Brown et al.[3]). Therefore, the first step to standardize safe doses is, in our opinion, to use the median life as a standard exposure time in Equation 1 or 2. Now another question arises: how does one find the median life for a given species? In fact, median lifetimes are unknown in most cases. Estimates can be obtained if sufficient data at zero dose level are available. To overcome this drawback and to go one step further in standardizing safe doses we introduce a concept of "tolerance in cumulative tumor hazard rate" which is:

$$\text{TOH} = H(T^*, x; \alpha, \beta)/H(T^*, 0; \alpha, \beta) \tag{3}$$

where $H(T^*, x; \alpha, \beta)$ = cumulative hazard rate at dose level x and exposure time T*

TOH = "tolerance" or permissible percentage increment in the spontaneous cumulative hazard rate proportion

Animal experiments are conducted at dosage levels high enough to induce tumors during short-term tests, while environmental safe residual doses must be low enough to induce very few or no tumors in humans for a long-term exposure. Results from laboratory animal studies are used to assess possible risks to humans. The scientific value of this assessment depends on the validity of three extrapolations: from high dosage to low dosage, from short-term experiment to long-term exposure, and from animal to man. Valid extrapolations should be based on statistical models grounded in biology of cancer and well fitted to empirical data, as well as efficient methods of estimation of parameters in these models.

In recent years, the multistage models which were introduced by Armitage and Doll,[4] and studied and utilized by many authors (e.g., Hartley and Sielken,[1] Brown et al.,[3] Armitage,[5] and Rockette[6]) have been widely used in carcinogenic risk assessment. Multistage models, based on the biology of cancer, provide a framework that facilitates understanding the statistical relationship between dose-time effects and tumor incidence.

Multistage models were originally used for describing the relationship between dose and response, without time dimension. However, just dose-response curves are no longer sufficient for studying long-term exposure of potential carcinogens; and more comprehensive approaches taking account of the progression of tumor through time are needed to describe tumor process and to make a full use of tumor data. Hartley and Sielken,[1] are the first researchers who introduce a time factor into the multistage models and use survival analysis theory to model carcinogenic experiments as accelerated life tests.

The multistage model with the time dimension, proposed by Hartley and Sielken,[1] is of the product form for the hazard rate h:

$$h(x, t; \alpha, \beta) = \left(\sum_{S=0}^{A} \alpha_s x^s \right) h_0(t; \beta) \tag{4}$$

where $h_0(t; \beta)$ is the spontaneous hazard rate, and a change in the dosage level x has a multiplicative effect on the spontaneous hazard rate by a factor $g(t; \alpha) = \Sigma \alpha_s x^s$.

This approach summarizes carcinogenicity data through a life regression model in which time to tumor has a life distribution that depends on the dosage level of a chemical. This is a proportional hazards model, and the form of the dose effect function g is specified as a polynomial by multistage theory of cancer. The proportional hazards model is a popular family of life regression model in biological applications, since the multiplicative effect of dosage on the hazard rate has a clear and intuitive biological meaning.

The multistage model describes a dose-time-response surface within one animal species and provides a useful tool for the first two extrapolations: from high dosage to low dosage and from short-term experiment to long-term exposure.

Under model assumption (Equation 4), TOH has an appealing property: independence of time T*.

In fact, from:

$$H(T^* x; \alpha, \beta) = \int_0^{T^*} \left(\sum_{S=0}^{a} \alpha_s x^s \right) h_0 (t; \beta) \, dt \qquad (5)$$

we have

$$TOH = \left(\sum_{S=0}^{A} \alpha_s x^s \right) / \alpha_0 \qquad (6)$$

We now make use of this property to standardize safe doses. It is easy to find the relationship between TOL and TOH which is:

$$TOH = 1 + \log(TOL)/\log[Q(T^*, 0; \alpha, \beta)] \qquad (7)$$

If T* is a median life, then $\log[Q(T^*, 0; \alpha, \beta)] = \log(0.5)$. Using this formula one can convert a tolerance TOL to a tolerance TOH, and then a safe dose is given by the solution of Equation 6. We call this dosage level standard safe dose of the carcinogenic substance to the given animal species. Notice that estimation of the unknown median life is no longer needed, and hence a safe dose is determined in a more certain way.

The following steps describe the method of standardizing safe doses:

1. Choose the test animal's median lifetime as exposure time for standardized safe doses (the median lifetime is usually unknown and need *not* to be estimated in our method).
2. Specify TOL, e.g., 0.9999 or 0.99999.
3. Convert TOL to TOH by:

$$TOH = 1 + \log(TOL)/\log(0.5) \qquad (8)$$

4. A safe dose is given by the solution of Equation 6.
5. Only α_i (neither β nor median lifetime) needs to be estimated in solving Equation 6. (Note that Hartley and Sielken's procedure needs to estimate α_i, β, and the median lifetime.)

Table 1.

TOL	TOH
$1 - 10^{-3}$	$1 + 1.44e - 3$
$1 - 10^{-4}$	$1 + 1.44e - 4$
$1 - 10^{-5}$	$1 + 1.44e - 5$
$1 - 10^{-6}$	$1 + 1.44e - 6$
$1 - 10^{-7}$	$1 + 1.44e - 7$
$1 - 10^{-8}$	$1 + 1.44e - 8$

Table 1 converts some commonly used TOL to TOH.

Another interesting point is that once a standard safe dose is obtained, one can determine the tumor risks caused by the carcinogenic substance at the given species life stages by the following formula:

$$R_i = 1 - \exp[(TOH - 1)\log(1 - i/100)] \tag{9}$$

where $R_i = [Q(T_i, 0; \alpha, \beta) - Q(T_i, x; \alpha, \beta)]/Q(T_i, 0; \alpha, \beta)$ denotes the risk, the percentage increment of tumor incidence caused by a carcinogenic substance, at the i th percentile of the life distribution, T_i. Table 2 gives risks of standard safe doses at five life stages for six permissible levels.

Table 2.

TOH	R_{10}	R_{20}	R_{30}	R_{40}	R_{50}
$1 + 1.44e - 3$	$1.52e - 4$	$3.213e - 4$	$5.135e - 4$	$7.353e - 4$	$9.976e - 4$
$1 + 1.44e - 4$	$1.52e - 5$	$3.213e - 5$	$5.135e - 5$	$7.353e - 5$	$9.976e - 5$
$1 + 1.44e - 5$	$1.52e - 6$	$3.213e - 6$	$5.135e - 6$	$7.353e - 6$	$9.976e - 6$
$1 + 1.44e - 6$	$1.52e - 7$	$3.213e - 7$	$5.135e - 7$	$7.353e - 7$	$9.976e - 7$
$1 + 1.44e - 7$	$1.52e - 8$	$3.213e - 8$	$5.135e - 8$	$7.353e - 8$	$9.976e - 8$
$1 + 1.44e - 8$	$1.52e - 9$	$3.213e - 9$	$5.135e - 9$	$7.353e - 9$	$9.976e - 9$

The following example illustrates the use of standard safe doses.

Example 1

Brown et al.[1] conducted a study of the prevalence of skin cancer among 40,421 persons consuming arsenic-contaminated drinking water and predicted, using the Hartley and Sielken multistage model with time, that an American male (female) would have a risk (percentage increment) of skin cancer of $R = 2.6 \times 10^{-3}$ (1.2×10^{-3}) if exposed to 1 μg/kg/day for a median life span in the United States. Using TOL = 1 - R and Equations 8 and 9, we can determine the skin cancer risks at different life stages of American people.

Skin cancer risks at five life stages of American people are listed in Table 3. Here $R_i = [Q[T_i, 0; \alpha, \beta) - Q(T_i, x; \alpha, \beta)]/Q(T_i, 0; \alpha, \beta)$ denotes

Table 3.

	R_{20}	R_{30}	R_{40}	R_{50}	R_{60}
Male	8.38e − 4	1.34e − 3	1.92e − 3	2.60e − 3	3.44e − 3
Female	3.86e − 4	6.18e − 4	8.84e − 4	1.20e − 3	1.59e − 3

the percentage increment of skin cancer caused by inorganic arsenic at the i th percentile of the life distribution of American people, T_i; and R_{50} is the risk at the median life span in the United States, 76.2 years.

Our discussion so far is about the formulation of standardized safe dose. From a statistical inference point of view, it also has an advantage that standard safe dose can be estimated in the absence of knowledge of the baseline time-to-tumor distribution, while the usual safe dose estimation requires the assumption on the background tumor incidence distribution. Notice that the standard safe dose estimate is found by solving Equation 6 which involves only parameter α. Therefore a procedure of estimating standard safe doses based on a distribution-free approach, (e.g., Cox's[7] partial likelihood method) can be developed. Research on this procedure is in progress.

COMPETING RISKS

Kodell et al.[2] point out that if crude experimental proportions of animals with tumors from chronic bioassays for carcinogenicity are used for low-dose extrapolation in a risk analysis, different dose-specific patterns of mortality due to competing risks can bias results. Therefore estimation of standard safe doses must be adjusted for competing risks. Basu and Klein[8] proposed a model for estimating safe doses under competing risks. In this section we consider a special case of this model. Suppose that an animal is exposed to dose x of a carcinogen, where the animal is at risk to m independent diseases or types of tumor with onset times (T_1, T_2, . . . ,T_m). For each of the m diseases we assume that the hazard rate of the time to occurrence of the i th type of disease is:

$$ h_i(x, t; \alpha_i, \beta_i) = \left(\sum_{S=0}^{A} \alpha_{si} x^s \right) h_{0i}(t; \beta_i) \qquad (10) $$

To obtain estimates of α_i and β_i we select d dose levels, $x_1, x_2, \ldots, x_d$ to test the carcinogen. At dose x_d we put N_d animals on test and keep them on test until their death from natural cause or until they are sacrificed. For each animal a necropsy is performed to determine which diseases are present on removal from the study. The information contributed to the likelihood of disease *i* comes in three categories.

Category 1. The disease was present on removal, and the onset time is known or estimated by the pathologist to be $\tau_i \leq t$. The contribution to the likelihood is the probability density function $f_i(\tau_i, x; \alpha_i, \beta_i)$.

Category 2. The disease is not present on removal from the study at time t. The contribution to the likelihood is the survival function $Q_i(t, x; \alpha_i, \beta_i)$.

Category 3. The disease is present on removal at time T, but the time of tumor onset is unknown. The contribution to the likelihood is the cumulative distribution function $F_i(T, x; \alpha_i, \beta_i)$.

To find the maximum likelihood estimators of α_i and β_i, note that the likelihood of interest is the product of d likelihoods, one for each dose level. For the dth likelihood an animal contributes m terms, one per disease, of the form determined by the above categories. After writing the total likelihood and arranging terms, one can show that the total likelihood is the product of m likelihoods each of the form of Hartley and Sielken's likelihood (their Equation 17). Hence, in the multiple disease setting one collects data as indicated above and performs m separate analyses, one analysis for each disease. We illustrate this procedure by the following example.

Example 2

A two-risk competing model was simulated. One risk can be considered as a tumor and another as a second disease. Onset times were simulated from a Weibull distribution with shape parameter 3. The scale parameters at the various doses follow $g_i(x; \alpha_i) = \Sigma \alpha_{is} x^s$, i = 1, 2, with the values $(\alpha_{10}, \alpha_{11}, \alpha_{12}) = (0.0145, 0.00725, 0.003625)$ and $(\alpha_{20}, \alpha_{21}, \alpha_{22}) = (0.052, 0.026, 0.013)$, respectively. Five dose levels, 0, 6.1, 12.2, 18.2, and 24.3, were used. Death times were simulated from another Weibull population with shape parameter 3 and scale parameter 1.

At each dose level 120 observations were generated. For each observation the tumor and onset times of the other disease were generated from the above model. If the disease time was greater than the death time, then observation on that disease was placed in category 2. If the tumor or disease onset occurred prior to death, a fair coin was flipped to determine if the onset time was known and hence the observation is in category 1; or if the onset time was unknown and hence in category 3. Following Hartley and Sielken,[1] the total sample of 120 observations at each of the five dose levels was subdivided into six subsamples of 20

Table 4.

Group	Tumor Xhat	Disease Xhat
1	3.9768	0.7618
2	2.2978	0.5761
3	0.0061	0.0013
4	2.7950	0.0065
5	0.0719	0.7986
6	0.0046	0.0010
Est. stand. safe dose	0.0153	0.0032
Lower 95% CI	0.0013	0.00024
True stand. safe dose	0.0198	0.0055

observations at each dose level so that confidence intervals could be obtained.

To find the maximum likelihood estimates of parameters we assume that $h(t; \beta) = \Sigma\beta_s t^s$. Solutions of the likelihood equations are obtained by the following iterative scheme for each subgroup. At the initial state we consider each of the five dose levels separately. Using $\alpha_{i0} = 1$ and $\alpha_{i1} = \alpha_{i2} = 0$ a reduced gradient method is used to find those estimates of β_s, $s = 1, \ldots 8$, which maximize the contribution to the disease likelihood by those observations at dose x_i $i = 1, \ldots, 5$.

Once these initial estimates are obtained, a two-step iteration scheme is used. On alternating steps the likelihood is maximized by a reduced gradient method, with respect to either α_i or β_i using the corresponding α_i or β_i from the previous half step. This procedure is continued until the change in the relative size of the likelihood is at most 0.0001.

To estimate x_i we first solve safe doses for each of the six groups using the estimates of α_{ij}, β_{ij}. Let x_{jg} be the resulting estimator. One estimate of x_i is $(\Pi x_{jg})^{1/6}$, the geometric mean. For median lifetime $T^* = 0.885$ and $TOL(T^*) = 0.9999$, the estimated safe doses and the lower 95% confidence interval are reported in Table 4.

We plan to develop a user-friendly software for practical oncologists to calculate standard safe doses under competing risks.

In this chapter, estimation of standard safe doses within one animal species is discussed. Another extrapolation—from laboratory animal to man—might be more important, yet more difficult. More comprehensive models incorporating other factors such as body weight, daily food amount, age, and life expectancy are required for analyzing interspecies data. However, the concept and procedure developed for estimating standardized safe doses shall provide a framework that enables us to be in a better position to transfer this knowledge to the species extrapolation.

ACKNOWLEDGMENT

This work was sponsored in part by the U.S. Environmental Protection Agency under cooperative agreement No. CR818945, and, in part, by the U.S. Air Force Office of Scientific Research, Air Force System Command, USAF, under grant number AFOSR F49620-92-J-0371.

REFERENCES

1. Hartley, H. O. and Sielken, R. L., Jr. "Estimation of Safe Doses in carcinogenic experiments," *Biometrics*, 33:1–30 (1977).
2. Kodell, R. I., Gaylor, D. W., and Chen, J. J. "Standardized tumor rates for chronic bioassays," *Biometrics*, 42:867–873 (1986).
3. Brown, K. G., Boyle, K. E., Chen, C. W., and Gibb, H. J. "A Dose-Response analysis of skin cancer from inorganic arsenic in drinking water," *Risk Anal.*, 9(4):519–528 (1989).
4. Armitage, P. and Doll, R. "Stochastic models for carcinogenesis," In *Proceedings of the Fourth Berkeley Symposium on Mathematical Statistics and Probability*, 4:19–38 (1961).
5. Armitage, P. "Multistage models of carcinogenesis," *Environ. Health Persp.*, 63:195–201 (1985).
6. Rockette, H. E. "Statistical issues in carcinogenic risk assessment," *Environ. Health Persp.*, 90:223–227 (1991).
7. Cox, D. R. "Partial likelihood," *Biometrika*, 62:269–276 (1975).
8. Basu, A. P. and Klein, J. P. "A model for life testing and for estimating safe dose levels," In *Statistical Theory and Data Analysis*, K. Matusita, Ed. (Amsterdam, Netherlands: Elsevier Science Publishers B.V., 1985), pp. 69–80.

CHAPTER 21

Where Next? Adaptive Measurement Site Selection for Area Remediation

H.T. David and Seongmo Yoo

ABSTRACT

We seek efficient adaptive measurement site selection. The methodology calls for spatial interpolation, adaptive measurement site selection, and a stop-decision rule. Regarding spatial interpolation, we choose Hardy's multiquadric-biharmonic method (MQ-B) for which spatial interpolation is based on distance. Adaptive measurement site selection, assuming no information other than the coordinates of, and measurements for, the sites selected so far, is based on three ideas: adaptively attaining nearly uniformly dispersed measurement sites; selecting next a measurement site which, together with some earlier smaller set of measurement sites, would have yielded a map near the current map; and avoiding use of measurement sites on the boundary of the area of interest. Regarding the stop-decision rule, we base terminal remediation decisions on an estimate of the discrepancy between the actual contamination map and the current sample map. This estimated discrepancy not only approximates the actual current discrepancy, but also will fail to converge with increasing measurement site number whenever contamination is granular enough to make spatial interpolation unreliable. Examples and illustrations are included.

INTRODUCTION

Assessing the state of the environment includes, in particular, determining the level and extent of contamination at sites (called areas in the present chapter) where such contamination is known to have occurred.

This contamination may be clearly granular as would, for example, be the case if it were due to widely dispersed contaminated equipment, with little possibility of leaching.

Or it may have occurred in a manner making possible either granular or relatively smooth distribution of the contamination. Windblown surface contamination, which is the contamination mechanism at several sites in this country, provides a surface illustration of this latter sort of contamination. Leaching of contamination into the ground from a central area, forming plumes of unknown extent, provides a subsurface illustration.

It seems generally agreed on that, for the type of contamination of the previous paragraph, a good way to determine the level and extent of the contamination is to attempt to map it. There seems also to be agreement that the mapping and the drawing of conclusions from it should, if at all possible, proceed according to a reasonably inclusive plan drawn up at the start. This is what we address in this chapter.

Consider, in particular, that surface contamination levels are to be mapped so that spatial interpolation based on measurements at sampled sites will be called for, yielding a contamination map. Such spatial interpolation is not necessarily restricted to direct interpolation of measured pollution levels. It may, for example, pertain to interpolating the values of ratios of pollution level to an observed explanatory variable; or it may pertain to interpolating the discrepancies between actual measured pollution levels and prior expert estimates of these pollution levels. In any event, it is required in the first instance to choose a spatial interpolation method for interpolating measurements Y_i at selected sites s_i to all sites s of the area of interest. Other authors' experiences[1-5] and also our own have led us to choose Hardy's multiquadric-biharmonic (MQ-B) method, which is rather astatistical in nature and is sketched below. Also required is a method for successively selecting the measurement sites s_i. Prior work on this problem of site selection appears limited to just a few papers[6-9] formulated in a kriging context and thus requiring the idea of a relevant correlogram or variogram. Consistent with our in part astatistical point of view above, we have derived instead an essentially deterministic selection method described below that aims at nearly uniformly dispersed sampling sites. Our method also incorporates a certain back-to-the-future feature, which judges potential new sites according to how well they would have contributed to predicting the current map, had they been added to an earlier, smaller site set.

The next section details our ideas concerning the three steps of an inclusive plan, or strategy, for area remediation. The following two sections explain our proposals regarding the first and second steps. The last section, consisting in the main of six figures, shows our application of our methodology to five areas, of which three are actual contaminated areas and two are synthetic. Figure 1 and 2 describe these areas. Figure 3 compares our proposal for step 1 to the basing of successive maps on uniformly dispersed sites. The latter provides a suitable and demanding benchmark; suitable, because it is recognized[10] to be an efficient site selection method; demanding, because it cannot be implemented sequentially. Figure 4 pertains to step 2 in the following way: in the next-to-last section below, we propose a stop-decision rule based on a certain map-discrepancy index $\hat{\delta}_o(m)$ intended to estimate the discrepancy $\delta_o(m)$ between the current map based on m site measurements, and the actual map of the area. Figures 4–6 show how $\hat{\delta}_o(m)$ behaves with increasing m, relative to $\delta_o(m)$, when our methodology is applied to our five data sets.

REMEDIATION STRATEGY

We see the development of a remediation strategy as involving the following steps:

1. Formulating a strategy for successively selecting contamination measurement sites — The strategy proposed here calls for successively developing empirical site contamination maps, based on the contamination measurements made at the successively selected sites and a site selection rule based in part on these empirical maps.
2. Formulating a stop-decision rule — The decision is *either* to base selective remediation on the last developed empirical contamination map *or* to judge area contamination as too granular for effective selective remediation.
3. Objectively evaluating the effectiveness of the remediation strategy — This is discussed more fully below.

In this chapter, we make rather specific proposals concerning steps 1 and 2, and apply these proposals to both actual and synthetic data sets. These applications show our proposals to be reasonable ones. As already alluded to in the Introduction, our proposals concerning steps 1 and 2 are rather astatistical in nature; This astatistical point of view needs to be judged in light of the following considerations:

Developing effective steps 1 and 2 of a remediation strategy seems to us to be the main order of business; and this does not necessarily call for a stochastic framework, as will be illustrated below.

Step 3 seems to us to be the secondary order of business, and to perhaps be done best with a stochastic model. We use the term perhaps because it is possible to pragmatically evaluate a remediation strategy *without* a stochastic model, by witnessing the performance of the strategy and possible competitors on both actual and synthetic test contamination areas. This approach has, for example, been adopted,[1-5] in verify-

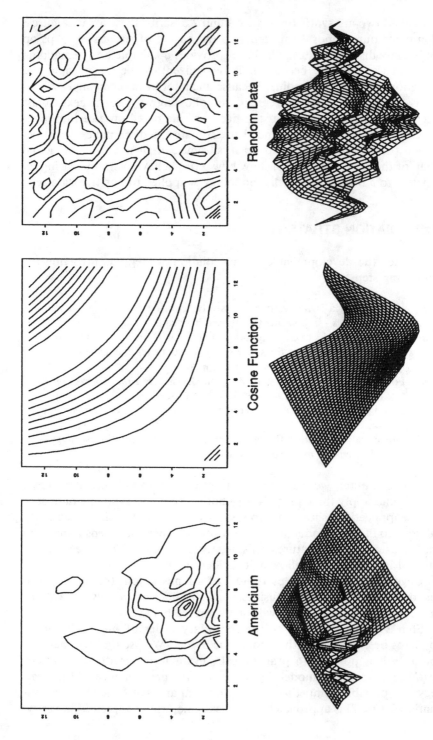

FIGURE 1. Contours and perspective plots for three data sets on regular grids.

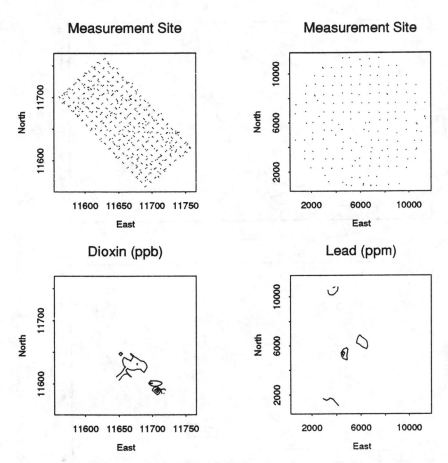

FIGURE 2. Measurement sites and contour plots for two data sets on irregular grids.

ing the effectiveness of Hardy's multiquadric mapping methods, and has the advantage of avoiding the danger of any possible false sense of security that a probabilistic figure of merit might entail.

Despite our to some extent astatistical point of view, we feel appropriately placed in these proceedings: *first*, because our presentation concentrates just on the above main order of business, leaving the door open to reenter the statistics world at the (third) evaluation step; *second*, because questioning and debate always are salutary.

Step 1 : Contamination Measurement Site Selection

Two well-known methods of spatial interpolation are Matheron's kriging[11] methods;[12] and Hardy's multiquadric methods,[12] in particular, his

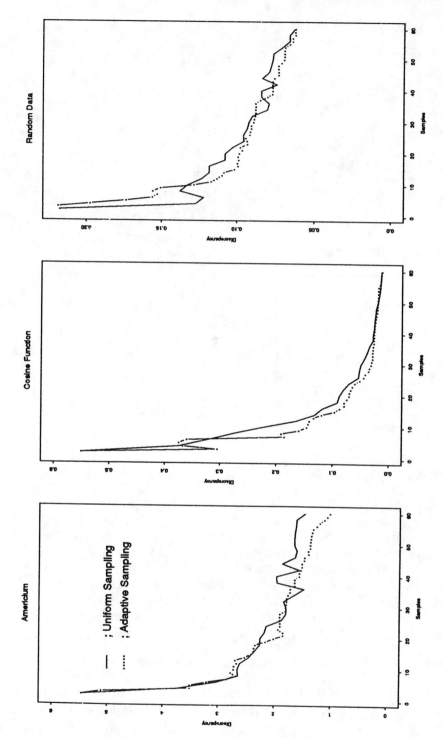

FIGURE 3. Plots of actual discrepancy index, under uniform and adaptive sampling, for the three data sets of Figure 1.

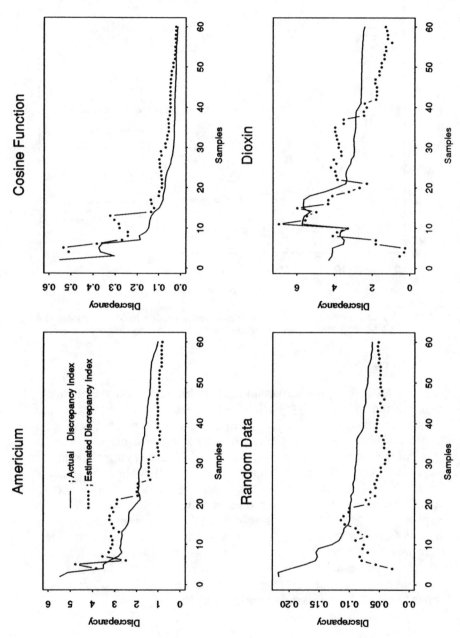

FIGURE 4. Plots of actual and estimated discrepancy indices under adaptive sampling.

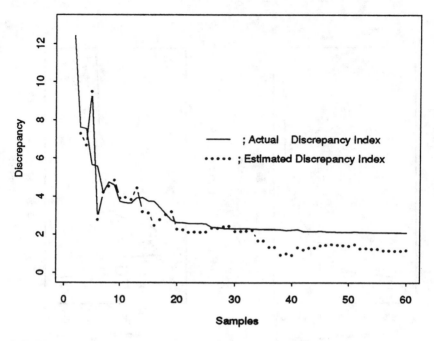

FIGURE 5.   Plot of actual and estimated discrepancy indices for dioxin data under adaptive sampling when sampling starts near high-contamination zones.

method MQ-B which is the spatial interpolation method that is incorporated into our methodology. As has become recognized,[5,13] most spatial interpolation methods are formally equivalent. Indeed, with regard to the aforementioned two methods, the MQ-B method may be seen as substituting distance itself for the function of distance that is the semivariogram of kriging. It is others' experience[1-5] (as well as our own) that it is not disadvantageous for purposes of accurate spatial interpolation to forego (as does Hardy) using an estimated spatial covariance structure in the process of spatial interpolation itself. (Note, however, that as indicated in the previous section this is not to say that stochastic considerations might not play a role in the evaluation of mapping methods, in particular, and remediation strategies, in general.)

The constructed map $Y^{(m)}(s)$ is linear in the responses Y_i at the m sampled sites s_i, both in kriging and in MQ-B:

$$Y^{(m)}(s) = \sum_{i=1}^{m} \alpha_i(s) Y_i. \tag{1}$$

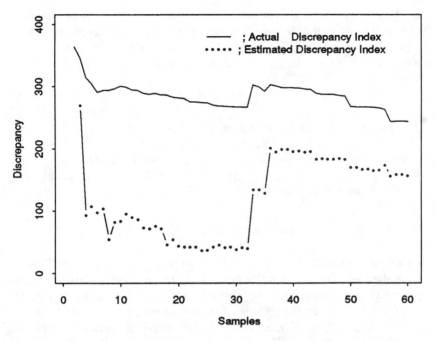

FIGURE 6. Plot of actual and estimated discrepancy index for lead data under adaptive sampling.

Here the coefficients $\alpha_i(s)$ are derived in terms of certain matrix operations, with nonzero and nonunit matrix elements that are distances $||s_i,s||$ and $||s_i,s_j||$ in the case of the MQ-B method, and are the corresponding semivariogram values in the case of kriging.

We see our methodology as adaptive in the sense that measurement sites are to be chosen sequentially. For purposes of the present discussion, we suppose no information available for this sequential site selection other than the coordinates of—and measurements for—the sites selected so far.

Three principles guide our method for choosing the next measurement site. The first principle is to seek an accommodation between basing the spatial interpolation on uniformly dispersed measurement sites,[10] and the adaptive nature of our methodology. This accommodation is essentially implemented by choosing next a site that maximizes the least distance to sites measured so far:

$$min_{i=1,\ldots,m}||s_i,s_{m+1}|| \geq min_{i=1,\ldots,m}||s_i,s||.$$

The second principle is to select next a site which, together with some earlier smaller set of measurement sites, would have yielded a map close to the current map. Of course, contamination levels $Y(\sigma)$ at candidate sites σ are not available when the next site is to be chosen. Estimates of $Y(\sigma)$ are therefore required using Equation 1. The third principle is to avoid measurement sites at or near the boundary of the area A of interest.

The adaptive measurement site selection methodology that we propose incorporates the above three principles: define the discrepancy $\delta(Y(\cdot), Y'(\cdot))$ between a map $Y(\cdot)$ and a map $Y'(\cdot)$ as $\delta(Y(\cdot), Y'(\cdot)) = \int_A | Y(s) - Y'(s)|ds$. Suppose that at a certain stage, m sites $s_1, \ldots, s_m$ have been selected, with corresponding contamination measurements $Y_1, \ldots, Y_m$. Define $\mu \equiv max_{s \in A} \, min_{i=1,\ldots,m}||s_i,s||$; and with p near 1, let the set T consist of sites s with $min_{i=1,\ldots,m}||s_i,s||$ between $p\mu$ and μ:

$$T \equiv \{s : p\mu \leq min_{i=1,\ldots,m}||s_i,s|| \leq \mu\}$$

Using randomization if necessary, identify a small subset of T as the candidate set, from which the next site is to be selected. That selection proceeds as follows: let σ be a site in the subset, and let $\delta_j(\cdot)$ be the discrepancy between the (Hardy) map constructed on the basis of (s_1, Y_1), $\ldots, (s_j, Y_j), \ldots, (s_m, Y_m)$, and the (Hardy) map constructed by replacing (s_j, Y_j) by $(\sigma, Y^{(m)}(\sigma))$. The next site s_{m+1} is the site σ minimizing $max_{j=1,\ldots,m} [\delta_j(\sigma)]$.

Step 2 : Stop-Decision Rule

The second principle underlying our site selection methodology counted on the idea that the future will be related to the present in much the same way that the present is related to the past. That idea is retained in the stop-decision rule that we propose: ideally, we would want to stop sampling when $\delta_o(m) \equiv \delta(Y^{(m)}(\cdot), Y^o(\cdot))$ is small, where $Y^o(\cdot)$ is the actual contamination map. $\delta_o(m)$ is, of course, not at hand; however we propose estimating it by $\hat{\delta}_o(m) \equiv \delta(Y^{(m/2)}(\cdot), Y^{(m)}(\cdot))$, and basing our decision making on this function of m. Actually for reasons that we do not as yet understand it appears (see Figures 4–6 of the next section) that $\hat{\delta}_o(m)$ is close to $\delta_o(m)$, so that $\hat{\delta}_o(m)$ can be relied on not only to qualitatively determine whether convergence is taking place, but also to gauge the actual current discrepancy between $Y^{(m)}(\cdot)$ and $Y^o(\cdot)$. Figure 6 reveals, in particular, that the lack of convergence of $\hat{\delta}_o(m)$, with m can be expected to reflect lack of convergence of $\delta_o(m)$, which in turn will be associated with the type of high granularity for which spatial interpolation is unreliable and which in the absence of special information requires remediating the entire area as a whole. Detailed analysis of the behavior of $\hat{\delta}_o(m)$, relative to $\delta_o(m)$ pertains to step 3, which is not addressed in this chapter.

NUMERICAL ILLUSTRATIONS

We illustrate our methodology with several data sets. Figure 1 shows three data sets on a regular grid. The first of the three is americium data from the Rocky Flats plant, Colorado; the second is a synthetic set generated by a certain periodic version of the function xy; the third data set is obtained from independent uniform variables, by averaging neighboring values. Figure 2 shows two actual data sets for dioxin and lead, respectively, on irregular grids. The dioxin data serve to illustrate the fact that our methodology applies equally well to irregular grids, and the lead data illustrate the granular situation mentioned at the end of the previous section.

Figure 3 shows—for the three data sets of Figure 1—simultaneous plots of the actual discrepancy index $\delta_o(m)$, computed both from successive uniform grids (solid line) and by our methodology (dotted line).

Figure 4 shows the good correspondence between the actual discrepancy index $\delta_o(m)$ and the estimated discrepancy index $\hat{\delta}_o(m)$, for four of our data sets.

Figure 5 illustrates the fact that the actual discrepancy index $\delta_o(m)$ and estimated discrepancy index $\hat{\delta}_o(m)$ will agree better, under our methodology, in the case of granular data if sampling begins near zones of high contamination.

Figure 6 illustrates the fact that the estimated discrepancy index $\hat{\delta}_o(m)$ (as also the actual discrepancy index $\delta_o(m)$), will show little tendency to converge under extreme granularity.

ACKNOWLEDGMENTS

We are indebted to R. L. Hardy for introducing us to the multiquadric method; to Igi Litaor of the Rocky Flats plant, Colorado, for valuable comments and for providing us the americium data; and to Evan J. Englund of the Environmental Monitoring Systems Laboratory, Las Vegas, for providing us the dioxin and lead data. Hardy and Litaor, in addition, provided valuable references. We would like to acknowledge as well the help of Michael L. Thompson of the Department of Agronomy, Iowa State University. Funding for this work was provided by the Office of Technology Development, Office of Environmental Restoration and Waste Management, U.S. Department of Energy. Ames Laboratory is operated by Iowa State University for the U.S. Department of Energy under Contract No. W–7405–ENG–82.

REFERENCES

1. Hardy, R.L. "Theory and Applications of the Multiquadric-Biharmonic Method: 20 Years of Discovery," *Comput. Math. Appl.* 19 (8/9): 163–208 (1990).
2. Hein, G.W. and Lenze, K. "Zur Genauigkeit und Wirtschaftlichkeit verschiedener Interpolations-und Prädiktions-Methoden," *Z. für Vermessungswes.* 104: 492–505 (1979).
3. Franke, R. "A Critical Comparison of Some Methods for Interpolation of Scattered Data," *Technical Report NPS* 53-79-003. Naval Postgraduate School. Monterey, CA. (1979).
4. Hardy, R.L. "Comparative Studies of Multiquadric Equations and Other Methods of Interpolation and Approximation," Technical Papers *Am. Congr. Surveying and Mapping 41st Annual Meet.,* Washington, D.C. (1981).
5. Sirayanone, S. "Comparative Studies of Kriging, Multiquadric-Biharmonic, and Other Methods for Solving Mineral Resource Problems," PhD Dissertation, Iowa State University, Ames, IA (1989).
6. McBratney, A.B., Webster, R., and Burgess, T.M. "The Design of Optimal Sampling Schemes for Local Estimation and Mapping of Regionalized Variables-I: Theory and Method, *Comput. Geosci.* 7: 331–334 (1981).
7. McBratney, A.B. and Webster R. "The Design of Optimal Sampling Schemes for Local Estimation and Mapping of Regionalized Variables. II. Program and Examples, *Comput. Geosci.* 7: 335–365 (1981).
8. Yfantis, E.A. and Flatman, G.T. "On Sampling Nonstationary Spatial Autocorrelated Data," *Comput. Geosci.* 14: 667–685 (1988).
9. Cressie, Noel, Gotway, C.A., and Grondona, M.O. "Spatial Prediction from Networks," *Chemometrics and Intelligent Lab. Syst.,* 7: 251–271 (1990).
10. Ripley, B.D. *Spatial Statistics* (New York, NY: John Wiley & Sons, Inc., 1981), p. 25.
11. Matheron, G. "Principles of Geostatistics," *Econ. Geol.* 58: 1246–1266 (1963).
12. Hardy, R.L. "Multiquadric Equations of Topography and Other Irregular Surfaces," *J. Geophys. Res.* 76: 1905–1915 (1971).
13. Hardy, R.L. "Kriging, Collocation, and Biharmonic Models for Applications in the Earth Science," Technical Papers. *Am. Congr. Surveying and Mapping 43rd Annual Meet.,* Washington, D.C. (1984).

CHAPTER 22

The Center for Environmental Statistics: Interim Status and Vision of Products

Brand Niemann, Carroll Curtis, and Eleanor Leonard

ABSTRACT

An interim framework for presenting "the facts and figures and the discipline" of environmental statistics using an innovative information system is described. The interim frameworks include the following concepts/methods and products: 1. Metadata standards: guides and databases published; 2. Classification of indicators: Sourcebook of Environmental Indicator Fact Sheets for the Public; 3. Spatial metadata and statistics: Atlas of Selected Environmental Databases and Cross-Media Issues; 4. Data integration: annual reports on environmental problems areas and national and regional goals to the public; 5. Information system: Infobases on agency servers and the Internet; and 6. Strategic quality planning: Master environmental data infobase for EPA planners and managers. Application of the interim frameworks are illustrated with sample products that have been produced or are being produced. The applications of these sample products for the different clients of the EPA Center of Environmental Statistics (the Center) are also discussed.

INTRODUCTION

The word statistics is defined in two ways.[1] Statistics (plural) are graphs, charts, numbers, and tables of numbers; or in other words, facts and figures. Statistics (singular) is a body of methods for reaching conclusions about data, or in other words, a discipline. In the words of the prominent statistician Lincoln Moses, "the quality and credibility of a statistic, indeed its very meaning depends on the process that produced it."

Applying the above definitions, environmental statistics should be the facts and figures and the discipline that attempt to characterize the condition of the world around us. However, according to another prominent statistician,[2] the state of the art of environmental statistics is less advanced than for other disciplines because "environmental statistics are impossible, are unavoidable, are of bad quality; and require special answers and resources!" This has given rise to the situation where it has been said that "environmental statistics, to be sure, are not always reliable, but we have nothing better and we must make as much of them as we can."

Attempts to improve the state of the art for environmental statistics have given rise to a new discipline, envirometrics, which has been defined[3] as the science of applying statistical tools and methods to real-world problems which should include the certification of training and products of applied statisticians with real-world data. Envirometricians need to be good communicators to policymakers ("poor data pollutes the environment of the mind") and follow the "keep it simple statistics" (KISS) approach.

The purpose of this chapter is to describe an interim framework for presenting "the facts and figures and the discipline" of environmental statistics using an innovative information system. The application of the interim frameworks are illustrated with sample products that have been produced or are being produced. The applications of these sample products for the different clients of the EPA Center of Environmental Statistics are also discussed.

FRAMEWORK AND INTERIM PRODUCTS

Background

The Office of Strategic Planning and Environmental Data (OSPED) is within the Office of Policy, Planning, and Evaluation (OPPE). The development of environmental goals and measures for strategic planning and the development of methods and reports on environmental data and statistics are the major functions of OSPED.

The EPA is actively engaged in strategic quality planning. The goal is full implementation of "How It All Fits" by fiscal year 1995 as shown in Figure 1. Strategic goals are "what" we want to achieve and total quality management (TQM) is "how" we achieve those goals. TQM is based on continuous improvement from a customer focus, teamwork, and performance measurement.

The EPA Science Advisory Board has recommended using comparative risk analysis to identify priorities. The scientists' ranking of envir-

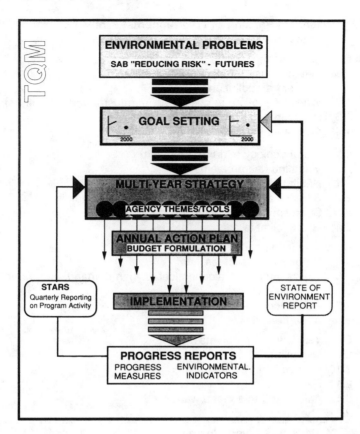

FIGURE 1. Schematic diagram of "how it all fits."

onmental problems for the EPA Science Advisory Board[4] is shown in Table 1. The agency's 10 strategic themes represent tools for reducing risk.[5] The EPA strategic theme 2: science/data—improving the EPA knowledge base includes: (1) improving the scientific underpinnings of EPA policies (a customer-oriented goal of providing the right information to those who need it); and (2) enhancing our capability to use environmental indicators, statistics, and activity measures to evaluate the success of programs and to report the results of our activities to our stakeholders. The strategic themes help integrate actions across media-specific boundaries. A number of cross-media initiatives were used to formulate the agency's fiscal year 1994 budget. The agency is developing a set of national goals to focus its efforts to reduce risk.[6]

The EPA strategic quality planning and performance reporting is as follows: (1) multiyear strategic plans that chart the steps and tools to

**Table 1.   Scientists Ranking of Environmental Problems for the EPA
Science Advisory Board, September 1990**

1. Risks to human health
    Relatively high-risk problems
    1.1   Ambient air pollution
    1.2   Worker exposure to chemicals in industry and agricultural
    1.3   Pollution indoors
    1.4   Pollutants in drinking water

2. Risks to natural ecology and human health
    Relatively high-risk problems
    2.1   Habitat alteration and destruction
    2.2   Species extinction and loss of biological diversity
    2.3   Stratospheric ozone depletion
    2.4   Global climate change

    Relatively medium-risk problems
    2.5   Herbicides/pesticides
    2.6   Toxics, nutrients, biochemical oxygen demand, and turbidity in
           surface waters
    2.7   Acid deposition
    2.8   Airborne toxics

    Relatively low-risk problems
    2.9   Oil spills
    2.10 Groundwater pollution
    2.11 Radionuclides
    2.12 Acid runoff to surface waters
    2.13 Thermal pollution

Source: EPA Science Advisory Board.[4]

reduce risk; (2) annual action plans that will link the long-term planning and budgeting process; (3) performance measures (input, output, and outcome) that will provide feedback on program effectiveness and adjustments to the strategies; and (4) progress reports (internal and external) that will document achievement of the proposed national environmental goals.

The EPA is also analyzing trends and exploring the use of scenarios. Experts have advised the EPA to take advantage of four interconnected trends, namely: (1) economic growth and environmental quality are complimentary goals, (2) the locus of power for environmental decision making is shifting, (3) attitudinal issues are likely to pose the most difficult challenges, and (4) consensus building may be more important than technological or administrative solutions.

The EPA center is developing a series of environmental statistics reports (ESRs) and electronic infobases leading eventually to a state of the environment report (SER). A recent survey[7] has found the number of SERs and ESRs have multiplied worldwide. This survey also explains that ESRs are largely numerical whereas SERs have substantial explanatory text and have some difficulties as follows: the interpretation changes as the monitoring base evolves; the databases are usually inadequate and data quality poor or nonexistent; the reports are expensive and unattractive to the public; there is a lack of public trust in the independence of the producer; and the reports have conflicting aims and political constraints. One of the best examples of a combined ESR-SER is the recent report from the United Nations Statistical Commission and the Economic Commission for Europe[8] which includes a part devoted to standard statistics, time series, and indicators on several aspects of the environment; and a part devoted to a statistical monograph on a complex environmental issue (agriculture) which attempts to describe the linkages between actual environmental effects and the physical, social, and economic characteristics of agricultural production.

The EPA environmental statistics report and infobase series are being organized around the following set of guiding principles: (1) statistics are data and the process that produced them (metadata); (2) metadata standards need to be developed and metadata organized; (3) simple classifications of indicators and priority environmental problem areas are helpful in organizing reports; (4) state of the environment frameworks are helpful in deciding what to measure and in writing the "data story"; (5) indicators are trends that provide a basis for action; (6) maps present special problems of design and interpretation; and (7) infobases are very effective at managing and distributing free forms of information such as text, graphics, and databases. Table 2 provides a comparison of the priority environmental problem areas and the proposed national environmental goals.

The following subsections provide an overview of six individual frameworks involved in producing environmental statistics reports and infobases and in providing them to the EPA and the public. Table 3 provides an overview of the concepts or processes on the left-hand side and the interim products on the right-hand side of the table, respectively.

Metadata Standards: Guides and Databases Published

Statistics are data and the process that produces them (metadata). Statistics has also been referred to as the science of uncertainty; thus metadata should characterize the uncertainty about the data. The goal of quality environmental statistics reporting is to combine the data and the metadata so that users have the results as well as the caveats about the results together in one place to guide their interpretation and application of the results.

Table 2. Comparison of ESID Priority Environmental Problem Areas and the Proposed National Environmental Goals

Environmental problem areas	National environmental goals
Global warming from carbon dioxide (2.4)	Global climate change (2.4)
Stratospheric ozone depletion (2.3)	Protection of the ozone layer (2.3)
Ground-level ozone trends (1.1)	Clean air (1.1, 2.8, 2.12)
Acid deposition trends (2.7) Loss of wetlands (2.1) Loss of songbirds (2.2) Loss and recovery of forests (2.1) Lead contamination in biota (2.1) Degradation of shellfish waters (2.6)	Ecological protection (2.1, 2.2)
Surface water quality trends (1.4, 2.6, 2.12)	Clean water (1.4, 2.12) Safe drinking water (2.10)
	Clean up of contaminated sites (2.6)
	Prevention of ongoing releases of harmfull toxic chemicals (2.6)
Pesticide usage trends (1.2, 2.5)	Safe food (2.11)
Chesapeake Bay region Great Lakes region Gulf of Mexico region Mexican-United States border region	
	Improved understanding of the environment Safe indoor environment (1.3) Worker safety (1.2) Prevention of oil spills and chemical accidents (2.9)

Note: Parenthetical numbers refer to Table 1.

Table 3. Overview of Framework and Interim Products

Concepts/Process	Product
Metadata standards	Guides and databases published
Classification of indicators	Sourcebook of Environmental Indicator Fact Sheets for the Public
Spatial metadata and statistics	Atlas of Selected Environmental Databases and Cross-Media Issues
Data integration	Annual reports on environmental problem areas and national and regional goals to the public
Information system	Infobases on agency servers and the Internet
Strategic quality planning	Master environmental data infobase for EPA planners and managers

The following interim hierarchy of metadata standards are suggested:

Level 1: Data from recognized statistical programs, i.e., programs that were selected for the Guide to Selected National Environmental Statistics in the U.S. Government (the Guide).[9]
Example: The Guide to Selected National Environmental Statistics in the U.S. Government Infobase (1993).

Level 2: Microevaluation of summary statistics — complete the Data Quality Profile "short form" by interview.
Example: "Atlas of Selected Global Indicators of Sustainable Development" (1992).[10]

Level 3: Detailed evaluation of summary statistics and selected source data with major policy significance — complete the Data Quality Profile "long form" by interview and work with the actual databases.
Example: "Draft Data Set Profiles for the "Environmental Progress Report," May 1992 draft for discussion."[11]

Level 4: Complete review and publication of selected key databases — the data certification process.
Example: Interagency Working Group on Data Management for Global Change (IWGDMGC) Inter-Agency Reference (InterRef) CD-ROM.[12]

The latter refers to an interagency effort that the Center participates in and supports that will periodically publish high-quality metadata and data on compact computer disks with large storage capacity. Initially the InterRef is focusing on packaging the high-quality metadata and data that has already undergone peer review and needs to be made more accessible to users in the global change research community.

The Center is asking its National Advisory Committee on Environmental Policy and Technology (NACEPT), the EPA Statistical Policy Advisory Committee (SPAC), and its American Statistical Association (ASA) Review Committee for comments on the proposed hierarcy of metadata standards and the following suggested structure of metadata information systems like the Guide:

- directory ("where the data are")
- quality/accuracy ("fitness for use")
- dictionary ("what the data are")
- samples ("the data actually exist")

A sample page from the Guide to Selected National Environmental Statistics in the U.S. Government (1993) is shown in Figure 2. The electronic version of the Guide contains links to about 400 tables, graphs, and maps of sample summary statistics provided by the statistical programs appearing in the print copy of the Guide.

The Guide to Selected National Environmental Statistics in the U.S. Government responds to the need to help analysts, decision makers, researchers, students, and others obtain policy-relevant environmental statistics and publications and locate experts who are knowledgeable about the data. The Guide is a reference to national level, time-series environmental statistics that are compiled and distributed by the U.S. government on a regular basis. It is a guide to statistical programs and the primary and secondary statistics they generate, not a guide to raw data or databases. As a starting point to learn more about various environmental statistical programs of the U.S. government, the Guide is not intended to supplant information that can be obtained directly from the government agencies. Furthermore, it is not an inclusive guide to U.S. governmental statistical programs, but ones that produce frequently sought after, national level statistics. The criteria for inclusion are the five items mentioned in the title, namely, selected, national, environmental, statistics, and U.S. government. The 1990 Annual Review of Information Science and Technology[13] cited the Guide as a good example of a more expansive multidisciplinary compilation of data sets that can be used to protect and improve the environment rather than one just to enforce standards. The statistical programs in the Guide are arranged by government department and agency. Each entry contains information about a separate statistical program (e.g., program purpose, data coverage and collection methods, geographic coverage, agency contacts, pertinent publications, and database access options). Information in the records was prepared and provided by government agencies in response to a questionnaire. The Guide also contains an index to over 150 key words and phrases that can be used to locate desired programs.

DEPARTMENT OF ENERGY

Carbon Dioxide Information Analysis Center

OFFICE:

Oak Ridge National Laboratory
Environmental Sciences Division

SUMMARY PROGRAM DESCRIPTION:

The objective of the Carbon Dioxide Information Analysis Center (CDIAC) is to compile, evaluate, and distribute information related to carbon dioxide (CO_2) in support of the Department of Energy's Carbon Dioxide Research Program (CDRP). To accomplish this objective, CDIAC identifies researchers' needs for data, models, and information; obtains, evaluates, and ensures the quality of the information; and works with other national and international data centers as well as with individual researchers to promote and facilitate the exchange of data. CDIAC supports the data and information needs of researchers studying the effects of increasing atmospheric CO_2 on climate, carbon cycle processes, and resources.

DATA COVERAGE:

Variables measured and analyzed include any CO_2–related or greenhouse gas–related parameter. Trend data include: atmospheric CO_2 and methane concentrations from surface monitoring sites and from ice cores; CO_2 emissions resulting from fossil fuel consumption and cement production; historical land use data in Southeast Asia; long-term temperature and precipitation, cloudiness, and sunshine records for the United States; global and hemispheric temperature anomalies; dust veil indices; umbral/penumbral ratios; and radiocarbon data from oceanographic cruises.

COLLECTION METHODS:

Data sets that are archived and distributed by CDIAC have either been sent to CDIAC voluntarily by the collecting agency or researcher or have been sent to CDIAC as a result of contracts made by CDIAC. CDIAC identifies data sets critical to greenhouse and global warming issues by conducting surveys of researchers and users of CDIAC's data products; contacting researchers and agencies addressing global warming issues; attending scientific conferences and symposia; and soliciting suggestions from DOE managers. CDIAC does not impose format restrictions on individuals and agencies that archive data at CDIAC. CDIAC accepts the data in whatever form (i.e., hardcopy, dBASE files, LOTUS files, flat ASCII files) is most convenient for the contributor. Irrespective of the source, CDIAC reviews all data sent to CDIAC before documenting and distributing the data set. These reviews, which are often extensive, involve consultation with the contributing agency or researcher. CDIAC does not correct or distribute any data sets or computer models without the written consent of the contributing individual or agency.

COLLECTION FREQUENCY:

The frequency of data collection with the CDRP program ranges from hourly (e.g., atmospheric CO_2 concentrations) to decennial (e.g., land use changes in Southeast Asia).

GEOGRAPHIC COVERAGE:

Global.

CONTACT:

Robert M. Cushman, Director
Carbon Dioxide Information Analysis Center
Oak Ridge National Laboratory
P.O. Box 2008
Oak Ridge, TN 37831–6335
Phone: (615) 574–0390

FOR PUBLIC INQUIRIES:

Contact Sonja B. Jones at the address and phone numbers listed above.

PUBLICATIONS:

Boden, T.A., P. Kanciruk, and M.P. Farrell. 1990. Trends '90: A compendium of data on global change. ORNL/CDIAC–36. Oak Ridge, TN: Oak Ridge National Laboratory, Carbon Dioxide Information Analysis Center.

Boden, T.A., R.J. Sepanski, and F.W. Stoss (eds). 1991. Trends '91: A Compendium of Data on Global Change. ORNL/CDIAC–46. Oak Ridge, TN: Oak Ridge National Laboratory, Carbon Dioxide Information Analysis Center.

Burtis, M.D. (ed.). 1989. Carbon Dioxide Information Analysis Center Catalog of Databases and Reports. Environmental Sciences Division Publication No. 3477. Oak Ridge, TN: Oak Ridge National Laboratory, Carbon Dioxide Information Analysis Center.

Quinlan, F.T., T.R. Karl, and C.N. Williams, Jr. 1987. CDIAC Numeric Data Collection: United States Historic al Climatology Network (HCN) Serial Temperature and Precipitation Data. NDP–019. Oak Ridge, TN: Oak Ridge National Laboratory, Carbon Dioxide Information Analysis Center.

DATABASE(S):

All reports and data packages described in the above reports are available on request. For a complete listing and description of CDIAC databases, order "CDIAC Communications" from the contact listed above.

FIGURE 2.   Sample page from the guide to selected national environmental statistics in the U.S. government (1992).

In future editions, coverage of the Guide may be expanded to include regional and national spatial environmental databases; to provide more information on international, transnational, and global environmental data; and to include additional references to important health, ecological, and economic impacts including costs and damages. Environmental statistics gathered by private sources — nongovernmental organizations, corporations, research institutions, and national associations — may also be included. If possible, future editions will more clearly document the quality, completeness, and limitations of the data.

Classification of Indicators: Sourcebook of Environmental Indicator Fact Sheets for the Public

Simple classifications of indicators and a short list of priority environmental problem areas are helpful in organizing reports for the nonexpert and the public. The Norwegian Central Bureau of Statistics[14] and Environment Canada[15-18] have made considerable progress in producing environmental indicator reports and bulletins for the public, and we have decided to follow their methods and products in order to provide essential environmental information to the public.

An environmental indicator is a number that is meant to indicate the state or the development of important aspects of the environment. Usually an indicator will be presented as a set of numbers, for instance, a time series or a geographic cross section. Strictly speaking, the term indicator refers to a specific number along the time or space dimension. However, we will employ a less precise language and denote the whole set of numbers as an indicator.

An indicator without a unit of measurement is an index — several indicators weighted together to capture the total impact on an aspect of the state of the environment. A leading indicator to an environmental indicator is one that gives early warning of the development in the environmental indicator.

As the Norwegians, we focus on man-made (anthropogenic) impacts on the environment and do not attempt to develop indicators of sustainable development since we need to know a lot more to determine whether development is sustainable. In addition, we do not expect that a set of environmental indicators will give environmental experts new information; thus we consider the principal users of the environmental indicators to be nonexperts that are concerned about the environment.

A set of indicators should give information on development, environmental quality, and environmental policy (e.g., total area of protected national parks). While it is desirable to have the same set of indicators as used by other countries, there are yet no generally accepted standards for environmental indicators. The length of the time series is an important aspect in the choice of indicators as well as information on the effects of natural variability since we need to distinguish natural variations from changes in trends. Environment Canada has provided a list of eight criteria to assist in the selection of suitable indicators.[16]

Environmental problems are complex: the same problem may have several causes and the same stress may cause several problems. Unfortunately any classification scheme will cut problems that naturally belong together into pieces due to the many interconnections between environmental problems. Some possible indicator classification schemes are as follows:

- stresses by pollutant
- stresses by economic sector responsible for the pollutants
- media of recipients in the ecosystem (air, water, land, etc.)
- effects by ecosystems
- final effects (human welfare)

The Norwegians have reasoned that if nonexperts are the main audience, it is natural to choose a classification scheme as close as possible to the final effects (responses) and a short list of environmental problems that is easily recognizable by the nonexpert as follows, respectively:

- response indicators — main set as close as possible to final effects
- stress indicators — stresses consistent with indicators in the response indicator set
- structural indicators — relevant demographic and economic parameters related to the secondary (stress) indicator set
- speculative indicators — indicators with linkages to the stress indicators that are not well established

and

- global problems
- health problems including noise
- eutrophication
- forest and fish damage
- contamination
- recreation
- biodiversity

Interestingly, the Norwegians and others have found that different schemes for indicator classification generally give the same list of environmental problems.

A specific example of the Norwegian classification scheme is environmental effects or problems — climate change:

- response indicator—radiative forcing (watts per square meter)
- stress indicators—global CO_2 emissions (metric tons) and atmospheric CO_2 concentrations (parts per million volume)
- structural indicators—energy use and world population growth
- speculative indicator—change in mean global temperature (degrees C)

The Norwegians plan to have their indicator data accompanied by commentaries on the development over time of the indicators, to possibly compare them to similar indicators from other countries, and to include a list of potential indicators in addition to the main list.

The Canadian State of the Environment Reporting program[15,16] has developed a preliminary set of some 50 indicators covering 18 "issues" under five main headings:

- atmosphere
- water
- biota
- land
- natural economic resources

This classification is based on recipient and problem areas. The Canadian State of the Environment Reporting program has also begun to publish a series of Indicator Bulletins of approximately two to five pages in length at regular intervals in print and electronic form. The first Indicator Bulletin on stratospheric ozone depletion will be followed by others on climate change, urban air quality, and wildlife.

Work on a series of Environmental Indicator Fact Sheets has also begun with the selection of 11 priority environmental problem areas and four regional "geographic targets." The selection was based on scientists' ranking of environmental problems for the EPA Science Advisory Board in September 1990[4] (shown previously in Table 1) and other uses and clients. Tufte's[19,20] principles include that excellence in statistical graphics consists of complex ideas communicated with clarity, precision, and efficiency, and the suggested use of "small multiples." They are being employed in the display of the statistical graphics and tables in the Indicator Fact Sheets to enforce comparisons among the different displays and to thereby see whether qualitative relationships might exist among them. The global carbon dioxide emissions were selected as an example of an Environmental Indicator Fact Sheet (Figure 3).

Spatial Metadata and Statistics: Atlas of Selected Environmental Databases and Cross-Media Issues

Spatial metadata and statistics are both "young sciences";[21,22] and integrating statistical analysis and cartographic methods, including animation, for environmental data is a relatively recent activity.[23-27] The term cross-media, as used here, refers to combining by overlay or superposition separate spatial data themes such as populated places and toxic waste sites, sulfur emissions and acid deposition patterns, etc. An example of a cross-media issue is the spatial relationship between sulfur dioxide emissions and wet sulfate depositions in eastern North America.[23] Another recent development of interest to spatial data display and analysis is fractal interpolation,[28] which is a new technology for making complicated curves and fitting experimental data in ways that resemble natural phenomena like clouds. With iterated function systems (IFS) for computing fractals, geometrically complex graphs of continuous functions can be constructed to pass through specific data points. Fractal interpolation has been implemented recently in the Spatial Analysis System (SPANS) GIS which the EPA Center is using. The Center has also used the GisPlus software package to integrate spatial statistical analysis (kriging) and cartographic methods to estimate acid deposition values in the eastern United States on a congressional district basis to answer questions from Congress.

The Federal Geographic Data Committee (FGDC) recently sponsored an information exchange to discuss the need for improved spatial metadata.[21] The result was the need for a set of essential metadata elements that could serve as an operational model and be consistent with the proposed Spatial Data Transfer Standard (SDTS) which is currently under review by the spatial data community. Some examples of the proposed spatial metadata elements are: (1) location (projection parameter, resolution, etc.); (2) lineage (data version, processing steps, etc.); (3) general (scale, dates, etc.); (4) dictionary (data set identifier, theme identifier, etc.); and (5) quality (positional accuracy, attribute accuracy, etc.).

The EPA Center is involved with developing a hierarchical ecological framework for North American state of the environment databases and reporting. Because of fundamentally different philosophies for ecological frameworks in the United States and Canada,[23,29,30] the EPA Center and Environment Canada have proposed a series of steps leading to a workshop of experts for producing a common ecological classification system for North America as follows: (1) identify the mapping issues that will arise in deriving an ecological map for both countries; (2) prepare a synoptic report on the currently used ecological frameworks used in both countries; (3) propose an action plan for a workshop forum of experts from both countries to discuss the issues; (4) prepare a series of maps showing ecological frameworks in both countries for the workshop; (5) provide a summary of currently available georeferenced environmental information and display examples mapped on the North American base map for the workshop; and (6) conduct the workshop and prepare the proceedings which would include North American maps for the consensus ecological framework.

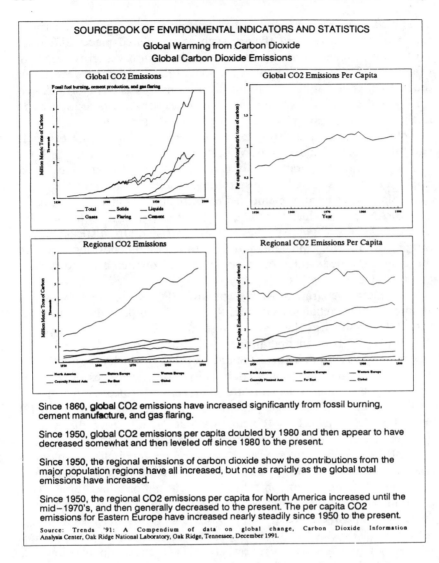

SOURCEBOOK OF ENVIRONMENTAL INDICATORS AND STATISTICS

Global Warming from Carbon Dioxide
Global Carbon Dioxide Emissions

Since 1860, global CO2 emissions have increased significantly from fossil burning, cement manufacture, and gas flaring.

Since 1950, global CO2 emissions per capita doubled by 1980 and then appear to have decreased somewhat and then leveled off since 1980 to the present.

Since 1950, the regional emissions of carbon dioxide show the contributions from the major population regions have all increased, but not as rapidly as the global total emissions have increased.

Since 1950, the regional CO2 emissions per capita for North America increased until the mid–1970's, and then generally decreased to the present. The per capita CO2 emissions for Eastern Europe have increased nearly steadily since 1950 to the present.

Source: Trends '91: A Compendium of data on global change, Carbon Dioxide Information Analysis Center, Oak Ridge National Laboratory, Oak Ridge, Tennessee, December 1991.

FIGURE 3. Sample pages from the Sourcebook of Environmental Indicator Fact Sheets for the Public—Global Carbon Dioxide Emissions.

Maps and spatial data analyses present special problems of design and interpretation. We are attempting to follow the principles of Monmonier[24] on the use of multiple maps and map animation or "atlas touring,"[25] and the cartographic principles of MacEachren[25] and Brennan[26] to improve the design and interpretation of the map and spatial analysis products of the EPA Center. We are also using cartographers who are

SOURCEBOOK OF ENVIRONMENTAL STATISTICS AND INDICATORS

Global Warming from Carbon Dioxide
United States Carbon Dioxide Emission Estimates

Year	Total	Solids	Liquids	Gases	Cement	Gas Flaring	Per capita emissions (metric tons of carbon)
1950	696.1	347.1	244.8	87.1	5.3	11.8	4.57
1951	716.7	334.5	262.2	102.7	5.7	11.7	4.63
1952	697.9	296.6	273.2	109.9	5.8	12.5	4.44
1953	714.5	294.3	286.6	115.5	6.1	11.9	4.46
1954	680.5	252.2	290.2	121.2	6.3	10.6	4.18
1955	746.0	283.3	313.3	130.8	7.2	11.4	4.50
1956	781.9	295.0	328.5	138.1	7.6	12.7	4.63
1957	775.1	282.7	325.8	147.6	7.1	11.9	4.51
1958	750.8	245.3	333.0	155.8	7.5	9.3	4.29
1959	781.4	251.5	343.5	169.9	8.1	8.4	4.40
1960	799.5	253.4	349.8	180.4	7.6	8.3	4.43
1961	801.9	245.0	354.1	187.4	7.7	7.7	4.37
1962	831.5	254.2	364.3	198.7	8.0	6.3	4.46
1963	875.6	272.5	378.8	210.3	8.4	5.6	4.63
1964	912.9	289.7	389.7	219.8	8.8	5.0	4.76
1965	948.3	301.1	405.6	228.0	8.9	4.7	4.88
1966	999.7	312.7	425.9	246.4	9.1	5.5	5.08
1967	1039.2	321.1	443.6	258.5	8.8	7.2	5.23
1968	1081.0	314.8	471.9	277.4	9.4	7.6	5.38
1969	1132.0	319.7	497.4	297.8	9.5	7.7	5.58
1970	1165.5	322.4	514.8	312.1	9.0	7.2	5.68
1971	1173.2	305.7	530.5	323.3	9.7	4.2	5.66
1972	1227.3	310.4	575.5	327.6	10.2	3.6	5.86
1973	1275.4	334.0	605.4	321.7	10.6	3.6	6.03
1974	1231.1	330.1	580.7	307.9	10.0	2.4	5.76
1975	1179.0	317.6	565.1	286.0	8.4	1.9	5.46
1976	1262.0	351.6	608.1	291.3	9.0	2.0	5.78
1977	1269.7	355.6	641.9	260.5	9.7	2.0	5.76
1978	1293.4	361.2	655.0	264.7	10.4	2.2	5.80
1979	1300.9	378.7	634.6	274.8	10.4	2.4	5.77
1980	1259.3	394.6	581.0	272.5	9.3	1.8	5.53
1981	1210.6	403.0	533.1	264.2	8.8	1.4	5.26
1982	1146.9	390.1	502.2	245.4	7.8	1.4	4.93
1983	1149.4	405.5	500.1	233.8	8.7	1.4	4.89
1984	1187.5	427.8	507.1	241.5	9.6	1.6	5.01
1985	1201.3	448.0	505.6	236.7	9.6	1.4	5.02
1986	1204.5	439.7	531.1	222.6	9.7	1.4	4.99
1987	1253.8	463.3	545.3	233.8	9.6	1.8	5.15
1988	1313.8	491.4	566.3	244.6	9.5	2.1	5.35
1989	1328.9	498.4	566.5	252.4	9.5	2.1	5.37

FIGURE 3. Continued.

specialists in displaying and communicating geographic information in environmental problem areas.

Work on the Atlas of Selected Environmental Databases and Cross-Media Issues (the Atlas) began with the state level databases provided by the federal statistical programs that are already part of the Guide infobase. In addition, databases for other geographic scales and resolutions have and are being acquired for other sections of the Atlas, namely,

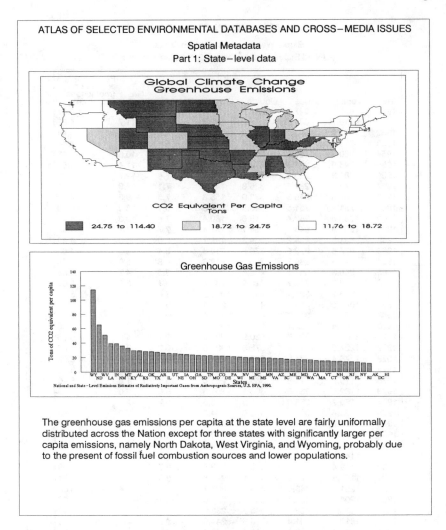

FIGURE 4.   Sample pages from the Atlas of Selected Environmental Databases and Cross-Media Issues—carbon dioxide emissions by state.

world country level from the World Resources Institute,[31] national point and county level, regional and state level, and cross-media issues. The Atlas format adopted consists of the following: (1) a thematic map with clear legend; (2) the rank order or frequency distribution of the thematic values; (3) a word description of the result the graphic shows along with caveats; (4) the source of the database and explanation of the map classification scheme and map projection; (5) a table of the data used in the map; (6) descriptive statistics of the data and a microevaluation of the

ATLAS OF SELECTED ENVIRONMENTAL DATABASES AND CROSS-MEDIA ISSUES

Spatial Metadata
Part 1: State-level data

Agency: Environmental Protection Agency

National and State-Level Emission Estimates of Radiatively Important Gases from Anthropogenic Sources (1990).

Office:

Air Quality Planning and Standards

Summary Program Description:

These emissions were estimated from source activity levels and emission factors.

The 1990 Census state-level populations were used to calculate the per capita values.

Map Classification Scheme:

The data were displayed thematically based on the national state average. This usually means that the national average range has the fewest or no states within that range. This approach helps the reader to see which states fall above or below the national average. However, outlier state values are not accounted for and may distort the graphical representation. Outliers are readily seen in the rank order bar graph on the accompaning page with the map itself. This methodology was adapted from Monmonier, 1991, "How to Lie with Maps", University of Chicago Press, Chicago 60637.

Map Projection:

Albers equal-area conic

Data: Greenhouse Gas Emissions (1990)

STATE	CO_2PERCAP	STATE	CO_2PERCAP
AL	29.25	NE	24.75
AK		NV	19.89
AZ	18.37	NH	14.62
AR	26.04	NJ	13.88
CA	15.65	NM	35.96
CO	21.89	NY	12.23
CT	14.63	NC	19.49
DE	21.53	ND	65.22
DC		OH	23.51
FL	13.57	OK	27.99
GA	22.93	OR	13.89
HI		PA	21.22
ID	17.02	RI	11.76
IL	25.74	SC	17.44
IN	39.07	SD	22.55
IA	24.16	TN	22.52
KS	28.38	TX	27.26
KY	29.41	UT	25.33
LA	39.23	VT	15.17
ME	17.11	VA	18.72
MD	16.85	WA	16.51
MA	15.27	WV	51.14
MI	19.65	WI	20.58
MN	19.30	WY	114.40
MS	19.48		
MO	21.93		
MT	32.83		

Descriptive statistics:

Count:	48
Maximum:	114.4
Minimum:	11.76
Range:	102.64
Sum:	1205.32
Mean:	25
Variance:	272
Std Dev.:	16
C.V.:	0.65

FIGURE 4. Continued.

database behind the table and the graphic. An example of the Atlas format is shown in Figure 4 (carbon dioxide emissions by state map).

An example of a cross-media issue map that superimposes the 1989 National Priority List (NPL) sites (highest priority Superfund sites for cleanup action) over counties with above average minority populations is shown in Figure 5. In this figure, we choose to show only the subset of counties (29) containing 20% or more minority populations based on the

Environmental Equity

Displayed counties represent counties with a combined Black and Hispanic population percentage greater than or equal to 20 and contains 3 or more NPL sites.

Data Source: USEPA, Office of Emergency and Remedial Response, Hazardous Site Evaluation Division, National Priority List.

Graphic: EPA Center for Environmental Statistics.

FIGURE 5. Sample map from the Atlas of Selected Environmental Databases and Cross-Media Issues—environmental equity (see text for explanation and statistics).

1990 census of population and three or more NPL sites. The national county average minority (Black and Hispanic) population is about 20% and the total NPL sites in 1989 was 1302. The total population as of December 31, 1989 was 249,840,300 while the total Black and Hispanic populations were 31,028,599 and 20,853,279, respectively. The number of counties identified by this procedure was far less than identified in an earlier analysis.[32]

Data Integration: Annual Reports on Environmental Problems Areas and National and Regional Goals to the Public

There are different types of data integration and one needs a structure for effectively displaying the results to nonexperts. One could employ so-called metaanalysis to combine and interpret different data sets into a common display and search for possible multivariate relationships between two or more variables, or one could pull together a wide collection of databases into a common database management system and display them in a consistent manner as the Guide infobase. True data integration, of course, is very difficult due to the inherent differences in the spatial and temporal scales and accuracies of almost all environmental databases. One of the best examples is the problem being encountered with integrating conventional EPA databases collected primarily for regulatory purposes with those from the Ecological Monitoring and Assessment program (EMAP) collected for statistical assessment purposes.

The EPA center clients need credible and effective data integration in annual reports on environmental problems areas and on progress toward national and regional goals in the form of a good "data story" that nonexperts can understand and use to take effective action. Through the use of "secondary" environmental data from a variety of sources, the EPA center will assess quality, organize, integrate, and develop descriptive statistics (i.e., statistics which can be used to describe the state of the environment, or the status of some measure in terms of trend). Work on an environmental statistics report series has begun based on the large number of database examples provided by the federal statistical programs in the Guide. The basic format of the data story consists of the following: (1) a statistical graphic, (2) a word description of the result the graphic shows with caveats, (3) the source of the database, (4) a table of the data used in the graphic, and (5) a microevaluation of the database behind the table and the graphic.

The Organization for Economic Cooperation and Development (OECD)[33] has employed a so-called pressure-state-response framework for organizing environmental statistics and information into a comprehensive story. The EPA center is experimenting with a slight modification of the basic pressure-state-response framework as a data integration structure by simply expanding each topic into two subtopics for a total of six broad categories, namely:

1. Pressures
   a. basic pressures (e.g., demographic, technological, political, etc.)
   b. actual pressures (e.g., release of pollutants, natural resource harvesting, land-use changes, etc.)
2. state (of the environment)
   a. status and trends (e.g., concentrations of contaminants, indoor and work environments, etc.)
   b. health and welfare impacts (e.g., changes in exposure, damages to resources, etc.)
3. responses
   a. short-term (e.g., environmental regulations, land-use planning, etc.)
   b. long-term (e.g., mitigation measures, planned consumption of resources, etc.)

The success of any environmental statistics framework and data integration structure still depends on the effective and efficient display of information,[19,20] which in turn depends on the innovative uses of state-of-the-art information system technology. Global warming from carbon dioxide was selected as a priority environmental problem area to illustrate the use of the data integration structure. Figure 6 shows a list of some of the issues under each of the six broad categories, and Figure 7 shows a selected graphic for each category. The data story is as follows:

Certain gases capture heat from the sun-warmed earth and hold some of this warmth in the atmosphere in what is known as the greenhouse effect. The principal greenhouse gases are carbon dioxide (CO_2), methane (CH_5), nitrous oxide (N_3O), chlorofluorocarbons (CFC-11 and CFC-12), and ozone (O_3) and other trace gases. Carbon dioxide is the principal greenhouse gas; and its sources are combustion of fossil fuels such as coal, oil, and natural gas; deforestation; and soil destruction. The "Basic

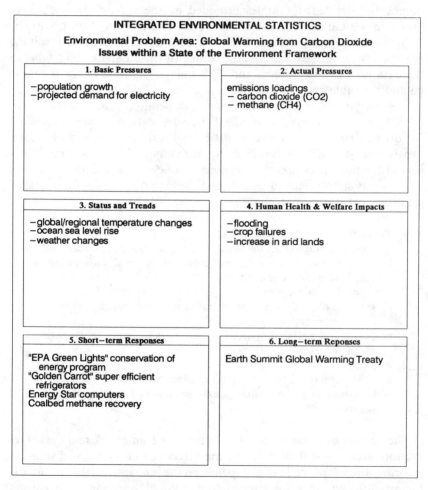

INTEGRATED ENVIRONMENTAL STATISTICS

Environmental Problem Area: Global Warming from Carbon Dioxide
Issues within a State of the Environment Framework

1. Basic Pressures	2. Actual Pressures
−population growth −projected demand for electricity	emissions loadings − carbon dioxide (CO2) − methane (CH4)

3. Status and Trends	4. Human Health & Welfare Impacts
−global/regional temperature changes −ocean sea level rise −weather changes	−flooding −crop failures −increase in arid lands

5. Short−term Responses	6. Long−term Reponses

"EPA Green Lights" conservation of
   energy program
"Golden Carrot" super efficient
   refrigerators
Energy Star computers
Coalbed methane recovery
	Earth Summit Global Warming Treaty

FIGURE 6. Sample of Integrated Environmental Statistics Fact Sheets: Global Warming from Carbon Dioxide—selected issues in a state of the environment framework.

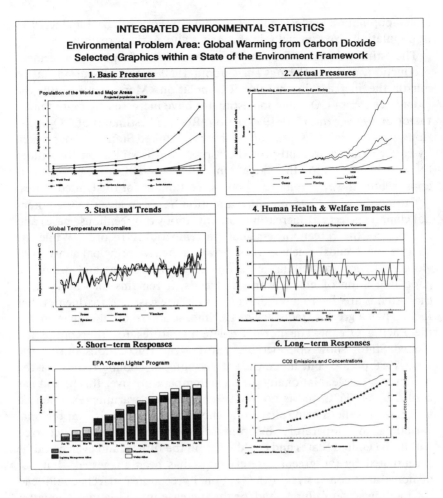

FIGURE 7. Sample of Integrated Environmental Statistics Fact Sheets: Global Warming from Carbon Dioxide—selected graphics in a state of the environment framework.

Pressure" (see box 1) for the generation of carbon dioxide emissions to the atmosphere is population growth and demand for electricity. The carbon dioxide emissions are the "Actual Pressure" (see box 2).

Time series of growth in world population size provide two crucial indices of population dynamics: the rate of growth over time and the absolute changes of population size over a given time period. The latter is often more pertinent as far as resource use and environmental impacts are concerned. The time-series data indicate that while the rate of growth appears to have peaked during the 1950–1985 period, the brunt of the global population explosion is yet to come in terms of absolute increases in population size especially in the Asian and African regions.

The estimated global CO_2 emissions from fossil fuel burning, cement production, and gas flarings since about 1860 to the present appear to mirror the increase in the global population. More significantly the global per capita CO_2 emission estimates have increased by more than a factor of 10 over the 1860–1990 period. Regional estimates of CO_2 emissions and per capita CO_2 emissions show the United States was overtaken by eastern Europe recently for the former, but is still about five times that of the world for the latter. Only the eastern European region is approaching the per capita CO_2 emission estimates for the United States.

During the past 20 years, much attention has been given to the potential climatic effects of increasing concentrations of greenhouse gases and to isolating the "greenhouse signal" in temperature records. While the "Status and Trends" (see box 3) as shown in the five principal analyses of global temperature anomalies are in general agreement and provide a good indication of global warming trends, it remains unclear whether these trends are due to a buildup of greenhouse gases or are due to other factors like a greater frequency of El Nino/southern oscillation events in the Pacific Ocean. The normalized temperatures for the continental United States show considerable year-to-year fluctuations during the 1900–1988 period, but no consistent warming or cooling trends for the nation as a whole. Nationally the mean temperature over this period has cooled by –0.06°C and is not consistent with the warming experienced across the northern hemisphere as a whole. Certainly significant regional variations in temperature and other climatic variables have occurred within the United States and could occur in the future, and even could be accentuated by the greenhouse effect. It has been suggested—but not demonstrated conclusively—that regional greenhouse effects could lead to flooding, crop failures, and/or the increase in arid lands. The latter could be possible "Human Health and Welfare Impacts" (see box 4) from global warming.

While the U.S. contribution to estimated global CO_2 emissions has decreased from about 40% in 1950 to about 20% in 1989, its per capita CO_2 emissions estimates have increased slightly over the period and are among the highest in the world. In response to world pressure, the United States has agreed to limit CO_2 emissions to present levels by some future date. A cap on future CO_2 emissions would constitute a "Long-Term" (see box 6) response. A composite graph of global CO_2 emissions and concentrations shows that the U.S. CO_2 emissions have increased slowly during the past 40 years compared to those for the rest of the world, which is responsible for the steady increase in global CO_2 concentrations.

The EPA has initiated several programs designed to reduce greenhouse gas emissions and demand for electricity in the United States and reduce deforestation in the world. The EPA Green Lights is a innovative program that encourages major U.S. corporations and states and local governments to install energy-efficient lighting. Lighting accounts for 20–25% of electricity used annually in the United States and lighting for industry, stores, offices, and warehouses represents 80–90% of total lighting electricity use. If energy-efficient lighting were used everywhere it were profitable, the electricity required for lighting would be cut by approximately 50% and aggregate national electricity demand would be reduced by 10%. This reduction is estimated to free $186 billion per year from rate payer bills for useful investment and lower annual CO_2 emissions by about 232 million tons (4% of the national total), the equivalent of 42 million cars. The EPA Green Lights is one example of a "Short-Term" (see box 5) response to reducing global warming from CO_2 emissions.

Information System: Infobases on Agency Servers and the Internet

The EPA Center performed a formal information system requirements analysis and developed an interim architecture assisted by experts in the fields of statistics, document design and engineering, and computer systems among others.[34] An information systems architecture is a description of all functional activities to be performed to achieve the desired mission; the elements needed to perform the functions, including all resources (hardware, software, facilities, data, and people); and the performance levels of those system elements. An architecture includes information on the technologies, interfaces, and locations of functions. It is considered an evolving description of an approach to achieving a desired mission. Some critical criteria for tests of the effectiveness of the EPA Center data and information systems were defined as follows:

- Are needed data readily available, preferably in a computer-usable format, at low cost?
- Are available data sets described well enough (metadata) so that researchers anywhere can find and use them?
- Can data taken by one researcher, or one agency, be used readily by others?
- Will data taken by this generation of scientists and statisticians be usable by the next generation?

It became clear that a critical role for the EPA Center should be to unite or reunite the data with its metadata. Metadata is information about the data, in other words, a description of the methods used to develop the databases and the discipline used to prepare the statistics. As a minimum, statistical metadata should include the following components, especially when reporting summary (aggregated) statistics:

- statistical microdata — measured values or variables of individual objects
- preparation process — coding, editing, and data entry
- aggregation process — weighting, models, etc.
- statistical macrodata — estimated values of characteristics, parameters, populations, etc.

The critical question then is what descriptions of macrodata in a statistical database are needed in order to help potential users of the databases judge the usefulness of the macrodata for different purposes?

By implementing the state of the art in electronic publishing technology on desktop computers and workstations, the EPA Center is able to eliminate traditional "cut and paste" of documents containing graphics and images and even create on-line documents that can be read like E-MAIL. The documents can literally be interfaces to databases where the tables and statistical graphics based on those databases are updated automatically when the underlying databases are changed or updated. The documents are being assembled in layers with links to the underlying databases so that each layer can stand alone in hard copy form, yet be linked electronically together in a popular and attractive format at a low cost which makes possible wide distribution.

As a first step toward implementing the state of the art in electronic publishing technology, the EPA Center has developed a series of Folio Views infobases that can be used on both stand-alone PCs and the agency superservers. Folio Views is already available on the agency superservers as part of the on-line help system for Novell Netware and is rapidly becoming an EPA and national standard for text retrieval and hypertext applications. An infobase is a repository of electronic information that is analogous to a database, but is more effective at managing the free forms of information such as textual metadata and "hyperlinking" to tables, graphs, and maps. A partial list of the infobases that the EPA center is producing itself or assisting others to produce is shown in Table 4. The work with infobases of guides to data and metadata supports the Interagency Working Group on Data Management for Global Change's Master Directory,[35] Interagency Reference CD-ROM, and Wide-Area Information Servers[36] activities, the Office of Management and Budget's Government-Wide Information Inventory/Locator System,[37] and the Intergovernmental Task Force on Monitoring of Water Quality, as well as the EPA program offices and EPA Master Directory efforts. The Guide infobase of the EPA center is also available on the

Table 4. Partial List of Infobases Completed or In Process

File name	Document title	Source	Status
GES.NFO	Guide to Selected National Environmental Statistics in the U.S. Government (1993)	EPA	Revised 1/93
WATER.NFO	Office of Water Compendium	EPA	Revised 1/93
21STCENT.NFO	Proceedings of an International Forum	Environment Canada	February 1993
MSTR.NFO	Global Change Master Directory	NASA	To be updated
TRENDS91.NFO	Carbon Dioxide Trends	CDIAC	September 1992
LEGACY.NFO	EPA Progress Report Securing Our Legacy	EPA	October 1992
ITFM.NFO	Task Force on Water Quality Monitoring	ITFM	November 1992 draft
FGDCMAN.NFO	Manual of Federal Geographic Data Products	FGDC	December 1992 draft
GMPO.NFO	Gulf of Mexico Program Support	GMPO	December 1992 draft
SQP.NFO	Strategic Quality Planning	OSPED	November 1992 draft
BIODIV.NFO	Biodiversity and Ecological Directory	EPA	October 1992 draft

Internet in compressed file format that can be downloaded and decompressed to a PC runtime file.

More recently, the center has been asked to support the Interagency Gulf of Mexico program and the upgrade of the EPA Quarterly Progress Report system. It is clear that regional case studies of the integration of progress measures and environmental indicators are critically needed to design a more comprehensive reporting system for the EPA. It is obvious that to accomplish national goals, the goals would have to also be accomplished at the local, state, and regional levels as well, so a logical first step is to compare and integrate the proposed national goals with those of the Gulf of Mexico program (see Table 5). The menu for the prototype Gulf of Mexico program office infobase is shown in Figure 8 which will be installed on the Gulf of Mexico program office bulletin board system soon to provide most near-term information management and distribution needs of the program. An Environmental Statistics Report on the Gulf of Mexico program support also serves as a user's guide to the infobase.[38]

Strategic Quality Planning: Master Environmental Data Infobase for EPA Planners and Managers

Strategic planning is an ongoing, systematic process aimed at articulating organization goals over the long- and short-term, and deciding on priorities for most efffectively achieving those goals. The new vision of the EPA, or "target that beckons," is one in which data integration and science have a major role in policy decisions and that progress measures and environmental indicators provide feedback to the planning process. One critical mission of strategic planning is to identify the tools to be used to make the vision a reality. Certainly one of the key tools that has been identified is for access to the vast storehouse of quantitative environmental data for a variety of users and purposes. In support of that vision, the EPA center is developing a Master Environmental Data Infobase for EPA planners and managers as shown schematically in Figure 9. Initially the EPA center has attempted to integrate the activities and products within the Office of Strategic Planning and Environmental Data in an infobase with an interface that follows the structure in Figure 1 as shown in Figure 10. An environmental statistics report on the support of strategic quality planning and environmental data by the EPA center also serves as a user's guide to the infobase.[39] This infobase may be used to support the EPA inreach and outreach efforts as part of the proposed national environmental goals effort and will be used as a prototype to build the master infobase for EPA planners and managers that provides them with the best access and analytical capabilities.

NEXT STEPS AND CHALLENGES

The production of products with credible and useful environmental statistics is the goal of the EPA Center for Environmental Statistics. The first steps have been to conceptualize and produce some interim products. The next steps are to complete those interim products and have them evaluated by various clients and our review committees. One major challenge will be to effectively integrate the scientific expertise, the statistical expertise, and the document engineering expertise needed to pro-

Table 5. Comparison Between Proposed National Environmental Goals and the Nine Environmental Challenges of the Gulf of Mexico Program

National environmental goals	The nine challenges for the Gulf of Mexico program
Global climate change (2.4)	
Protection of the ozone layer (2.3)	
Clean air (1.1, 2.8, 2.12)	
Ecological protection (2.1, 2.2)	Significantly reduce the rate of loss of coastal wetlands Improve and expand coastal habitats that support migratory birds, fish, and other living resources Achieve an increase in Gulf Coast seagrass beds
Clean water (1.4, 2.12)	Ensure that all Gulf beaches are safe for swimming and recreational uses
Safe drinking water (2.10)	Protect human health and food supply by reducing input of nutrients, toxic substances, and pathogens to the Gulf
Clean up of contaminated sites (2.6)	
Prevention of ongoing releases of harmful toxic chemicals (2.6)	
Safe food (2.11)	Increase Gulf shellfish beds available for safe harvesting by 10%
Improved understanding of the environment	Expand public education/outreach tailored for each Gulf Coast county and parish Reduce by at least 10% the amount of trash on beaches Enhance the sustainability of Gulf commercial and recreational fisheries.
Safe indoor environment (1.3) Worker safety (1.2) Prevention of oil spills and chemical accidents (2.9)	

Note: Parenthetical numbers refer to Table 1.

```
^Press Enter for Table of Contents
^Press Enter for Photograph

              Gulf of Mexico Program

Mission:   Develop a comprehensive strategy to protect and enhance
           the environmental quality of the Gulf of Mexico.

           A Prototype Infobase Developed by the
          EPA Center for Environmental Statistics
                         for the
                Gulf of Mexico Program
           U.S. Environmental Protection Agency

                    November 1992

Hit Space Bar to begin random searches

Photograph of Gulf of Mexico

                  Table of Contents

     ^The Gulf of Mexico - A Vast and Valuable Resource
     ^Priority Environmental Problem Areas
     ^Organization
     ^Summaries of Subcommittee Reports (draft)
     ^Metadata Information System Components
     ^Sourcebook of Environmental Statistics
     ^Atlas of Selected Spatial Databases
     ^Integrated Environmental Statistics
     ^Proposed National Environmental Goals - EPA's Measurable
     Environmental Goals Project
     ^Nine Five-Year Environmental Challenges
```

FIGURE 8.   Menu for the Gulf of Mexico program office infobase.

duce sourcebook, atlas, and integrated statistics data stories that will be useful to a broad range of interests and clients. Another major challenge is to address the contributions of natural variabilities and delayed responses in environmental systems as they affect the interpretation of status and trends and the planning of environmental protection activities. Probably the greatest challenge will be to quantify the links between the elements in the state of the environment framework so that the EPA center products will support quantitative strategic quality planning.

ACKNOWLEDGMENTS

The authors acknowledge the guidance and support of Elisabeth LaRoe, Arthur Koines, James Morant, and N. Phillip Ross as well as the staff of the Center for Environmental Statistics and ViGYAN in the production of this chapter.

Design for a Data Retrieval and Information Acquisition System

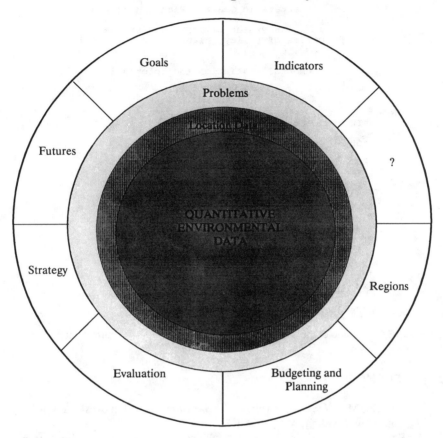

How does the user select, access, retrieve, format, manipulate, and use data, charts, graphs, maps, other visuals, and textual information available through a computer to solve a problem?

FIGURE 9. Schematic diagram of relationships between environmental data and strategic planning.

DISCLAIMER

This chapter has been produced as part of the authors' work for the U.S. Environmental Protection Agency, but has not received formal agency review so that no formal endorsement of its contents or commercial software products mentioned should be inferred.

```
Press ENTER for Menu

            "EPA... Preserving Our Future Today"

          Strategic Quality Planning at EPA

            An Prototype Infobase for the
       Office of Policy, Planning and Evaluation

                    by the
       Environmental Statistics and Information Division

                 December 1992

Menu
       ^Introduction - "How It All Fits"

       ^Other Infobases

       ^State of the Environment

       ^Futures

       ^Environmental Problems: Risk Based Priorities

       ^Goal Setting

       ^Implementation

       ^Performance Results

       ^Progress Reports

       ^Feedbacks
```

FIGURE 10.   Menu of the strategic quality planning and environmental data infobase.

REFERENCES

1. Lincoln, M., *Think and Explain with Statistics*, Addision-Wesley Publishing Company, Reading, MA, 1986.
2. Kahnert, A., Statistics Division, UN Economic Commission for Europe, Eighth Annual EPA Statistics Conference, 1992.
3. Hunter, S., President-elect, American Statistical Association, Eighth Annual EPA Statistics Conference, March 12, 1992.
4. EPA Science Advisory Board, Reducing Risk: Setting Priorities and Strategies for Environmental Protection, September 1990.
5. Preserving Our Future Today: Strategies and Framework, EPA 230-R-92-010, September 1992.
6. Setting Environmental Goals for the Nation, A Proposal for Discussion and Review, U.S. Environmental Protection Agency, draft December 23, 1992, 50 pp.
7. Parker, J. and Hope, C. The State of the Environment—A Survey of Reports from Around the World, Environment, Volume 34, Number 1, January/February 1992.
8. The Environment in Europe and North America: Annotated Statistics, United Nations Publication No. E.92.II.E.14, New York, 1992, 366 pp.

9. Guide to Selected National Environmental Statistics in the U.S. Government, U.S. EPA, 230-R-92–003, Center for Environmental Statistics, Office of Policy, Planning and Evaluation (PM-223), Washington, D.C., April 1992 (available in hardcopy and electronic forms).

10. Atlas of Selected Global Indicators of Sustainable Development, draft prepared for the WRI-Hosted Workshop on Global Indicators of Sustainable Development, by the U.S. EPA Environmental Statistics and Information Division, December 7, 1992.

11. Environmental Progress Report, draft for discussion, Environmental Results Branch, Office of Strategic Planning and Environmental Data, Office of Policy, Planning, and Evaluation, U.S. EPA, May 1992.

12. Committee on Earth and Environmental Sciences, The U.S. Global Change Data and Information Management Program Plan, Federal Coordinating Council for Science, Engineering, and Technology. Office of Science and Technology Policy, Washington, D.C., 1992.

13. Annual Review of Information Science and Technology (ARIST), Volume 25, 1990.

14. Alfsen, K. H. et al., Environmental Indicators – Discussion Paper, No. 71, Central Bureau of Statistics, Oslo, Norway, April, 1992, 76 pp.

15. A Report on Canada's Progress Towards a National Set of Environmental Indicators, A State of the Environment Report, No. 91-1, Environment Canada, Ottawa, Canada, January 1991, 98 pp.

16. Technical Supplement to a Report on Canada's Progress Towards A National Set of Environmental Indicators, Technical Report Series No. 20, Environment Canada, Ottawa, Canada, 1991, 82 pp.

17. Environmental Indicator Bulletin, Stratospheric Ozone Depletion, SOE Bulletin No. 92-1, Environment Canada, Ottawa, Canada, 1992, 4 pp.

18. Technical Supplement to the Environmental Indicator Bulletin on Stratospheric Ozone Depletion, Technical Supplement 92-1, Environment Canada, Ottawa, Canada, 1992, 8 pp.

19. Tufte, E., *Envisioning Information*, Graphics Press, Cheshire, CT, 1990.

20. Tufte, E., *The Visual Display of Quantitative Information*, Graphics Press, Cheshire, CT, 1983.

21. Information Exchange Forum on Spatial Metadata, Sponsored by the Federal Geographic Data Committee, U.S. Geological Survey, June 16–18, 1992, 236 pp.

22. Cressie, N. A. C., *Statistics for Spatial Data*, John Wiley & Sons, Inc., New York, 1991.

23. Carr, D. B., Olsen, A. R., and White, D., Hexagon Mosaic Maps for Display of Univariate and Bivariate Geographical Data, *Cartography and Geographic Information Systems*, Vol. 19, No. 4, pp. 228–236, 271, 1992.

24. Monmonier, M., *How to Lie With Maps*, University of Chicago Press, Chicago, IL, 1991.

25. Monmonier, M., Authoring Graphic Scripts: Experience and Principles, *Cartography and Geographic Information Systems*, Vol. 19, No. 4, pp. 247–260, 272, 1992.

26. MacEachren, A., Primer on Cartographic Symbolization and Design for GIS, prepared for the U.S. Environmental Protection Agency Office of Information and Resource Management, National GIS Program, 1993, 83 pp.

27. Brennan M. S., Some Animation Possibilities with Arc/Info Macros Using Environmental Data, unpublished manuscript prepared for the U.S. EPA, 1993.

28. Barnsley, M., *Fractuals Everywhere*, Academic Press, Boston, MA, 1988.

29. Rizzo, B. and Wiken, E., Assessing the Sensitivity of Canada's Ecosystems to Climate Change, *Climate Change*, Vol. 21, pp. 37–55, 1992.

30. Omernik, J., Personal Communication to Robert Shipman, ViGYAN, Falls Church, VA, January 11, 1993.

31. World Resources 1992–93, World Resources Institute in collaboration with the United Nations Development Programme and the United Nations Environment Programme, 1992, 400 pp.

32. Commission for Racial Justice, United Church of Christ, Toxic Wastes and Race in the United States, A National Report on the Racial and Socio-Economic Characteristics of Communities with Hazardous Waste Sites, 105 Madison Avenue, New York, NY 10016, 1987, 20 pp.

33. The State of the Environment, Organization for Economic Cooperation & Development (OECD), Paris, France, 1991, 297 pp.

34. EPA Office of Information Resources Management, Requirements Analysis for the Center for Environmental Statistics, June 1991.

35. Committee on Earth and Environmental Sciences, The U.S. Global Change Data and Information Management Program Plan, Federal Coordinating Council for Science, Engineering, and Technology, Office of Science and Technology Policy, Washington, D.C., 1992.

36. Christian, E. J. and Gauslin, T. L., Mechanisms to Access Information About Spatial Data, *Environmental Statistics, Assessment, and Forecasting*, Cothern, C. R. and Ross, P. N., Eds., Lewis Publishers, Boca Raton, FL, 1994.

37. McClure, C. R. et al., Identifying and Describing Federal Information Inventory/Locator Systems: Design for Networked-Based Locators, Final Report to Office of Management and Budget-Office of Information and Regulatory Affairs, National Archive and Records Center-Center for Electronic Information, and General Services Administration-Regulatory and Information Service Center, School of Information Studies, Syracuse University, Syracuse, NY, August 25, 1992.

38. Environmental Statistics and Information Division, Environmental Statistics Report Series: Gulf of Mexico Program Support, draft report prepared for the Gulf of Mexico Symposium, December 10–13, 1992.

39. Environmental Statistics and Information Division, Environmental Statistics Report Series: Strategic Quality Planning Support for Priority Environmental Problem Areas and the Proposed National Environmental Goals, draft report prepared for the Office of Strategic Planning and Environmental Data, November 1992.

SECTION VII

Environmental Statistics: Where Do We Go From Here?

N. Phillip Ross and C. Richard Cothern

CONCLUSION

Present Situation and Needs

Decision makers and risk managers need sound, valid information on environmental trends and conditions at both the macro- and microlevels of resolution. The big picture should be presented first, followed by detail descriptions including the uncertainties. This information is not readily accessible and in many instances, it is not available because it was never collected. The United States is one of a few countries in the world which does not report environmental information in a consistent and continuous manner. U.S. environmental data collection is decentralized and is not coordinated in an effective manner. In the United States, if we ask the question, "What is the state of our environment?," the answer can only be given in bits and pieces presenting both the public and the decision maker with a patchwork of information which is incomplete and in many instances lacking known quality and validity.

Generally speaking, environmental processes can be characterized as: complex, multifaceted, nonlinear, chaotic, and dynamic. Complexity can be seen in such diverse areas as global warming, species diversity, and synergistic and antagonistic interaction of chemical pollution. The chaotic elements are the instabilities in environmental phenomena introduced by the activities of humans from automobiles, power plants, large cities, industry, and other activities. One example of nonlinearity is thresholds such as those in dose-response curves for health endpoints. The study of such environmental processes as those described above must involve many different disciplines and skills. Technical, scientific (biological, physical, social), and engineering skills are needed involving such major disciplines as: biology (including such specialties as microbi-

405

ology, toxicology, medicine, physiology, pathology), chemistry (environmental, analytical), physics (environmental, health), civil engineering, environmental engineering, economics, political science, law, and communications.

Our ability to identify potentially toxic chemicals in the environment has increased; our knowledge of the short- and long-term health effects of human exposure has improved as well as our understanding of how pollutants may affect human health and the health of our ecosystems. Unfortunately, these effects are difficult to assess. Incidence rates are very low, and exposure to toxins is often over long periods at low ambient levels. We can no longer rely on our visual or olfactory perceptions to guide us in determining the nature and extent of the threats posed to the health of our environment; we must rely on information provided from carefully designed and implemented studies.

A major problem that we face in understanding our environment and assessing risk to both human as well as ecosystems, is the presence of major gaps in our information base. The ability to assess the state of the environment and risks inherent in that state of the environment require that we have more complete information of better quality. The collection of environmental data is an extremely expensive and difficult enterprise. One cannot simply take a survey of the flora and fauna in our national ecosystems and ask them how they feel or whether they are being exposed to harmful levels of pollutants. The challenge is to develop methods of measurement, to determine appropriate placement of monitoring sites (sampling design), and to assure the quality and consistency of the data collected over time. As we learn more and more about the environment and its affects on human health, welfare, and ecosystems, the more we recognize the need for comprehensive information on trends and conditions. Much of our national environmental information needs will require establishment of extensive monitoring systems which must provide data representative of both space and time. We need to develop better statistical approaches which minimize the number of observations and maximize the amount of information. This can only happen if the users of information articulate more clearly what they want to know, and what level of accuracy and precision the data must reflect.

The uncertainty associated with environmental data and risk assessment has always been a concern for environmental scientists and decision makers. In the past, major environmental problems were easily identified and approaches to their correction in many instances were simply a matter of cleaning up, or the abatement of the pollution source. In the 1960s and 1970s some of the rivers in our nation were on fire, and human health was visibly affected by episodes of acute air pollution in major urban areas. The risks associated with these problems "appeared" to be self-evident. The solutions also "appeared" to be self-evident: clean up the toxic wastes in the rivers and stop the sources of emissions of pollutants into our air. Over the last 25 years we have learned that the problems and their solutions are not quite that simple.

Where Do We Go From Here?

Bureau of Environmental Statistics

Where do we go from here? We need a systematic approach toward collecting information both at the micro- and macrolevels. There are initiatives being discussed at the federal level to better organize the manner in which the federal government collects environmental data. Congress has proposed the establishment of a federal bureau of environmental statistics as part of its efforts to elevate the Environmental Protection Agency to a cabinet level department. This bureau would provide a focal point for environmental data integration, analysis, and state of environment reporting of national environmental conditions and trends. If successful, the bureau will provide the public and decision makers with environmental statistics of known quality at the macrolevel of resolution.

Statistical design and analysis of environmental data are the keys to dealing with the probabilistic character of the natural processes and their responses to man-made perturbations and stresses. Although the current tools and methods that statistical analysis provides are powerful, there are areas where it can be improved and areas where the current techniques have not been fully applied. However, the application requires a broader view than is currently used; and this requires the creation of a new kind of research organizational structure that can deal with the breadth and complexities of environmental reactions, interactions, and interrelationships involving contaminants, physical effects, and other phenomena that impinge on the health of the human and ecological environment. The new kind of structure should involve a coordination of the many centers conducting environmental statistical research.

Now we need to turn to the more complex problems by moving beyond the current approaches into new frontiers of knowledge that involve deeper understandings of the world in which we live. What appears to be needed is more sophisticated statistical theory to address environmental problems of today and tomorrow; better tools; and more broadly trained statisticians, scientists, and information specialists.

Conventional research designs in the area of environmental statistics are often inadequate to describe the environment adequately. Thus needs emerge in three general areas of statistical analysis which can achieve a better understanding of the state of the environment: the investigation and study of environmental processes, the training of personnel to do this work, and the communication of the findings to those involved in decision making regarding the environment.

The future of environmental statistics research holds some very challenging problems. As we have mentioned, the collection of environmental data is difficult and expensive. Over the last 30–40 years, large quantities of environmental data have been collected by a variety of institutions and organizations, both in the United States and throughout the world. Unfortunately, there has been little or no coordination of these efforts, and as a result we are a nation that is environmentally data rich and information poor. Besides collecting and coordinating these data, the challenge to the environmental statistician is to develop new methods which will allow analysts to mine this rich source in a manner that produces defensible and meaningful information. The data sets that are available often provide information for spatially and temporally distinct sets of measures. For example, the Great Lakes provides an interesting microcosm where there is a tremendous amount of data collected over a number years by many different organizations; however, very little use can be made of these data in providing a holistic picture of the state of the environment for the Great Lakes region. A branch of environmental/ ecological statistics is developing which deals with the problems of "encountered" data, (i.e., data which were collected for other purposes, but which one finds may have value for his own purposes). Encountered data techniques vary with the particular situation. The statistician takes on the role of bringing together varied disciplines to focus on the problem of integrating the disparate sets of found data. Much research will be needed for developing methods to map these disparate data sets into single comprehensive sets from which meaningful information can be derived. These mapped sets or "synthetic databases" will require very detailed documentation (i.e., metafiles which describe how the information was collected and analyzed) to allow multiple users to access the databases and to use the "synthesized" estimates. It is critical to the integrity and validity of the analyses that users of these data and the resulting information understand the process by which the data bases were integrated and the intricacies of the models used for the analysis.

Global Concerns and Communication

Environmental concerns are no longer being looked at from the provincial perspective of the individual state or locality; environmental problems do not recognize political borders and indeed what happens in Los Angeles, CA can affect people living in other parts of the world. For example, in the Great Lakes, pesticide residues were found in the fatty tissue of fish; these pesticides are not used in the north, but are used in the southern areas around the Gulf of Mexico. There are many hypotheses that could explain this. Most probably the residues are showing up because of air transport of the pesticides. In order to address these problems, it will be necessary to collect environmental data on a global scale. A potential area for further investigation is remote sensing. Much of the technology and analytical models associated with the use of remote sensing for gathering information have been associated with the defense activities of the United States. With the lessening of world tensions, the potential for focusing these sophisticated technologies for assessing the state of the environment is becoming a reality.

The Rio Conference gave new emphasis to the global implications of environmental concerns. With rapid industrialization taking place in third world countries, we are beginning to see dramatic changes in the environmental landscape. These changes have both short- and long-term effects. If we want to be in a position to make rational judgments about dealing with these potential problems, we must think how by collecting more harmonized data we can see the interrelationships between the stresses we place on the environment and the global responses they cause.

In Europe, the European Economic Community (EEC) has put a great deal of emphasis on the environmental concerns inherent in a free trade atmosphere. The EEC has established an environmental agency, and one of the first actions taken by that agency was initiation of the development of a common environmental information system for the 12 member countries. All member countries will require baseline information on the state of the environment in their individual nation. Information will be needed on the trends in emissions of pollutants to the environment, the use of toxic chemicals, and other measures of environmental stress and response.

Informing the Public

One of the challenges in environmental statistics in the future is to communicate this information in a manner that the public can understand and will be able to appreciate the uncertainties associated with the data collection, integration, and analysis. More research is needed into the methods of multivariate data visualization. We need to add to the recent advances in visual methods of presenting this complex information along with indicators of the quality of the information. As people become better educated, they will want to know more about their own environment as well as the impacts their actions will have on the future environment. In the United States, the public has the "right to know"; and it is the responsibility of environmental statisticians to provide information to the public in an accessible manner. The consequences of this more open access to information will require that we develop better techniques for protecting the confidentiality of data and maintaining the integrity of the databases made available to the public. With the availability of sophisticated computer software for managing and analyzing data, there is the inherent danger of easy misuse. A good example is the evolution of software known as geographic information systems (GIS). GIS software is capable of providing very polished geographic pictures which purport to relate the presence or absence of environmental pollution in some specified geographic area. It is possible to take a measurement from a single air monitoring site in a county and input this value into a mathematical dispersion model, and the result will be a color coding of the entire county. The user of this graphic may or may not know how valid the dispersion model is or how accurate and reliable the data measurements were—all the user sees is the very pretty graphic showing the county colored red. There is a need to develop procedures and safeguards, and to educate users so that they will not react to anomalies of the analytical method as being real problems.

DISCLAIMER

The thoughts and ideas expressed in this chapter are those of the authors and are not necessarily those of the U.S. Environmental Protection Agency.

Index